1001 Questions Answered About Astronomy

1001 Questions Answered About Astronomy

JAMES PICKERING

Fully Revised and brought up to date

by PATRICK MOORE

LUTTERWORTH PRESS

GUILDFORD AND LONDON

Revised edition first published in Great Britain, 1960
Second revised edition, 1965
Third revised edition, 1969
Fourth revised edition, 1972
Fifth revised edition, 1975

ISBN 0 7188 0609 3

PRINTED IN GREAT BRITAIN BY
LOWE AND BRYDONE (PRINTERS) LTD, THETFORD, NORFOLK

To
M. D. P.
who suffers long and is kind

If the stars should appear one night in a thousand years, how would men believe and adore; and preserve for many generations the remembrance of the city of God which had been shown!

RALPH WALDO EMERSON

CONTENTS

LIST OF PHOTOGRAPHS

(*between pages* 148-9)

1001 Questions Answered About Astronomy

I. THE SUN

1. What is the Sun? The Sun is a star.

2. What is the Sun made of? The Sun is a sphere of gas. In it can be found evidence of about ⅔ of the elements that are known to exist. Since these elements known to be upon the Sun are all light, it is more than likely that the rest of the known elements, which are heavier, could be found nearer the centre of the Sun, which would be the natural place for them to be because of their greater density.

3. How big is the Sun? The diameter of the Sun is 865,370 miles. That is 109 times the diameter of the Earth. The Sun's volume is 1,300,000 times that of the Earth, but its mass, determined by its gravitational effect upon the Earth and the other planets, is only 333,-434 times the mass of the Earth.

4. How far from the Earth is the Sun? The distance has recently been re-measured. The accepted figure is now 92,956,000 miles, though there is naturally some uncertainty to a few hundreds of miles. A good round figure of 93,000,000 miles is probably as safe to use as any. This is the Astronomical Unit, abbreviated AU.

5. How does the Sun compare with the Earth in size, volume, density, etc.? It would perhaps be dramatic to set up two columns for this answer. In the first column, the Earth would be considered as unity in many of its dimensions. The comparative figures for the Sun are in the second column.

	Earth	*Sun*
Diameter	7,927 miles (equatorial)	865,370 miles
	7,900 miles (polar)	Sun is not oblate
Volume	1	1,300,000
Density	1	0.26

	Earth	Sun
Specific gravity	5.52	1.41
Surface gravity	1	28
Velocity of escape	6.95 mi/sec	383 mi/sec
Surface temperature	32° F. (average)	10,300° F.
Interior temperature	5,000° F.	35,000,000° F.
Mass	1	333,434

6. What are the apex and the antapex of the Sun's way? The apex of the Sun's way is the point in space toward which the Sun is moving. The Sun is moving, with a velocity of 12 miles a second, in the direction of the eastern portion of the constellation of Hercules. It is difficult to determine precisely the point toward which the Sun is heading, but the consensus of astronomical opinion is that it is located somewhere within a triangle on the celestial sphere whose apices are Alpha (α), Lyrae (Vega), star 111 Herculis, and Delta (δ) Herculis, about 18 hours right ascension, 34° north declination.

The antapex of the Sun's way is the point from which the Sun is moving and is opposite to the apex. It is in the constellation of Columba, 6 hours right ascension, declination 34° south.

7. How was the apex of the Sun's way determined? The position of the apex of the Sun's way was determined by a study of the proper motions of stars. All the stars in the region near the bright star, Vega, seemed to be moving apart from one another. Sir William Herschel reasoned that this general opening up of the stars here could be the result of the Sun's real motion toward this part of the sky. He pursued this idea and confirmed it by careful measurements of the apparent motions of many stars in that neighbourhood. Other astronomers have since calculated this point and are in general agreement about its location. Herschel's son, Sir John Herschel, located the antapex of the Sun's way in the southern skies during his stay in South Africa. He found that the stars in Columba at a point diametrically opposite to those in Lyra and Hercules, seemed to be closing in on each other, as they would appear to do if the Sun were travelling away from them.

8. How hot is the Sun? The temperature of the surface of the Sun is about 10,300° F. From this figure, the Sun's temperature rises very rapidly toward the centre of the Sun where it is probably between 35,000,000° F. and 50,000,000° F.

9. What is the density of the Sun? The mean density of the Sun is 0.26, with the density of the earth considered as unity.

10. How much of the mass of the solar system is in the Sun? The Sun contains 99.86% of the mass of the solar system. This is 999/1,000.

11. Does the Sun rotate upon its axis? Yes, but not uniformly. The Sun is not a solid; it is a sphere of gas. Its rotation, as learned from observation of the apparent travel of sunspots across the visible disk of the Sun, varies, being most rapid near the equator of the Sun and slowest near its poles. At its equator, the average rotational period of the Sun is 24.65 days. Halfway between the equator and the poles, the period is 27.50 days, and at the poles it is 34 days.

12. Does the Sun have an atmosphere? In a sense, the Sun is all atmosphere, since it is composed of materials in a gaseous state. There is, however, a definite central sphere to the Sun. This central sphere is surrounded by two levels or shells of gases, and these two levels are called the Sun's atmosphere.

13. What is the central body of the Sun called? The central body of the Sun is the photosphere, or sphere of light. This is the Sun that we see.

14. What are the two layers of the solar atmosphere? Directly above the photosphere is the "chromosphere" or sphere of colour. The bottom region of the chromosphere is called the "reversing layer." The outermost atmosphere of the Sun, above the chromosphere, is the corona.

15. Why is the lowest part of the chromosphere called the reversing layer? It is in the reversing layer that the energy created in the photosphere is, to some extent, absorbed by the cooler gases of the reversing layer. This absorption creates the darker lines in the solar spectrum. These lines, if they could be received without interference from the photosphere, would be brighter than the rest of the spectrum. The lowest portion of the chromosphere, by absorption, reverses the lines, making them dark lines instead of bright emission lines.

16. If the photosphere of the Sun is made of gas and we can see it, why can't we see right through it? The photosphere is opaque because it contains a vast quantity of a special type of hydrogen, each atom of which has acquired an additional neutron. This hydrogen is called "deuterium," and it occurs because of the tremendous energies at work in the superheated solar interior. (See question 1001.)

17. What is the chromosphere? The chromosphere is the first of the two layers of atmosphere that surround the photosphere. *Chromosphere* means the sphere of colour, and it is called this because it has a light pink colour. The chromosphere extends for several thousand miles above the photosphere.

18. When can the chromosphere be seen? The chromosphere is visible during a total eclipse of the Sun. At such times, the bulk of the Moon passes between the Earth and the Sun, and the chromosphere, which is otherwise overwhelmed by the brilliance of the photosphere, can be seen. At other times, the chromosphere may be seen by means of a spectrohelioscope which, by means of permitting the light from one element only to pass to the eye, allows the outer atmosphere of the Sun to be seen whenever the Sun itself is visible.

19. What is the corona? The corona is the outermost portion of the Sun's atmosphere. It is overwhelmed by the brilliance of the Sun and can be seen only when the bright central part of the Sun is covered by the Moon during a total solar eclipse or by means of a specialized instrument, the coronagraph. Then it appears as an irregular, pearly-white halo around the black silhouette of the Moon. When man first began to notice the various phenomena which accompany a total eclipse, the corona was thought to be an emanation from the Moon which was made visible by the light of the Sun shining through it. It was also considered to be an effect of the Earth's atmosphere caused by the unusual conditions of an eclipse.

The corona is composed of extremely tenuous gases that are expelled from the Sun constantly. Its appearance varies greatly with each eclipse and is influenced by the condition of the surface of the Sun. If there are relatively few sunspots, the corona will be almost even and regular, extending farther from the Sun over the Sun's equatorial regions and remaining closer at the Sun's poles. The corona is made up of countless fine filaments of bright gases which, when the Sun is quiet, seem almost to be combed out. When there are a number of spots on the Sun, however, the corona is evidently expelled with great force. Its streamers extend outward for millions of miles and it reflects vividly the turbulence on the surface of the Sun below it.

20. How hot is the corona? Spectral analysis of the light from the outer regions of the corona indicates that temperatures there, a million miles from the surface of the Sun, are on the order of more than a million degrees Fahrenheit. The gas in these regions is, however, extremely rarefied.

21. How big is the corona? The rays of the corona extend for many millions of miles into space and, upon a few occasions, it is believed that coronal streamers have reached the Earth, and perhaps beyond.

22. When can the corona be seen? The corona is visible only at times of total solar eclipse or by means of a special instrument called the coronagraph. Then the corona appears as a faint and delicate halo of fine rays radiating outward unevenly from the Sun. The light from the corona during an eclipse is about half as bright as the full Moon.

23. Does the corona always have the same shape? No. Photographs and direct observations of the corona taken at the time of eclipses show that the corona is constantly changing. During periods of high sunspot activity, the corona will be more uniform about the Sun, but the general outline, although regular, will extend farther from the Sun's surface than at periods of solar quiescence. In minimal sunspot periods, the corona will be much more irregular in shape, with extremely long streamers extending out from the equatorial regions of the Sun, but with very slight activity from the Sun's poles.

24. What are prominences? Prominences are clouds of bright hydrogen which are expelled from the Sun and which often rise to tremendous distances from its surface. They look like flames, but they are not flames because the process that produces them is not combustion. They can be seen only under the same conditions that make the chromosphere visible because they are part of the chromosphere. They sometimes extend to the distance of a solar diameter—almost a million miles—and take a variety of forms. Their action is just as often descending as ascending. The material of which they are composed seems to be drawn into the Sun, possibly into a sunspot group. Some of the prominences are in violent motion, either away from or toward the Sun. Others remain quiescent above the surface of the Sun, like luminous clouds, and still others seem to form in space above the surface of the Sun and shoot downwards. There is evidently tremendous magnetic force involved in their formation and behaviour, but there is still much to be learned about them.

25. What is the cause of solar prominences? The most spectacular of the prominences seem to originate from sunspots. It is possible that they are hydrogen and other light gas blown out of the chromosphere by the overwhelming energy from the photosphere which is permitted to escape through sunspots. The indrawing action which is often seen in prominences is also characteristic of sunspots whose centres show a downward action while the expulsive effect occurs around the edges of sunspots.

26. How big are prominences? Prominences come in a wide range of sizes and shapes. Some are relatively small, slender flickerings; some are broad, low clouds; some are twisting, flamelike shapes that shoot upward from the Sun for as many as 100,000 miles in a few hours; some take the forms of great arches, spiralling upward from two apparent sources to break away many hundreds of thousands of miles out in space. The largest arch prominences may extend as far as a million miles above the surface of the Sun.

27. What are spicules on the Sun? Spicules are fine spurtings of glowing gas that shoot up from the surface of the Sun near the Sun's poles. They are short-lived; each one lasts for only a few seconds.

28. What are sunspots? Sunspots are dark areas in the photosphere, the bright visible part of the Sun. Sunspots have been called storms on the Sun, but this is a questionable description. Terrestrial storms move through the Earth's atmosphere. Sunspots do not travel across the face of the Sun, but are carried around by the Sun's rotation. They are centres of great turbulence with tremendously strong magnetic fields. Sunspots appear dark because they are cooler than the general bright surface of the Sun, but if an electric arc could be placed in an apparently dark sunspot, it would appear dark by comparison. Sunspots appear to penetrate the photosphere to some depth and to provide a vent for some of the forces created in the tremendous pressures and high temperatures that exist there.

29. How big may sunspots be? Sunspots come in all sizes. The smallest are a few hundred miles in diameter and appear tiny as seen from the Earth. There are probably sunspots which are so small that they cannot be seen, even through a telescope. On the other hand, there are spots whose total area is several thousand million square miles. Such a spot would have a diameter of anywhere from 50,000 to 100,000 miles. The largest spot of which there is any record appeared in April, 1947, and had an area of 6 thousand million square miles. A spot larger than 25,000 miles in diameter can be seen, with a protective filter, without a telescope.

30. Are sunspots black? No. They appear dark only by contrast with the intensely bright general surface of the Sun. If it were possible to place an electric arc within the area of a sunspot, the electric arc itself would appear as a tiny black dot in the sunspot.

31. Are sunspots cooler than the rest of the Sun? Yes. Since the gases in the areas of sunspots are radiating less energy than the gases of the general solar surface, the temperature within the area of a sunspot is about 2,000° lower than the temperature of the normal surface of the Sun.

32. How long do sunspots last? As might be expected, the larger sunspots last longer than the smaller. Sunspots which are of about the minimum diameter that can be seen last from one to four days.

Less than 10% of the sunspots observed at any given time last long enough to be carried completely around the Sun by the Sun's rotation. Still fewer last longer than one rotation period—about four weeks—and the endurance record at the moment is held by an enormous sunspot which appeared in 1840 and which remained visible for 18 months.

33. Are there always spots on the Sun? No. There have been times when no spots were discernible, but such times are rare. The cycle of sunspots is one of the most interesting studies in astronomy. A German scientist, Schwabe, discovered the cycle of sunspot frequency in 1843. The sunspot cycle is not rigidly regular but, over the years, an average period of 11.3 years passes between one minimum of sunspot activity and the next. The high points of sunspot frequency, however, fluctuate widely. A maximum may occur soon after a minimum or soon before the following minimum. There have been intervals between maxima of 7.3 years, and there have been intervals as long as 17 years between the high points of sunspot frequency. There are occasional periods of weeks at the time of a sunspot minimum when no spots at all can be seen on the Sun.

34. Do sunspots appear all over the Sun? No. Sunspots are generally confined to a region between the Sun's equator and the poles of the axis of the Sun's rotation. They are rarely seen on the equator and never at the poles. At the beginning of a sunspot maximum period, they appear nearer the poles than the equator, but as the cycle progresses, successive spots appear at lower latitudes. The final spots of one cycle may linger near the equator as the new spots of another cycle are making their appearance in high latitudes.

35. Do sunspots have any effect upon the Earth? Sunspots, or phenomena connected with sunspots, have a very definite effect upon the Earth. The tremendous bursts of energy which produce flares and prominences and which are associated more often than not with sunspots, expel particles of matter of subatomic size for millions of miles into space and often strike the atmosphere of the Earth. When such energy bursts reach the Earth, they upset the structure of the atmosphere by causing magnetic storms of greater or less intensity.

The delicate electronic instruments of mankind which depend upon the integrity of the atmospheric structure are often adversely affected: radio-telephone service is interrupted; teletype machines transmit gibberish; radios misbehave; television sets perform poorly; currents of electricity may be set up that sometimes interfere with the operation of the ordinary telegraph system. The arrival of the actual particles, usually two or three days after the outbreak of a flare, is believed to be responsible for auroral activity. This is probably caused by an ionization of particles of gas in the upper regions of the Earth's atmosphere by the impact of the particles expelled from the Sun. It has also been found that a very slight increase in the temperature of the air is experienced during sunspot maxima.

There is a possibility of a connection between sunspot cycles and the weather, though most astronomers tend to be sceptical.

36. What is meant by the word *gauss* in connection with sunspots? A gauss is a measure of the strength of a magnetic field. A single loop of wire with a radius of 0.628 cm., through which flows a current of 1 ampère, possesses a field of 1 gauss at its centre. The magnetic fields of sunspots sometimes amount to 2,500 to 3,000 gauss over vast areas. Sunspots may be considered to be tremendous electromagnets.

37. How does sunspot activity affect radio? The long waves of energy that carry radio programmes are reflected from the ionosphere, the upper region of the Earth's atmosphere, and are returned to the Earth. For this reason, they can be received by radio sets at any direction and at a reasonable distance from the point of origin. When the ionosphere is disturbed or disrupted by the bombardment of energy or particles from the Sun, it loses this power of reflection, or the reflective pattern is disrupted, and the radio waves, instead of being returned to Earth, pass through the ionosphere and go out into space. Sometimes, during partial disruption, freakish effects are caused. Local radio stations, such as police transmitters, which generally have a very short range, have been clearly heard thousands of miles away.

38. What are flocculi? Flocculi are bright clouds, usually made of calcium, which are found on the Sun near sunspots and which are sometimes called "plages."

39. What are faculae? Faculae are bright mottlings, larger than the flocculi. They are usually associated with sunspots, and are known to be projections of solar prominences against the surface of the Sun. The word is Latin for "torches."

40. What are granulations on the Sun? Granulations are areas of brightness 500 to 600 miles in diameter which are all over the surface of the Sun. These bright nuclei are usually elliptical in shape and seem to be surrounded by darker areas or boundaries. They resemble, superficially, grains of rice and are so called. They are in constant turbulence and motion. The life of an individual rice grain is only a few minutes. It breaks up and disappears and a new one forms in the vicinity of the first. The whole surface of the Sun is made up of islands of brightness which are appearing, disappearing and changing their shape constantly.

41. Why is the limb of the Sun darker than the centre of the Sun's disk? The light from the limb of the Sun, the edge of the Sun's disk, must pass through a much greater thickness of overlying gases than does the light from the centre of the Sun's disk. This light also comes from a much higher and therefore cooler region of the Sun than does the light from the centre of the disk. For these reasons, the limb of the Sun is less bright than the centre of the disk.

42. How much energy is put out by the Sun? The total energy radiated by the Sun can be found by multiplying the solar constant by the number of square centimetres in the surface of a hypothetical sphere whose radius is the Earth's mean distance from the Sun—a sphere whose diameter is about 186 million miles. Since there is practically nothing between the Earth and the Sun to intercept or absorb this radiation, the energy which strikes this hypothetical surface is the same as the energy which leaves the surface of the Sun. It amounts to 500,000,000,000,000,000,000 horse power, or about 70,000 horse power per square yard of the Sun's surface.

43. How much of the Sun's energy does the Earth receive? The Earth receives only about one part in 2 billion of the energy that is produced by the Sun. This tiny fraction amounts to 4,690,000 horse power per square mile of the Earth's surface.

44. What is the Gegenschein? The Gegenschein is also known as the counterglow, which is a good translation of its German name. It is a broad area located opposite to the Sun in the very narrow and almost imperceptible band of the zodiacal light. The Gegenschein is probably produced by the same factors that are responsible for the zodiacal light, that is, the faint reflection of sunlight from the swarm of tiny meteoric particles which are in space within the boundaries of the solar system. The Gegenschein is elliptical, about 10° in its greatest diameter, and is very difficult to see.

45. On what date does the Sun rise at latitude 66½ degrees north and on what date does it set there? Latitude 66½ degrees north is the Arctic Circle. Here, the Sun rises on or about March 21—the time and date vary slightly from year to year—and sets about September 21. There is, however, twilight for several weeks both before and after these two dates.

46. Some films show big flares and explosions on the Sun. Why cannot we hear the sound of these explosions? Sound is the effect of a wave motion of air upon our ears. If there is no air to carry this motion from its source to us, we cannot hear any sound. There is no air between us and the Sun, outside of the Earth's atmosphere which extends possibly 2,000 miles out from the Earth's surface; hence there is no medium which can carry to us any sound that these prominences might produce.

47. What is the best way to look at the Sun through a telescope? Don't! Remember that the rays of the Sun, focused on a piece of paper with an ordinary hand magnifying glass will set the paper on fire. Telescopes, or even field glasses, are complicated and more powerful magnifying glasses. Do not put your eye in the position of the piece of paper! The safest way to look at the Sun is to place a piece of white cardboard beyond the eyepiece of your telescope and project the Sun's image upon it. The eyepiece of the telescope can be racked in or out and the piece of cardboard moved back and forth until the image you get is sharp and clear. The telescope will serve as a projection instrument, the Sun can be seen very well and no harm will result. The light of the Sun, intensified by a telescope, can do permanent injury to your eye.

48. What is a "mock Sun"? Another name for a "mock Sun" is a "Sun dog." Sun dogs are caused by the refraction of the Sun's light by tiny particles of ice suspended high in the atmosphere. Spurious images of the Sun are caused by this refraction, which are placed at intervals of 90° around a faint circle of light which has the Sun at its centre.

II. THE EARTH AND THE MOON

49. How big is the Earth? The equatorial diameter of the Earth is 7,927 miles. The Earth is not a perfect sphere. It is slightly flattened at the poles, which makes it an oblate spheroid. Those two words mean "shaped somewhat like a sphere, but flat at the poles." This oblateness makes a difference of only 27 miles between the Earth's diameter measured at the equator and the diameter at the poles, which is 7,900 miles. The flattening, then, is only one part in 297. The mass of the Earth is $\frac{1}{332{,}000}$ of the mass of the Sun. The weight of the Earth, in terms of its own gravitational pull upon its substance, may be translated into the figure 66 followed by 20 zeros if calculated in tons. This is six thousand, six hundred million, million, million tons.

50. What causes the flattening at the poles of the Earth? The Earth's oblateness was caused by its relatively rapid rotation. This motion, millions of years ago when the Earth was not yet as rigid as it is today, forced more of its substance into the region of its equator and made its equatorial diameter slightly larger than its polar diameter.

51. How many degrees away from a perpendicular to the ecliptic or to the Earth's path about the Sun, is the Earth's axis tilted? 23°26′44″.8.

52. By how much does the Earth's path around the Sun deviate from a straight line? The Earth's path deviates from a straight line by about 1 inch in 9 seconds of travel. This is 1 inch for every 166.5 miles, or about 1 part in 10,550,000.

53. How fast does the Earth travel about the Sun? The Earth's average speed is 18.52 miles per second. The small eccentricity of the Earth's orbit makes the variation in its speed very slight, but there is a difference in the length of the seasons because of this variation. The number of days between the beginning of spring and the beginning of autumn, in the northern hemisphere, is 186, but the number of days between the beginning of autumn and the beginning of spring is 179 days. These conditions are reversed, of course, in the southern hemisphere, where the winter season occurs during the northern summer.

54. What is an anomalistic year? a tropical year? The anomalistic year is the time that elapses between two successive passages of the Earth through its perihelion point—that point in its orbit where it is closest to the Sun. That point is reached by the Earth early in January each year. The anomalistic year contains 365 days, 6 hours, 13 minutes, 53.0 seconds. The tropical year is the year of the seasons which we know as the calendar year. It is the time that elapses between two successive passages of the vernal equinox by the Sun. The tropical year contains 365 days, 5 hours, 48 minutes, 46.0 seconds. The tropical year is shorter by 25 minutes, 7 seconds than the anomalistic year. The difference is caused by the motion eastward of the line of apsides of the Earth's orbit, which lengthens the anomalistic year, and by precession, which shortens the tropical year.

55. Is there any time during the year when the axis of the Earth is at right angles to the Sun? Yes. There are two occasions during the Earth's revolution about the Sun when it is standing at right angles to the Sun. The poles of the Earth are then equidistant from the Sun and a line through the centre of the Earth passing through the Earth's equator is at right angles to the Sun. These times occur when the revolution of the Earth brings it to the points that we call the equinoxes and the Sun is overhead at the equator. The Earth reaches these points each year about March 21 and September 21.

56. Why are the Tropics of Cancer and Capricorn so called? The Tropics of Cancer and Capricorn are two imaginary lines upon the Earth's surface which parallel the equator 23½° north and 23½° south of the equator, respectively. They mark the most northerly and southerly parts of the Earth ever to see the Sun directly overhead.

When man first became aware of these limits of the Sun's apparent march up and down the sky during the year, the Sun appeared to stand before the stars in the constellation of Cancer when it had reached its most northern aspect, and to be in front of the stars of Capricorn at its southern limit.

Since that time, the precession of the equinoxes has moved the meridian lines upon which the Sun stands at such times, westward among the stars. The two hour lines, called the solsticial colures, are now found passing through the constellations of Gemini and Sagittarius, and more modern and fitting names for the two tropics would be the Tropic of Gemini and the Tropic of Sagittarius.

57. What is the ecliptic? The ecliptic is the path of the Earth about the Sun during the year. From the Earth, the Sun appears to move through the heavens, and the path it seems to take is a great circle inclined 23°26′45″ from the Earth's equator. The polar axis of the Earth is tilted by this amount away from the vertical of the Earth's orbit about the Sun. Because the tilt, or inclination of the Earth's axis, is constant both in amount of tilt and in the direction in which

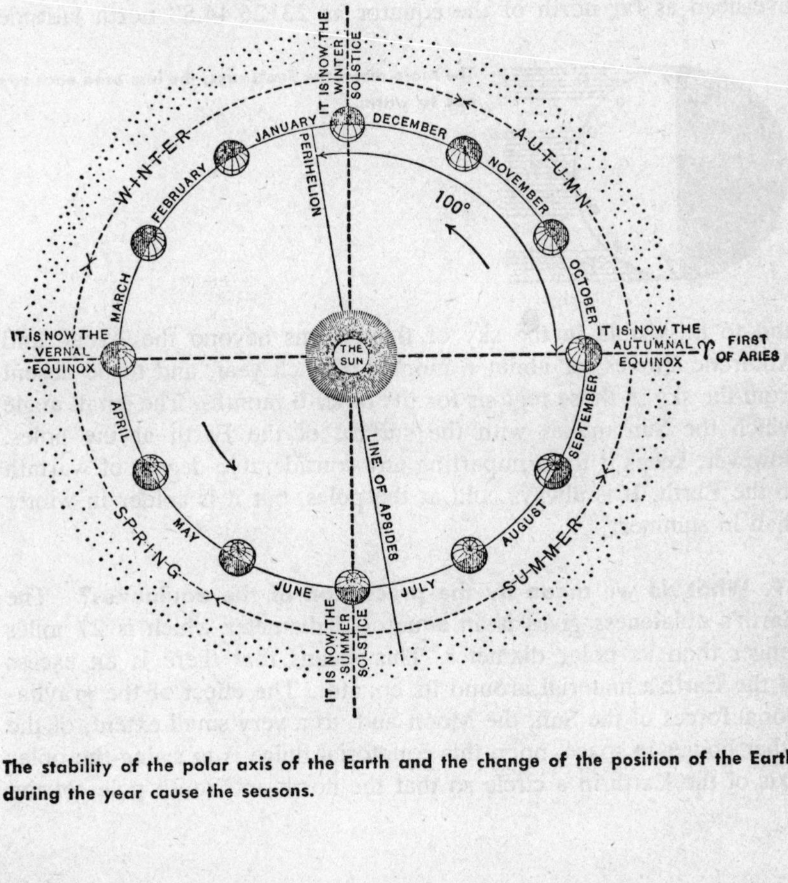

The stability of the polar axis of the Earth and the change of the position of the Earth during the year cause the seasons.

it is tilted, the Sun appears to move in a path which carries it about 23½° above the equator in the northern summer and 23½° below the equator in the northern winter. The points in the celestial sphere at which the ecliptic crosses the equator are the equinoxes and the points in the celestial sphere at which the Sun is either farthest above or below the equator are the solstices. The vernal equinox occurs every year about the 21st of March; the summer solstice, about the 21st of June; the autumnal equinox, about September 21 and the winter solstice, about December 21.

58. Does the Earth have seasons at the north and south poles? Yes. The difference between seasons at the Earth's poles is more one of light than of temperature. The inclination of the Earth's axis from the plane of its orbit about the Sun causes the Sun to appear overhead as far north of the equator as 23°26′44.8″ north latitude

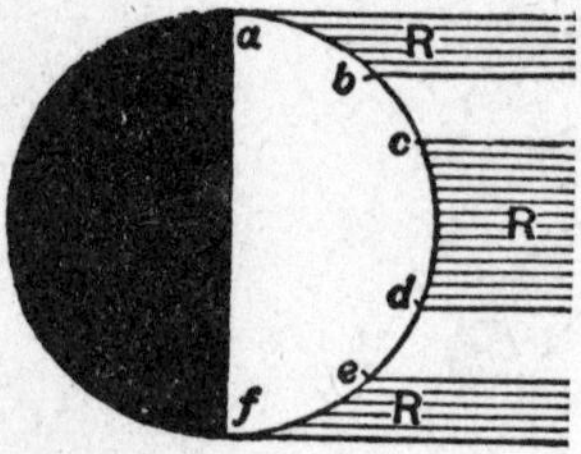

The more direct the Sun's rays, the less area each ray has to warm.

and to be visible in the sky of the regions beyond the Arctic and Antarctic Circles for about 6 months of each year, and to be absent from the sky in those regions for the other 6 months. The small angle which the Sun makes with the surface of the Earth at the poles, however, keeps it from imparting any considerable degree of warmth to the Earth. It is always cold at the poles, but it is colder in winter than in summer.

59. What do we mean by the precession of the equinoxes? The Earth's oblateness gives it an equatorial diameter which is 27 miles longer than its polar diameter. This means that there is an excess of the Earth's material around its equator. The effect of the gravitational forces of the Sun, the Moon and, to a very small extent, of the other bodies in space, upon this equatorial bulge is to swing the polar axis of the Earth in a circle so that the north and south poles of the

Earth describe enormous circles against the stars. A complete cycle of this motion takes 28,500 years to complete. Its result is a slow circling of the stars as they are seen from the Earth. The Pole star changes, because of this motion, very slowly. Within historical times, the Earth has had three north stars. The position of the equinoxes and the solstices, those points at which the Sun appears to cross the Earth's equator or to stand farthest from it, are also shifted westward very slowly. This shift of the Equinoxes is called precession. The precession must be accounted for in every precise measurement of star positions. Every star chart is referred to a particular date which is called an epoch, so that the user may know by how much to alter the given positions of the stars and other objects from those given in the chart.

60. Are summers warmer in the Southern than in the Northern Hemisphere? Yes. The Earth is at perihelion in January, during the southern summer, and is about 3 million miles closer to the Sun than it is in July during the northern summer. This difference, although it is less than 4% of the Sun's mean distance from the Earth, is sufficient to make the summers warmer and the winters colder in the Southern Hemisphere than in the Northern Hemisphere. The variation in temperatures between the two Hemispheres would be even greater if it were not for the fact that a vastly larger per cent of the area of the Southern Hemisphere is covered with water, which has a tempering effect on the temperature.

61. Where is the International Date Line and what is the reason for it? The International Date Line is the 180th meridian of longitude, which runs north and south at about the middle of the Pacific Ocean. It marks the point where one civil day ends and the next begins upon the Earth. It veers a bit from the actual meridian in some places to avoid cutting through a geographical location, such as a group of islands or the eastern tip of Siberia. Travellers crossing the International Date Line from east to west lose a day, while those travelling from west to east gain a day. Time on the Earth is not kept by the Sun. If it were, every town and village would have its own individual time, differing from every other town or village which was a bit to the east or west of it. This condition was true, to some extent, up until the late nineteenth century, when the various time zones now in effect were introduced. These time zones are areas running north and south which are roughly 15° in width, because the Earth's rate of rotation is 15° an hour. The boundaries of the time zones do not lie exactly upon the meridians, but are altered to avoid cutting across the middle of geographical entities, like states or territories. Everywhere within one time zone, civil time is the same, and is known as standard time. Across North America there are 8 time zones, in each of which time is later by an hour or, in some cases, half an hour, than in the zone next east. The time zones are Newfoundland, 3 hours 30 minutes later than Greenwich Time; Atlantic, 4 hours later; Eastern, 5 hours later; Central, 6 hours later; Mountain, 7 hours later; Pacific, 8 hours later; Yukon, 9 hours later; and Alaskan, 10 hours later.

62. What is the equation of time? First, let us understand that, relative to the Earth, the Sun does not move. Any motion that we

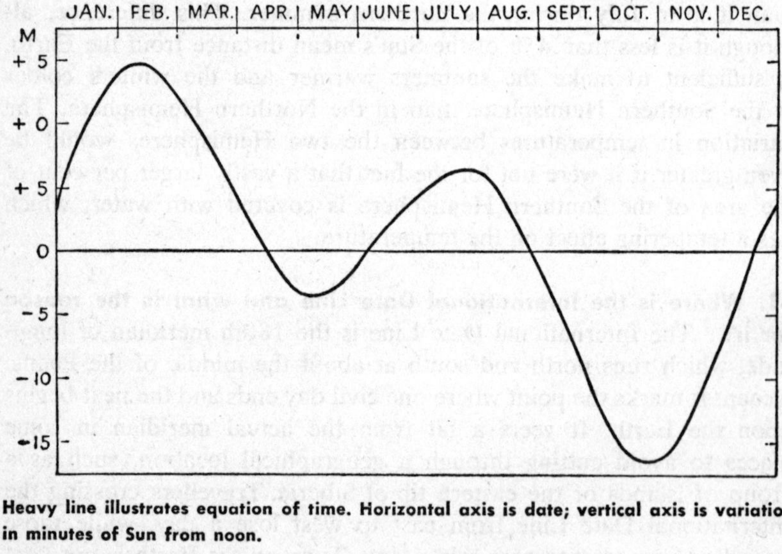

Heavy line illustrates equation of time. Horizontal axis is date; vertical axis is variation in minutes of Sun from noon.

see on the part of the Sun is really a motion of the Earth, shifting our viewpoint of the Sun and making it appear to move. The revolution of the Earth about the Sun, which gives us the year, is not uniform. The Earth follows the laws which govern the motions of bodies in space and moves more rapidly in its orbit when it is closer to the Sun, in January, than it does in July, when it is farther from the Sun. When the Earth moves faster, the Sun appears to move faster, and when the Earth moves more slowly, the Sun appears to move more slowly. Added to this, the Sun seems to travel not along the equator, which is at right angles to the meridian and the various hour lines which mark the progress of the Sun, but along the ecliptic, which is not at right angles to the hour lines. Hence, a projection to the equator of the motion of the Sun of 1° along the ecliptic would not result in an advance of 1° on the equator.

It would be awkward if man were to try to keep time by the Sun's apparent motion, because he would have to make an almost daily adjustment to compensate for the Earth's irregularities. Instead, man has devised a "mean Sun," which moves uniformly around the Earth through the year, *on the equator*. The mean Sun completes its annual journey in exactly the same time as does the real Sun, but the mean Sun travels evenly and uniformly. The difference in time between the real Sun, with its small losses and gains, and the steady progress of the mean Sun, is the equation of time. The real Sun is ahead of the mean Sun part of the time and part of the time it is behind the mean Sun. When the real Sun is ahead of the mean Sun, it is fast and crosses the meridian sooner than the mean Sun. When it is behind the mean Sun, it is slow, and crosses the meridian later than the mean Sun. On April 16, June 15, September 1 and December 26, the two Suns correspond. On February 14, the real Sun is about 14 minutes fast. It is about 3 minutes slow on May 16, 6 minutes fast on July 30 and about 17 minutes slow on November 2. The equation of time is usually given as "mean minus apparent" with a change in sign from negative to positive when the occasion warrants. This will translate apparent time into mean solar time. The equation of time must be used, for example, in reading sundials and in many delicate navigational and astronomical problems. The analemma, a lopsided figure 8 which is printed somewhere in the Pacific Ocean on most terrestrial globes, represents the equation of time.

63. Which of the two bodies, the Sun or the Moon, has the greater tidal effect upon the Earth? The Moon is responsible for about two-thirds of the tidal effect upon the Earth and the Sun for one-third. The Sun is, of course, tremendously more massive than the Moon, but it is also much farther away. Gravitational effect diminishes as the inverse square of the distance between the centres of bodies. Thus the Sun's vastly greater gravitational force is so lessened by its greater distance from the Earth that it is only half as great as the gravitational force of the Moon, as far as the Earth is concerned.

64. At what season of the year do spring tides occur? Spring tides occur at any season of the year when the Sun and the Moon happen to be either in conjunction or in opposition. The combined gravitational pull of the Sun and the Moon at such times causes unusually high and unusually low tides. These extreme tides are called spring tides—not from the word *spring* that is the name of a season, but from the word *spring* which means a jump. The moderate tides which generally occur when the Moon is in quadrature are called "neap" tides, from an Anglo-Saxon word meaning "scanty."

65. Does the sunlight that falls on the Earth have any weight? Yes. Light has appreciable weight—or pressure. (See question 1010.) The light pressure on the surface of the Earth is given as two pounds per square mile.

66. Is there such a word as *syzygy*, and what does it mean? *Syzygy* is a word used to describe the position of three bodies when they are more or less in line. When the Moon is either new or full, it is in syzygy, because at full it is on the opposite side of the Earth from the Sun, and when it is new it is between the Earth and the Sun. In either case the Earth, Moon and Sun are approximately in line.

67. How far from the Earth is the Moon? The Moon's distance from the Earth varies from 252,710 miles when the Moon is in apogee—farthest from the Earth—to 221,463 miles when the Moon is in perigee—nearest to the Earth. The mean distance between the Moon and the Earth is 238,857 miles. The measurements of the distances between the Earth and the Moon are considered accurate to one part in 300,000.

68. How big is the Moon? The diameter of the Moon is 2,160 miles. This is slightly more than one quarter of the diameter of the

Earth—the actual proportion is 0.273. The surface of the Moon is about $\frac{1}{14}$ of the surface of the Earth and its volume about $\frac{1}{49}$ of the Earth's volume. The mass of the Moon is $\frac{1}{81.5}$ of the mass of the Earth.

69. Is it scientific to say that the Moon revolves about the Earth? Both the Moon and the Earth revolve about the centre of gravity of the Earth-Moon system. Because the mass of the Earth is much greater than the mass of the Moon, the centre of gravity about which both revolve is located inside the Earth. The Earth's revolution about this point is so much smaller than the revolution of the Moon about the same point that it is correct to say that the Moon revolves about the Earth.

70. Where is the centre of gravity of the Earth-Moon system? The centre of gravity of the Earth-Moon system is 2,903 miles from the centre of the Earth, always on the side nearest the Moon. This means that the centre of gravity moves through the Earth about 1,060 miles beneath the surface of the Earth.

71. Does the Moon have a day and night? Yes. The Moon rotates upon its axis in 27.32 Earth-days. That is the length of the Moon's day, so that it has about 14 Earth-days of light and 14 Earth-days of darkness alternately at any one place upon its surface.

72. Are there a dawn and sunset on the Moon? There are sunrise and sunset on the Moon, but they would be without the spectacular effects that often accompany sunrise and sunset on the Earth. The glow in the sky, twilight, dawn clouds, sunset clouds and the like are all atmospheric effects and would be missing on the Moon, since the Moon has no atmosphere.

73. What is the extent of the Moon's influence on the Earth? The greatest effect that the Moon has upon the Earth is in the tides, which the gravitational pull of the Moon helps to raise in the waters of the Earth. In addition to that, there is a very small change in atmospheric pressure which is also caused by the gravitational pull of the Moon. Some slight changes in terrestrial magnetism are measureable, which are caused by changes in the position of the Moon with relation to the Earth.

74. What is the actual shape of the Moon's orbit? The Moon's orbit, like every orbit in space, is an ellipse. The eccentricity of this ellipse is, on the average, 1 part in 18. The close association of the Moon with the Earth, combined with the tremendous gravitational influence of the Sun, imparts many perturbations to the orbit of the Moon. The shape of its orbit may change so that the eccentricity is as small as 1 part in 23 or as great as 1 part in 15.

75. How fast does the Moon travel around the Earth? The Moon moves in its orbit at 2,287 miles per hour, on the average. It travels faster when it is in perigee—nearer the Earth—and slower when it is in apogee—farther from the Earth. The average angular velocity of the Moon as seen from the Earth—the distance across the sky which it seems to cover—is about 33′ of arc per hour. This is slightly more than the Moon's own diameter.

76. How long does it take the Moon to go around the Earth? If we time the Moon from the instant it leaves a given point among the stars until it returns again to that point, it will have taken 27 days, 7 hours, 43 minutes, 11.47 seconds. This is an average time, because the motion of the Moon is subject to many perturbations. (See question 994.) In fact, the time of the Moon from one point in the sky back to the same point may vary by as much as 7 hours. Because this period of time is in relation to the stars, it is called the "sidereal" month.

A more familiar and useful timing of the Moon's journey around the Earth is the interval between one new Moon and the next. This is the "synodical" month. The word *synodical* comes from two Greek words which mean "the same road," and the phrase refers to the conjunction, or apparent position in the same part of the heavens of two bodies. It takes the Moon, again on an average, 29 days, 12 hours, 44 minutes, 2.78 seconds to go from one phase to the same phase again. This period may vary by as much as 13 hours.

77. What is the cause of the regression of the Moon's nodes? The apparent path of the Moon about the Earth does not coincide with the apparent path of the Earth about the Sun—the ecliptic. The orbit of the Moon is inclined to the ecliptic by about 5°. For this reason, the orbit of the Moon crosses the ecliptic at two points each time the Moon revolves about the Earth. These points of intersection are the nodes of the Moon's orbit. These nodes move slowly westward around the Moon's orbit—backwards to the Moon's own motion—so that each node is about 1½° west of the previous node.

This regression of the Moon's nodes is caused by the gravitational influence of the Sun upon the Moon, and, to a lesser extent, of the Earth and the other planets. The gravitational power of the Sun tends to make the Moon move in a plane extending outward from the Sun's equator, but the Moon's own motion, in a slightly different plane, resists this force. Because of the variation in the Moon's distance from the Earth and from the Sun, the forces that work upon the Moon are not steady and constant during the month. Their net result, however, is to cause a slow westward movement of the nodes of the Moon's orbit.

78. What is the period required for the nodes of the Moon to make a complete circle around the sky? The length of time required for the nodes to regress completely around the ecliptic is 18.6 years. The nodes regress at an average rate of just under 20° a year.

79. If the Moon rotates, why do we always see the same side of it? We can see only one side of the Moon because it rotates in exactly the same length of time that it takes to revolve about the Earth. That combination of motions means that it always keeps the same side of itself toward the Earth. It is easier to understand this if you place a chair in the centre of a room. Now, always facing the chair, circle sideways about it. You will find, as you do this, that while you are always facing the chair, you have also faced, in succession, all four walls of the room. You must, therefore, have turned completely around—rotated—as you went once around the chair; you revolved, and yet you always faced the chair. This is what the Moon does in the case of the Earth.

80. How is it that we can see more than exactly one half of the Moon's surface if it always keeps the same side turned toward the Earth? We can, over a period of time, see more than one half of the

Moon's surface because of an apparent motion of the Moon which is called "libration" or "balancing." This is a sort of rocking of the Moon as seen from the Earth.

81. What is the Metonic cycle? The Metonic cycle is the number of months between successive recurrences of the Moon's phases upon the same date of the month. The Metonic cycle covers 235 lunations or complete changes of the Moon. This consumes 19 years, with variations of a day or two to compensate for leap years. The Metonic cycle is used in the determination of the date of Easter, which is fixed as falling on the first Sunday following the first full Moon after the vernal equinox. Meton was a Greek astronomer who lived in Athens about 430 B.C. He discovered the cycle of the Moon's changes by keeping meticulous records of the date of each step in the progress of the Moon through all its changes until the phases came back into step again with the calendar month.

82. Who first saw the Moon through a telescope? Probably the English astronomer Thomas Harriot, in 1609, but the first great lunar observer was Galileo.

83. Can an observer on the Moon see the Earth showing phases? Yes. The Earth, looking slightly less than four times as great in diameter as the Moon does to people on Earth, always remains in approximately the same position in the lunar sky. The Moon's revolution round the Earth makes the background of stars appear to slide slowly behind the Earth from left to right. When it is new moon on the Earth, a lunar observer sees full Earth. Nobody has yet observed this; but spectacular pictures of the Earth as seen from the Moon have been taken, the latest being those obtained from the camera on board the abandoned Lunar Rover of Apollo 16 in April 1972. These photographs showed a beautiful crescent Earth in the lunar sky.

84. Does the Earth rise and set for an observer on the Moon? No. Because the Moon always keeps the same side of itself toward the Earth, an observer on the Moon would always see the Earth in approximately the same position in the Moon sky. There would be slight changes, which would be the opposite of the Moon's libration changes, in which the Earth would change its position slightly through the Moon's night. The stars and the Sun would seem to slide around the skies behind the Earth. To someone on the side of the Moon away from the Earth, the Earth would never be visible, but every other object in the sky would be seen, at one time or another.

85. Would Earth's aurora borealis or aurora australis be visible from the Moon? Theoretically, yes. It is just a question of the amount of light and of eye-accommodation. The observer on the Moon would have to watch the dark of the Earth—its night side—from the dark of the Moon in order to see such a delicate illumination.

86. Is the dark part of the Moon the shadow of the Earth? No. It might be called the Moon's own shadow. It is the part of the Moon that is turned away from the Sun, so that the Sun's light cannot reach it. It is the Moon's night. The shadow of the Earth does strike the Moon at intervals, causing an eclipse of the Moon. (See question 405.)

87. Which way does the Moon really move in the sky? The Moon revolves around the Earth from west to east, in the same direction in which the Earth rotates upon its axis and in the same direction in which the Earth revolves about the Sun. It is the rotation of the Earth which makes the Moon appear to move from east to west during the night in exactly the opposite direction to that in which it is really moving. If the Moon is carefully observed with regard to the stars beyond it, it will easily be seen that it changes its position by 12.2° every day, moving to the east by that amount.

88. Why does the Moon rise later each night? The Moon is in motion in an orbit about the Earth, going from west to east, in the same direction in which the Earth is rotating. The Moon's daily motion is about 13°, so that each day while the Earth is spinning around once, the Moon is also moving to the east. Hence it is farther along, ahead of the spin of the Earth, each night. For that reason, the Earth must turn a little more than one complete revolution to catch up with the Moon every day.

89. How much later does the Moon rise each day than it did the day before? The average of the Moon's retardation—which is the official name for the difference between the rising of the Moon on one day and its rising on the next day—is 50.5 minutes. The Moon can be as long as 80 minutes later on one night than it was the night previous, or the time may differ by as little as 13 minutes. When the Moon is near the vernal equinox, its normal retardation in rising is small, for the Northern Hemisphere, but its retardation in setting is large. When the Moon is near the autumnal equinox, the reverse is true. The Moon rises late, but its setting is much nearer to the setting time of the previous day's Moon. In the Southern Hemisphere, these conditions are reversed.

90. Is there any place on Earth where the Moon does not set at any time? In latitudes north of 61°21′, the Moon may, at times, be seen to remain in the sky without setting for a day and a night. To satisfy the conditions, however, the Moon must be in that part of its orbit which will bring it farthest north in the sky. Its declination must be at the greatest possible extreme of 28°27′ north. To compensate for this, however, these same northern latitudes will go a day and a night without seeing the Moon at all, when the Moon is in the most southerly part of its orbit.

91. Why, for a succession of nights, does the harvest Moon rise approximately only thirty minutes later instead of the usual nearly full hour later each night? In September, the full Moon is in that part of its orbit which very nearly parallels the equator. Therefore the nightly eastward motion of the Moon—about 12°.6—does not carry it very much farther below the horizon, and it rises only a little later each night than it did the night before.

92. Does the Moon have seasons? Theoretically, yes. The Moon's equator is inclined 6°40.7′ from the plane of its orbit about the Earth. This would mean that the inclination of the Moon's equator from the plane of the ecliptic, or the Sun's apparent path, would vary from 28½° to 18½°. This would give it a varying season, but the actual effect, since there is no atmosphere upon the Moon, is negligible.

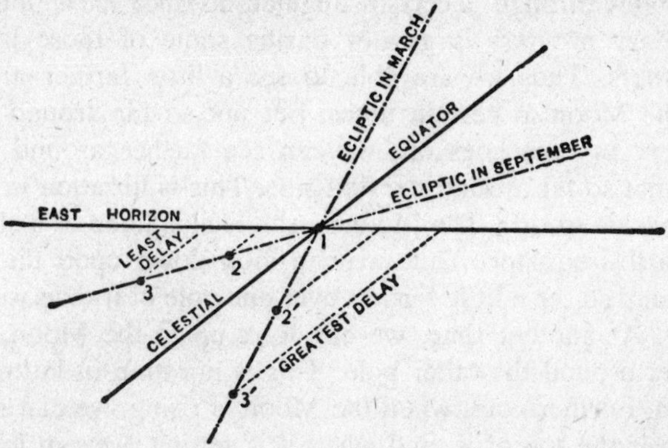

In September, the path of the Moon lies closest to the horizon, and its retardation is smallest. This is the season of the Harvest and the Hunter's Moons.

93. Would the skies seen above the Moon be filled with a very brilliant glow in the region of the Milky Way? Yes, at night. In fact, we should be able to see many stars in the Milky Way which we cannot see from the Earth. Our atmosphere deprives us of about 30% of what we should be able to see in the heavens.

94. Is the Moon actually silvery in colour? Various tests of the light reflected from the surface of the Moon compared with light reflected from a variety of substances found on the Earth show that the light from the Moon is most like that reflected from sandstone. That is a brownish gray in colour. Actually, the Moon is one of the poorest reflectors of light in the solar system.

95. Do we only see exactly one half of the Moon? From any one place at any one time, we see only one half of the Moon, but throughout the month we do not always see the identical area.

96. What causes libration on the Moon? If the Moon's orbit were circular and the Moon's speed around its orbit were uniform, we should never see more than one half of the Moon. The Moon's orbit is an ellipse, however, and for that reason the Moon moves more rapidly when it is nearer the Earth than it does when it is farther from the Earth. Its rotation, however, is uniform. This means that the

Moon rotates through a certain angular distance each hour, but its revolutionary progress is greater during some of those hours than during others. Thus we are able to see a little farther around one side of the Moon at certain times, but not so far around the other side. Then, as time goes on, we can see farther around the other side, but not so far around the first side. This is libration in longitude —or from side to side. The Moon's orbit is also tilted at a slight angle to the Earth's equator. Thus we can look down upon the Moon at one time and so see a little farther over one pole of it than we can over the other. At another time, we can look up at the Moon, and thus see farther around the other pole. This is libration in latitude, or up and down. Furthermore, when the Moon is rising, we can see a little farther over the top of it, and when it is setting, we can look a little farther over the top on the other side. Our natural satellite is not quite a perfect sphere. The diameter of the Moon, which is at right angles to the Earth, is a trifle longer than the diameter in the other direction. Its polar diameter is even shorter. This makes the Moon appear to wobble slightly as it goes around its orbit, and gives us a chance to see a bit more of its surface from time to time. These four bits of libration add about 9% to the half of the Moon that we could otherwise see.

97. How bright is the Moon? The full Moon is the second-brightest object in the heavens. The magnitude of the full Moon is −12.55. The magnitude of the Sun is −26.72. The Sun is, therefore, 400,000 times brighter than the full Moon.

98. How much of the sunlight that strikes it does the Moon reflect? The Moon reflects only 7% of the light of the Sun which strikes its surface. This is a measure of its albedo. The Moon is second to Mercury in being the poorest reflector in the solar system. The Earth's albedo is 39%.

99. If the whole sky were filled with full Moons, how much light would they give? About 1/5 of the light which is given by the Sun.

100. Is either half Moon half as bright as the full Moon? No. The quarter phases of the Moon, during which one half of the lighted side of the Moon can be seen, receive the light of the Sun at a smaller angle than does the full Moon. The quarter phases thus show shadows which drastically reduce the light from the visible surface of the Moon. There are no shadows on the full Moon. Even the tiny shadows cast by pebbles lying on the surface of the Moon contribute to the darkening effect in a Moon less than full. The quarter phases of the Moon, therefore, produce about 1/9 as much light as the full Moon.

101. Does the Moon have an atmosphere? There is no permanent atmosphere, as we know from results obtained from the equipment set up on the surface by the Apollo astronauts. The escape velocity of the Moon is so low—only 1.5 miles per second—that any former atmosphere has long since leaked away. There may be gaseous emissions from beneath the crust which produce localized temporary atmospheric conditions, and indeed this is extremely probable; but a general atmosphere is absent, and we are fully entitled to call the Moon "an airless world".

102. Did the Moon ever have an atmosphere? Probably not, in the sense in which we use the word. When the Moon was first formed, by whatever processes this happened, there must have been in it and around it a turmoil of superheated materials, including a vast quantity of gases. The mass of the Moon is so small, however, that the velocity of escape from its surface is low. Most gases, even when cooled down, have molecules which move at a greater speed than 1½ miles per second. The original gases of the Moon, travelling at vastly greater speeds, must have kept right on going out into space. By the time the Moon had cooled enough to be stable, all the gases had gone. Therefore, it might with truth be said that the Moon never did have an atmosphere.

103. What do we mean by "cusps" on the Moon? Cusps are the pointed ends of the crescent Moon. They always point away from the Sun.

104. Can a star ever be seen between the horns of the Moon? No. The horns or cusps of the Moon are the tips of its visible crescent. This crescent is the edge of the lighted portion of the Moon's surface. The circumference of the crescent continues around to make a circle between the horns which is the outline of the Moon's solid shape. There is no empty space between the horns of the Moon—it is all Moon, and no star farther away from the Earth than the Moon is—and all stars are—could possibly be seen between the horns.

105. The word *crescent* means growing. Why do we call the waning Moon a crescent? This is done simply by usage and custom. *Crescent* does mean growing, but it has come to be applied to the familiar curved and pointed shape, regardless of whether the size of the figure may be increasing or decreasing. The correct term for the increasing phases of the Moon is *waxing* and for the decreasing, *waning,* but these words do not tell us whether the Moon is crescent, quarter or gibbous. Sometimes the terms "crescent before new" and "crescent after new" are used to indicate clearly just which phase of the Moon is intended.

106. Why is it possible to see the dark part of the Moon sometimes inside the bright crescent Moon? Sunlight reflected from the bright surface of the Earth strikes the dark part of the Moon and is reflected from the Moon's surface back to the Earth, where it can sometimes be seen. This appearance, called "earthshine," is most frequently seen when the Moon is only a few days past new and appears as a slender crescent low to the west after sunset. It may also be seen at a corresponding time in the case of the waning crescent just before sunrise. It has poetically been called, in the "Ballad of Sir Patrick Spens," "the new moone wi' the auld moone in hir arme."

107. Why is the brightness of the Moon less after full Moon than at the corresponding phase before full Moon? The surface of the Moon which is lighted after it has passed its full phase contains a larger proportion of smooth area than the part which is lighted before the full phase. The great smooth region of the Oceanus Procellarum takes up a large part of the western side of the Moon and does not reflect as much light as the rougher regions which cover a large portion of the eastern—or new—side of the Moon. For this reason, the last-quarter Moon gives us less light than the first-quarter Moon.

108. What is the difference between the limb and the terminator of the Moon? The limb of the Moon is the outer edge of its visible disk. The terminator is the edge of the lighted portion of the Moon when it is in less than full phase. The terminator cuts across the visible body of the Moon and marks the edge of the lighted, or daytime portion and the beginning of the dark, or night-time portion. The

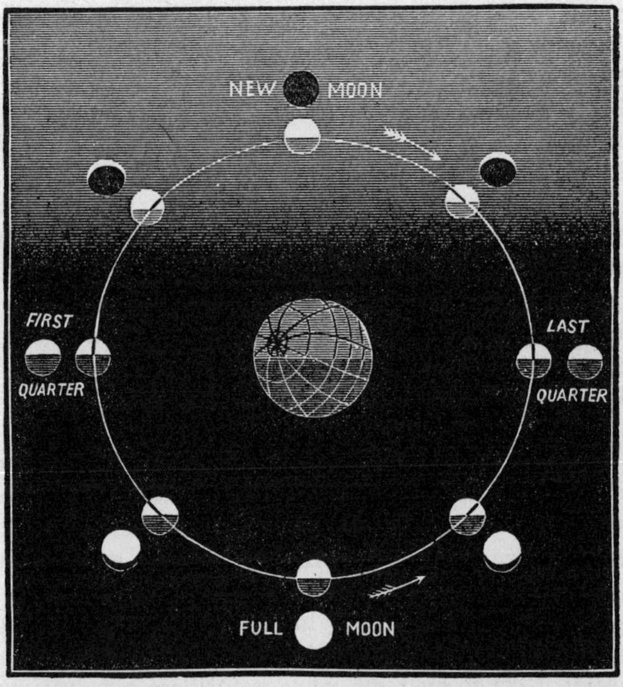

The outer figures show the appearance of the Moon as seen from the Earth at several points in its orbit about the Earth. The inner figures show that the Moon is always illuminated in the same manner.

terms *limb* and *terminator* are applied to any celestial body, although *terminator* can be correctly used only in the cases of bodies which show phases as seen from the Earth.

109. What is the cause of the phases of the Moon? One half of the Moon is always fully lighted by the Sun, but we are not always able to see all of the lighted side. At the time of the astronomical new Moon, we cannot see the Moon at all, because the Moon is then between the Earth and the Sun, and its dark side, away from the Sun, is turned toward the Earth. After a day or two the Moon has moved to the east, out of line between the Earth and the Sun, so that we can see the edge of the lighted half as a slender crescent. This crescent is called, by custom, the new moon.

In a little more than a week, the Moon's eastward motion around the Earth has brought the Moon into a position where we can see half of its lighted side. Since this visible lighted portion is a quarter of the Moon's surface, we call it the first quarter. After another interval of a little more than a week, the Moon has moved on around the Earth until it is opposite the Sun, rising as the Sun sets. We then see all of the lighted side. This is the full Moon.

The third-quarter Moon is caused by the Moon's continuing progress around the Earth, bringing it to the point where it is approaching the Sun, and we can again see half of its lighted side. This time, however, it is the other half from the one that could be seen at first quarter. The next interval of a little more than a week returns the Moon to its original position, almost in line between the Earth and the Sun, and produces another new Moon.

110. Why are the Moon phases seen from south of the equator opposite to the Moon phases as seen from north of the equator? To see the Moon from anywhere on Earth north of the equator, the observer must face generally south. The opposite would be true for an observer situated south of the equator; he would have to face north. Hence, west would be to his left and east to his right. He would thus see the new crescent Moon to his left with its terminator (see question 108) to his right, while a northern observer would see it in exactly the opposite manner. So with all the phases of the Moon. The full Moon, for example, which rises to the left of a northern observer facing south, would rise to the right of a southern observer facing north.

111. What is a gibbous Moon? When the Moon is between half and full, either waxing or waning, it is gibbous. *Gibbous* comes from an Italian word, *giboso,* which means "humpbacked."

112. From London, the Moon is always seen south of the zenith. (See question 342.) Where on the Earth would the Moon be seen overhead? The Moon's most northern declination, or celestial latitude, is 28½°, and its most southern is also 28½°. Therefore, it would be necessary to be somewhere on the Earth's surface between latitude 28½° north and 28½° south in order to be able to see the Moon occasionally overhead.

113. Why does the Moon seem to follow us when we are moving? The great distance between the Earth and all objects in space, including the Moon, makes it necessary for us to traverse great distances on the face of the Earth in order to change our viewpoint of the Moon. We can alter our viewpoint of anything upon the Earth by a very minor change of our position. Hence, as we watch the Moon while we are travelling over the surface of the Earth, the Moon seems to stand still, while local landmarks appear to approach us, pass us and recede into the distance behind us. The illusion thus created is that the Moon is moving in the same direction we are.

114. What caused the craters on the Moon? There is no general agreement as to the cause of the Moon's craters. The first and most obvious answer is that they had the same origin as terrestrial craters —that they were outlets for volcanic explosions from the interior of the Moon. If so, the Moon's craters must all be ancient, for there has been no evidence of violent volcanic activity there within the memory of man. The tremendous size of many of the craters casts doubt upon the volcanic theory of formation. The characteristic angle of slope of the walls of lunar craters is not like that of terrestrial craters, and there does not seem to be any difference, except in size, between the large craters on the Moon and the small ones. On the Moon, there are craters within craters; craters on top of mountains; craters in the mountainous walls of other craters. There are craters whose outlines have been almost obliterated by what appears to be smooth material which seems to have flowed and hardened over the original crater.

A second theory of the formation of lunar craters is that they are the scars left from the impacts of meteor bombardments that must have struck the Moon through the ages. Astronomers have been able to duplicate exactly the form and appearance of lunar craters, even to the central peaks, by dropping hard lumps of plaster into trays filled with finely powdered plaster. Depressions upon the Earth which were unquestionably formed by meteor impact ages ago resemble closely the craters on the Moon. The lunar craters, however, seem to have their openings in a plane corresponding with the surface of the Moon. If they had been made by meteorites, there should be some which were formed at an angle, for meteorites move in all directions, and some should have struck the Moon at angles smaller than 90°. The size of the lunar craters is easier to account for by the theory of meteor impact than by volcanic action, for a massive meteor, traveling at the velocities which these bodies assume in space, would have a tremendous store of energy due to mass and motion. When its motion was stopped, as it would be upon striking the Moon, that energy would be converted instantly into heat, and the meteoroid would explode like an artillery shell, forming a much larger hole than would be made if it had remained solid. It is probable that some combination of volcanic and meteoric action is responsible for lunar craters.

115. If the craters on the Moon were produced by the collision and explosion of meteorites, evidently the Earth must have been struck, too, at about the same time. Why doesn't the Earth have big craters, too? The Earth does have a few great depressions which closely resemble the craters of the Moon: the Meteor Crater in Arizona, near Winslow; the Chubb Crater, in Canada, and several others. Many astronomers believe that these were formed by the impact of meteorites. Most of these Earth craters are of comparatively recent origin. It is possible that there have been, in the past, many more such scars, but as the Earth has an atmosphere whose erosive action can, in time, level the highest mountains, such ancient craters would have been eliminated. The Moon has no atmosphere and no such processes of erosion are possible there. Any craters on the Moon, whatever their origin, would remain practically unchanged through time.

116. How many craters are there on the Moon? On the side of the Moon that is visible from the Earth about 30,000 craters can be seen telescopically. Photographs from space-probes and from the surface itself show vast numbers of extra craters which are too small to be seen from Earth, so that the full total is extremely high.

117. What is the largest crater on the Moon? The crater—or walled plain—Bailly, near the southern limb of the Moon, is the largest crater on the visible side of the Moon. It is about 180 miles in diameter. Larger craters exist on the Moon's far side.

118. What is the smallest crater on the Moon? There are many tiny craters much less than an inch in diameter. Of course, these pits cannot be seen from Earth, but they have been explored by the Apollo astronauts.

119. Which is the deepest crater on the Moon? Newton, a vast crater lying near the south pole of the Moon, is the deepest crater. Its walls rise 29,000 feet from the inside surface of the crater, and the centre of Newton is always in the shadow of these walls.

120. What are the two most conspicuous craters on the Moon? Tycho and Copernicus. Tycho, named for Tycho Brahe, the great Danish astronomer, is in the southern part of the Moon beyond the edge of the Mare Nubium. It is surrounded by craters and a very rough lunar surface. Tycho is very bright and has a tremendous system of rays that extend out from its edge for hundreds of miles in all directions. Tycho's rays are most conspicuous where they cross the darker surface of the Mare Nubium to the north. It is about 54 miles in diameter.

Copernicus is located a little west of the central line of the Moon on the border between the Mare Nubium and the Oceanus Procellarum. It is about the same size as Tycho—54 to 56 miles in diameter. Its ray system, which is not as extensive or as regular as that of Tycho, is most prominent at full moon. Both craters have central peaks. Copernicus was named for Nicolas Copernicus.

121. Who named the places on the Moon? Many men at many times. The first reasonably accurate map of the Moon was made by Hevelius, who named most of the prominent lunar features. In 1651, Giovanni Batista Riccioli published a map of the Moon in which he kept some of Hevelius' names, changed some and added many of his own. He was the first to name the craters for individuals. About 200 of Riccioli's names are still used. In almost every new map of the Moon, new names are added, but no longer are any of the older names changed.

122. Who names the craters now? The International Astronomical Union accepts suggestions for the naming of previously anonymous craters and acts upon these suggestions at one of its sessions. For example, one of the small craters bordering the great walled plain of Clavius, was recently named Porter, after the late Russell W. Porter, whose drawings of the 200″ Hale Reflector at Mount Palomar were of inestimable value in helping to build that instrument. Porter lived in Springfield, Vermont, and his home there, called "Stellafane," is now a gathering place and shrine for astronomers from everywhere.

123. Are there mountains on the Moon? There are many mountains on the Moon. There are great areas of the Moon's surface which are covered with closely packed mountain peaks, and there are several definite ranges of mountains. The ranges and many of the individual peaks have been named; the ranges, for the most part, after mountain ranges of the Earth, although some of the ranges were named as memorials to astronomical personalities. The recognized specific mountain ranges on the Moon include the Alps, the Altai Mountains, the Apennines, the Carpathians, the Caucasus, the Dörfel Mountains, the Haemus Mountains, the Harbinger Mountains, the Jura, the Riphaen Mountains, the Rook Mountains, and the Straight Range.

124. What caused the mountains of the Moon? Many of the great ranges are really the borders of the regular seas; thus the Lunar Apennines make up part of the border of the Mare Imbrium. They are thus rather different in nature from terrestrial mountain ranges such as the Himalayas. However, isolated peaks are very common on the Moon.

125. How high are the mountains on the Moon? The lunar mountains are relatively much higher than the mountains of the Earth, when the smaller diameter of the Moon is taken into account. There are many peaks on the Moon of more than 20,000 feet, and some peaks which rise to nearly 30,000 feet. In the same proportion, the highest peaks on the Earth would reach to about 100,000 feet. The Earth mountains, however, are subject to diminution of height because of weather erosion, but the virtual absence of atmosphere on the Moon does away with any erosion of that sort. There is no wind, rain, frost, snow or any agency which might wear down or soften the outlines of the lunar peaks. The only possible erosion of the mountains of the Moon might be a slight crumbling effect due to the expansion and contraction of the rock under the extremes of temperature that are found there.

126. Are there really seas and oceans on the Moon? No. The names of the "seas" and "oceans" on the Moon were given by men whose instruments were too crude to bring out detail. They assumed, by analogy with the Earth, that the dark areas of the Moon were covered with water. There has never been any trace of water upon the Moon within historical time.

127. Has the surface of the Moon undergone any change since it was first seen through a telescope? There have been no structural changes. It has been suggested that one object, Linné, changed from a crater into a small pit between 1843 and 1866, but the evidence is very slender, and not much faith can be placed in it. However, certain areas do show traces of activity. Transient red glows have been seen in the great crater Alphonsus, near the brilliant crater Aristarchus, and elsewhere. Their exact cause is not yet known, and they seem to leave no after-effects. Generally speaking, the face of the Moon is unchanging; any violent activity there belongs to the remote past. Neither is there any erosion on the Moon now, because of the absence of air and surface water. The Moon is a quiescent world.

128. What are the large dark regions on the Moon? The large dark areas are relatively smooth regions in which only a few scattered craters or other formations mar the general sweep of the lunar surface. In the early days of telescopic exploration, these places were considered to be covered with water and were named seas, bays, swamps, and so on. The largest of these is on the western side of the visible Moon and is the Oceanus Procellarum, the Ocean of Storms. This largest area borders on the north, with very little break, the Mare Imbrium, the Sea of Showers, and to the southeast, the Mare Nubium, the Sea of Clouds. Together, these three large, dark plains are the reason that the third-quarter Moon gives so much less light than the first-quarter Moon. The eastern side of the Moon is much more broken as to surface. There are six small "seas" on this side. The Mare Serenitatis, the Sea of Serenity; the Mare Tranquillitatis, the Sea of Tranquility; the Mare Foecunditatis, the Sea of Fertility; the Mare Crisium, the Sea of Crises, and the Mare Nectaris, the Nectar Sea, are all scattered over the eastern side of the Moon. At the extreme south is the Mare Frigoris, the Sea of Cold, an ill-defined, narrow dark area. Far to the northwest is a small sea, the Mare Humorum, the Sea of Humours. These regions are all relatively flat, with very few craters marking them. They look as though they had been formed by outflows of lava which, ages ago, escaped in some fashion from the then molten interior of the Moon and spread over the landscape. In some cases, craters along the edge of the seas have been half filled so that only a part of their mountain rim is visible. Perhaps a meteoroid of enormous size and mass, crashed into the Moon, broke through the satellite's crust and permitted the confined lava to escape and cool over the original surface.

129. What are the "rays" on the Moon? The rays on the Moon are tapering streaks of light areas which radiate from some of the larger and more conspicuous craters, such as Tycho and Copernicus. It has long been known that they are surface deposits, since they cast no shadows, and are best seen under high illumination; near the time of full moon they dominate the lunar scene. They were first investigated at close range in 1972, when the astronauts of Apollo 16, Young and Duke, landed near a small ray-crater in the area of Descartes. They found that the ray material was bright, and made up largely of breccia. Samples of the ray material were collected, and brought back to Earth for analysis. The origin of the lunar rays is still a matter for debate, but at least it is clear that they are among the youngest of all formations on the surface of the Moon.

130. Do astronomers believe that the surface of the Moon is covered with dust? Not nowadays. There used to be a popular theory that the lunar seas or maria were dust-filled, so that a space-craft landing there would simply sink out of sight, but this has now been finally disproved. Even before the flights of the space-probes, many lunar observers were doubtful about the dust theory, because it did not seem to fit the facts, but decisive proof was forthcoming only when the Russian vehicle Luna 9 actually landed there. Luna 9 came down in the vast Oceanus Procellarum (Ocean of Storms) and remained there without showing any tendency to sink into dust. This result was confirmed by later automatic space-probes, such as the U.S. Surveyors. The deep-dust theory became untenable. Had it been valid, manned landings on the Moon would have been extremely hazardous, and might have been impossible.

When Apollo 11 first came down on the Moon, in July 1969, and Astronauts Armstrong and Aldrin stepped out on to the surface, they found that their boots did not sink very far into the lunar surface. It now seems that the surface layer made up of unconsolidated material—known as the regolith—is at least several feet deep. On later space-voyages, some of the astronauts had difficulty in sinking core-tubes into the regolith to obtain samples; this was particularly the case with Apollo 15, which came down near the Hadley Rill, not far from the foothills of the Apennines.

It is very unlikely that any parts of the Moon are covered with deep dust, and as yet there is no evidence of any meteoritic dust. In all the areas so far examined, the surface is quite strong enough to bear the weight of a space-craft. This applies to the "sea" areas and also to the highlands—as we know from the results of the Apollo 16 flight to Descartes in April 1972.

131. How small an object can be seen on the Moon? With Earth-based telescopes, craters down to about a quarter of a mile in diameter. Photographs from space-probes show features only a few millimetres across.

132. Can we see things better on a full Moon than at other times? No. There are no shadows on a full Moon. Shadows give a three-dimensional effect to the Moon's surface and all the peaks and valleys, the craters and rifts, stand out much more distinctly. The surface of the full Moon looks flat and is much less interesting than the Moon's surface seen at any of the other phases.

133. Does the Moon have any effect on the weather? No.

134. Do the stars twinkle when seen from the Moon? No. The twinkling of the stars is caused by disturbances in the atmosphere of the Earth, which affect the light travelling through that atmosphere.

Upon the Moon, there is no atmosphere to disturb the light from the stars, so that they are steady and unchanging.

135. What causes tides? The gravitational attraction of the Sun and the Moon upon the Earth causes the tides.

136. Do the tides affect the motion of the Earth? Yes. The friction of the tides acts as a brake on the Earth's rotation, slowing it down measureably. The Earth is rotating about one second more slowly every 100,000 years.

137. Would the Earth raise tides on the Moon? It not only would, it does. The tide-raising force of the Earth on the Moon is 82 times the tide-raising force of the Moon on the Earth. If there were oceans and seas on the Moon, their tides would be prodigious. If the Moon were as rigid as the Earth, the distortion of the solid Moon because of the Earth's gravitational pull would be more than 6 feet. The Moon is probably less rigid than the Earth, so the distortion is probably much greater. We know that the diameter of the Moon in a line pointing toward the Earth is longer than the diameter at right angles to this line.

138. What will be the ultimate effect of tidal friction on the Earth? The Earth's rotation will be slowed down, theoretically, until the Earth and the Moon will circle each other slowly, each keeping the same side of itself constantly turned toward the other. The day and the month will then be equal in length and will be as long as 47 of our present days. At that point, the Sun's tide-raising force will upset the balance and reverse the circling of the Earth-Moon system. The day will decrease in length and the Moon will be drawn nearer to the Earth. When the Moon has approached the Earth to a point that brings it within the critical distance stipulated by Roche's Law (see question 352), it will be shattered to pieces by the Earth's tidal forces. The Moon's fragments will, most of them, form a ring around the Earth similar to the ring about the planet Saturn. This is scheduled to take place some thousands of millions of years after the Sun has lost its ability to warm and light the Earth, so don't worry about it; in any case we cannot tell whether the Earth itself will still exist.

139. Are there tides in the solid part of the Earth as well as in its waters? Yes, but these tides are measureable only with the most delicate instruments. The solid Earth is distorted about 4½ inches by the gravitational pull of the Sun and the Moon.

140. Just how does the tide-raising force operate? There is a difference between the force exerted by the Moon at the centre of the Earth and that exerted by the Moon at the surface of the Earth nearest the Moon. The force at the Earth's surface is greater than the force at the Earth's centre because the surface of the Earth directly under the Moon is nearer to the Moon than is the centre of the Earth. Thus, the Moon tries to pull the surface of the Earth away from the Earth's centre, and the centre of the Earth away from the side of the Earth that is farthest from the Moon. This effect results in the lifting of the waters of the Earth not only on the side of the Earth nearest the Moon but also on the side directly opposite. The tidal effect is always greatest along a straight line joining the centres of the Earth and the Moon, and is least along another line at right angles to this one.

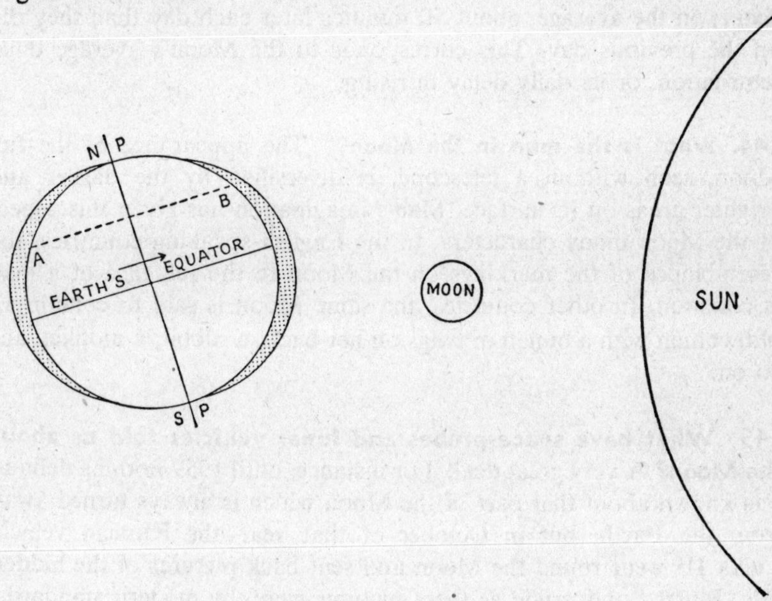

The Sun and Moon combine to produce tides on the Earth.

141. What effect does the Sun have upon the tides? When the Sun and the Moon are nearly in line with each other and with the Earth, as they are at new Moon and at full Moon, the Sun's gravitational force is added to that of the Moon and greater tides are experienced than at other times. At the quarter phases of the Moon, when the Moon is at quadrature, the forces of the Sun and the Moon are at right angles to each other, and moderate tides result. The greater tides—both higher high tides and lower low tides—are called "spring tides," from an Anglo-Saxon word which means a "leap." The moderate tides are called "neap" tides, from another Anglo-Saxon word meaning "scanty."

142. Why doesn't the tide sweep evenly about the Earth? It probably would if the continents didn't get in the way. It is the natural baffle of the land masses, the variations in depth of the waters and all the other irregularities of the Earth's surface that make the tidal effect irregular.

143. Do the tides keep time with the Moon? Yes. Successive tides occur, on the average, about 50 minutes later each day than they did on the previous day. This corresponds to the Moon's average daily retardation, or its daily delay in rising.

144. What is the man in the Moon? The appearance of the full Moon, seen without a telescope, is diversified by the darker and brighter areas on its surface. Man's imagination has given this aspect of the Moon many characters. In the English-speaking countries, the resemblance of the markings on the Moon to the full face of a man is common. In other countries, the same Moon is said to contain an old woman with a bunch of twigs on her back, a rabbit, a monkey and so on.

145. What have space-probes and lunar vehicles told us about the Moon? A very great deal! For instance, until 1959 nothing definite was known about that part of the Moon which is always turned away from the Earth, but in October of that year the Russian vehicle Lunik III went round the Moon and sent back pictures of the hidden side. Blurred and crude as these pictures were, by modern standards, they represented a tremendous technical achievement.

Great developments followed during the 1960s. The Russian Luna probes made soft landings on the Moon; so did the U.S. Surveyors, while the Orbiters circled the Moon and obtained detailed close-range photographs of the entire surface. Then, in 1969, came the first actual landings. On July 21 Neil Armstrong and Edwin Aldrin set foot on the bleak lunar plain known as the Mare Tranquillitatis or Sea of Tranquillity.

More space-flights to the Moon followed. Apollo 12 landed in November 1969, with Astronauts Conrad and Bean—who walked over to an old automatic soft-landing probe, Surveyor 3, and brought parts of it home for analysis. Apollo 13, in 1970, was nearly a disaster; on the outward journey an explosion crippled the spaceship, and it was a relief when the astronauts returned safely, though without landing on the Moon. The next Apollos (14, 15, 16 and 17 in 1971 and 1972) were completely successful. Apollo 17 carried the first official scientist to go to the Moon—Harrison ("Jack") Schmitt, a geologist.

The lunar samples brought back have made us alter all our ideas about the Moon. The rocks are essentially basaltic; there is no evidence of meteoritic material. The Moon's magnetic field is very weak indeed. There is no permanent atmosphere. However, the seismometers taken to the Moon and left there record periodical "moonquakes", which are due to internal disturbances, and are commonest when the Moon is near perigee—though of course they are very mild compared with terrestrial earthquakes.

We have learned more about the Moon in the past five years than we have been able to do in the previous five thousand. Nowadays we can examine lunar rocks in our laboratories, and we can study the photographs and the reports brought back by men who have actually been there. The permanent Lunar Base is no longer a wild dream. By the end of the century it may well exist.

III. MERCURY, VENUS AND MARS

146. Which of the planets is nearest the Sun? Mercury. Its mean distance from the Sun is 35,960,000 miles. The eccentricity of its orbit is large, however, so that at perihelion it is 28,600,000 miles, while at aphelion, it is 43,400,000 miles from the Sun. (See question 982.)

147. How fast does Mercury travel around the Sun? Mercury's speed in its orbit varies from 23 miles per second at aphelion—when it is farthest from the Sun—to 35 miles per second at perihelion—when it is nearest to the Sun.

148. How big is Mercury? Mercury is the smallest of the planets. Its diameter is 2,900 miles. Mercury has 0.06 of the volume of the Earth and 0.15 of the Earth's surface area. Mercury's mass is not known with great accuracy because Mercury has no satellites whose motion about Mercury would reveal its gravitational influence upon them and thus disclose its mass. It has probably about 1/24 of the mass of the Earth. It would take 8,000,000 Mercurys to make one Sun.

149. How long is a day on Mercury? 58½ Earth-days. It used to be thought that Mercury spun round on its axis in exactly the same time that it takes to go round the Sun (just under 88 Earth-days), but we now know that this is wrong.

Before the flight of Mariner 10, in 1974, we knew very little about the surface of Mercury. All that could be seen were darkish and lighter patches. Then came radar experiments, which indicated that there might be craters there; and Mariner 10 fully confirmed this. In February 1974 the probe by-passed Venus, and in March it by-passed Mercury, sending back detailed photographs. Mercury's surface proved to be very like that of the Moon, with craters of all sizes, though there were no really major "maria" comparable with the waterless seas of the Moon.

Mariner 10 made a second by-pass of Mercury in September 1974, and further photographs were obtained. We now have good maps of most of the surface, and we also know that Mercury is very hostile to all forms of life as we know it, so that there seems little chance of establishing a manned base there in the foreseeable future. Few of the craters have yet been named; one exception is a ray-crater which has been named Kuiper, in honour of the late Gerard Kuiper, a Dutch-American astronomer who played a major part in the planning of planetary probes.

Mercury, like Venus, has no known satellite.

150. How long is a year on Mercury? Mercury takes a few minutes less than 88 of our days to complete one trip around the Sun.

151. Is Mercury's motion about the Sun regular and even? No. Mercury follows the laws formulated by Johannes Kepler which govern the motions of all planets, and so it travels much faster when it is nearer the Sun than it does when it is in the part of its orbit which lies farther from the Sun.

152. What effect does this irregularity have upon Mercury's orbit? The orbit of Mercury, like all celestial orbits, is an ellipse. Hence, one axis of its orbit is greater than the other. Because of the large eccentricity of Mercury's orbit, the difference in length between the two axes is unusually great. The longer axis of Mercury's orbit swings slowly around the Sun, shifting slightly to the east, or advancing, with every revolution of the planet. This change takes place in the orbits of all planets, but it is unusually large in the case of Mercury.

153. What is the cause of the advance of the long axis of a planet's orbit? The increase in the speed of revolution that takes place when a planet is in perihelion—nearest the Sun—gives it an impetus that carries it slightly beyond the last turning place in its orbit. Hence, each turn that it makes is a little past the place where it made its turn the last time it revolved about the Sun. In the case of Mercury, the large eccentricity of its orbit makes this advance much greater than that of any of the other planets. Since this could not be accounted for by the laws of either Newton or Kepler, it gave rise to some strange speculation. The unusually large advance in this long axis of Mercury's orbit—the line of apsides—was thought to be the result of the gravitational force of an unknown planet lying between Mercury and the Sun. This hypothetical planet was given the name of Vulcan, but continued search failed to reveal any such planet. The suggestion was also made that a cloud of meteoric material might have such an effect upon the orbit of Mercury, but no trace of any such cloud has ever been found. A part of Einstein's theory of relativity stated that at high orbital speeds, such departure from an even orbital motion might be expected. Mercury's velocity in its orbit agreed very closely with that needed to account for the distortion of its orbit. It is now accepted that the slow revolution of Mercury's perihelion point is natural, caused by its high speed, and does not require any outside gravitational influence.

154. What is the temperature on Mercury? On the daylit side, at maximum, over 700° F. The temperature during night-time is extremely low.

155. Does Mercury have an atmosphere? In the accepted sense of the term, no. Mariner 10 results showed that there are very tenuous gases associated with the planet; but this does not make up an "atmosphere" which is permanent. This is not surprising, in view of the low escape velocity (2.6 miles/second).

156. When can Mercury be seen? In the spring or in the autumn—in April or October—at times of its greatest elongation, Mercury may be seen either in the evening sky shortly after sunset or in the morning before sunrise. At best, Mercury can never be more than 28° to one side or the other of the Sun, so that it is visible only for a short period of time—less than two hours—and very near to the horizon. Because of this, since it must be seen through the densest layer of the Earth's atmosphere, Mercury is often seen to twinkle and has a reddish appearance.

157. Why is Mercury more easily seen in the spring and the autumn than in winter or summer? Mercury always travels close to the ecliptic, the apparent path of the Sun in the sky. In the spring and in the autumn the ecliptic forms the largest angle with the horizon, while in the summer and in the winter the angle of the ecliptic with the horizon is smaller. The larger angle permits Mercury to be farther above the horizon during its greatest elongations which are east in March and April, making it an evening object, and west in September and October, when it is a morning object.

158. Was Mercury ever known by any other name? Yes. Because Mercury, like Venus, is seen with the unaided eye only for a brief time either before sunrise or after sunset, it was thought to be two different objects. Its evening appearance was called Mercury and it was called Apollo when it rose before the Sun. When its true nature became known, *Apollo* was dropped and *Mercury* retained, principally because its rapid motion was associated with swift-flying Mercury, who was the messenger of the gods.

159. How often is there a transit of Mercury? There are 13 transits of Mercury in one hundred years. There were transits of Mercury on May 6, 1957, November 7, 1960, and May 9, 1970. Mercury's orbit is inclined at an angle of slightly more than 7° from the ecliptic, the plane of the Earth's orbit, so that transits can occur only when Mercury reaches inferior conjunction at or near one of its nodes—the points at which the plane of its orbit crosses the plane of the Earth's orbit. The dates on which these crossings occur are May 8 and November 10 each year. Transits take place within 3 days of the May crossing or within 5 days of the November crossing. At all other times, Mercury's passage between the Earth and the Sun seems to us to be either above or below the face of the Sun. A transit of Mercury must be observed through a telescope, as Mercury has only $\frac{1}{150}$ the diameter of the Sun and is too small to be seen otherwise against the Sun's brilliant background.

160. How far is Venus from the Sun? The mean distance of Venus from the Sun is 67,275,000 miles. The orbit of Venus is so nearly circular that the difference between perihelion and aphelion is just about one million miles, as against a difference of more than 12 million miles in the case of Mercury and about 3 million in the case of the Earth.

161. How big is Venus? The diameter of Venus is 7,700 miles, just about 200 miles smaller than the diameter of the Earth. Venus is very nearly a twin to the Earth in other respects, for its mass is 82% of the mass of the Earth; its density is 89% of the Earth's density and its surface gravity is 86% of the surface gravity of the Earth.

162. How long is a day on Venus? Venus takes 243 days to make one rotation on its axis; this means that in theory, a "day" there is longer than a "year" (since Venus makes one revolution round the Sun in 224¾ days). Moreover, Venus spins from east to west, unlike the Earth or most of the other planets. To an observer on Venus, the Sun would appear to rise in the west and set in the east 118 Earth-days later —though, in fact, the dense atmosphere would hide the Sun permanently.

To make things even stranger, Mariner 10, the 1974 probe which by-passed Venus and sent back the first close-range pictures, confirmed that the upper atmosphere has a rotation period of only 4 days. The structure of the atmosphere must therefore be most peculiar.

163. How fast does Venus travel around the Sun? Venus revolves at an average speed of 22 miles per second. The orbit of Venus is so nearly circular that there is very little variation in its speed. Venus revolves about the Sun in 225 of our days.

164. How bright is Venus? Venus can be as bright as magnitude —4.4—twelve times as bright as Sirius, the brightest star. This peak brilliance occurs in those years when Venus is at perihelion near the end of December. At other times, it is still far brighter than any of the stars and is the third brightest object in the sky, exceeded only by the Sun and the Moon.

165. Does Venus have an atmosphere? Yes. The atmosphere of Venus is so dense that it hides the actual surface of the planet completely and permanently. During the 1930s the upper layers of the atmosphere were analysed spectroscopically, and were found to consist largely of carbon dioxide gas; there was no detectable free oxygen. Much later, in the 1950s, some water vapour was found. However, reliable information was not obtained until 1967, when the American vehicle Mariner V by-passed Venus, and a Russian vehicle actually landed there. It now appears that carbon dioxide is the main constituent of the atmosphere, and accounts for well over 95 per cent. of the total. Moreover, the atmosphere is much denser than might have been expected, and the Russians estimate its density, at the surface of Venus, as between 90 and 120 times that of the Earth's air at sea-level.

This will produce some surprising effects. The Sun and other celestial bodies could never be seen by an observer on Venus, and the whole planet would be an excessively gloomy place. It is also probable that light-rays would be bent sharply by the dense atmosphere, so that the observer's view of the surface would be grossly distorted. Altogether Venus seems to be a most unfriendly world, and there seems virtually no chance that any life exists there.

166. What can be seen on Venus with an ordinary telescope? Very little, apart from the characteristic phase. All we can make out are vague bright and duskier areas. It was only in 1974, with Mariner 10, that we were able to study the structure of Venus' upper clouds; the photographs showed the patterns very clearly indeed. However, nobody has yet seen the true surface of the planet.

167. How near to the Earth does Venus come? Nearer than any other of the planets. At inferior conjunction, when Venus is most directly between the Earth and the Sun, it is less than 25 million miles from the Earth. At superior conjunction, however, Venus can be as far as 160 million miles from the Earth.

168. Does Venus have phases? Yes. It was Galileo's discovery of the phases of Venus which helped to convince him of the correctness of the Copernican theory. When Venus is beyond the Sun as seen from the Earth, it is in its full phase. Its position then makes it very difficult to see, and the fact that it is then so far away from us makes its apparent diameter less than one sixth of what it is when Venus is nearest to us. When Venus is most favourably situated for seeing, it is always in crescent phase. The tremendous difference in apparent size and its position far to one side of the Sun, however, make it far brighter at such times than when it is full.

169. How often is there a transit of Venus? Transits of Venus occur in pairs, with an interval of 8 years between the transits of each pair, and with an interval of 105 to 122 years between the pairs. For example, a transit of Venus occurred in 1761 and another in 1769.

The next pair was in 1874 and 1882, and there will be another pair in 2004 and 2012.

Venus must be at or very near one of its nodes at such a time, and since the Sun too must lie in the same direction at the same time, the transits can occur only on June 7 or December 8, because those are the two dates when the Sun is in line with one or the other of the nodes of Venus' orbit. The transits of 1874 and 1882 took place in December. Those of 1761 and 1769 were in June, and those of 2004 and 2012 will also be in June.

170. Are there any theories as to what the surface of Venus may be like? Yes. The Russian Venera probes, which achieved soft landings and transmitted for some time, indicated that the surface temperature is of the order of 900 degrees Fahrenheit. This certainly seems to be approximately correct, and so Venus must be a fiercely hot, barren desert, with no trace of water or life. Whether men will ever be able to land there now seems questionable. In its way, Venus is just as hostile as Mercury, and considerably more so than the Moon.

171. Was Venus ever known by any other name? Yes. Because Venus is visible only for a brief interval either just before sunrise or just after sunset, astronomers of ancient times believed that Venus was two different objects. They called it Hesperus when it appeared in the evening sky and Phosphorus when it rose shortly before the Sun. When later information revealed that there was only one object, these names were dropped and the name of Venus was given to the planet.

172. How big is Mars? The diameter of Mars is 4,213 miles, a little more than half the diameter of the Earth. Mars, then, has 0.291 of the Earth's surface and 0.153 of its volume. The mass of Mars is 0.1078 of the mass of the Earth and $1/3{,}093{,}500$ of the mass of the Sun.

173. How far is Mars from the Earth? That depends entirely upon where Mars is in its orbit and in what part of the Earth's orbit our planet is. Mars can be as near to the Earth as slightly more than 34 million miles. The two planets are moving about the Sun in the same direction, but not at the same distance from the Sun nor at the same speed. The mean distance of Mars from the Sun is 141 million miles as compared with the Earth's mean distance from the Sun of 93 million miles, but the orbit of Mars is much more eccentric than is that of the Earth. This means that the difference between perihelion and aphelion in the case of Mars is much greater than with the Earth. With the Earth, that difference is three million miles, while with Mars, it is more than 24 million miles. Mars makes one revolution about the Sun in 687 of our days, against the Earth's year of 365¼ days. Hence Mars and the Earth are in opposition only once in 780 days. If an opposition occurs when Mars is at aphelion, with Mars being farthest from the Sun, it is also farther from the Earth, and its distance is about 62 million miles. When an opposition occurs with Mars at perihelion, the opposition is regarded as favourable, and the distance between the two planets could be reduced to just over 34 million miles.

174. What is the temperature on the surface of Mars? The Martian temperature ranges from well over 60° F. at noon on the Martian equator in midsummer to about —90° F. at the same place at night. The polar temperatures are considerably lower.

175. How fast does Mars travel about the Sun? Mars travels at 15 miles per second, the average between its perihelion velocity of 18 miles per second and its aphelion velocity of 14 miles per second. Mars takes 687 of our days to complete one revolution about the Sun.

176. How long is a day on Mars? A day on Mars is just a little longer than a day on the Earth. Mars rotates on its axis in 24 hours, 37 minutes, 22.58 seconds.

177. Does Mars have an atmosphere? Yes, but a very thin one. It used to be thought that the atmosphere was made up chiefly of nitrogen, and that the ground density was about the same as that of the Earth's air at 53,000 feet above sea-level. However, the Mariner probes of 1965, 1969 and 1971–2 have shown that the ground pressure is no more than 9 millibars anywhere on Mars, and that the main constituent is carbon dioxide. There is very little free oxygen or water-vapour.

178. Is there any water on Mars? Certainly no liquid water at all, at least on the surface; we cannot yet rule out the possibility of underground water supplies, though most astronomers regard this as rather improbable. The white polar caps seem to be made up chiefly of solid carbon dioxide, but there may be a certain amount of H_2O frost as well; when the cap is shrinking, a measurable quantity of water vapour is released into the atmosphere in the polar zone. However, we can certainly discount the possibility of seas or even major lakes on Mars at the present time, though they may once have existed there.

179. Does Mars have phases like the Moon or Venus? Not as seen from the Earth. When Mars is in quadrature, it may appear to be slightly gibbous. If it could be seen from Jupiter, or from any planet farther from the Sun than Mars is, Mars would show phases.

180. What is the surface of Mars like, and what are the famous Martian canals? It used to be thought that Mars must have a rather flat landscape, with no mountains of any real height. The canals were first described by Giovanni V. Schiaparelli in 1877; he saw them as straight, artificial-looking lines crossing the so-called Martian deserts. From 1895 to 1916 Percival Lowell studied the canals from his observatory at Flagstaff in Arizona, and regarded them as artificial. However, it became clear during the first part of our own century that conditions on Mars were unsuitable for advanced life.

The first successful Mars probe was Mariner IV, which by-passed the planet in 1965 and sent back some pictures which were remarkable by any standards. Instead of being flat, Mars proved to have a crater-scarred landscape, superficially not unlike that of the Moon. Astronomers in general had not expected anything of the kind. Mariner IV also proved that the atmosphere was much thinner than had been expected, and indicated that the polar caps were likely to be made up chiefly of carbon dioxide.

Next, in 1969, came two more probes, Mariners VI and VII, both of which were extremely successful, and sent back more detailed photographs. However, the most spectacular results of all were those of Mariner IX, which was launched in May 1971 and arrived in the vicinity of Mars in the following November. Instead of by-passing the planet and continuing round the Sun, it was put into a closed path around Mars, so becoming an extra satellite. During November and part of December little surface photography could be attempted, because Mars was experiencing one of its not infrequent dust-storms, and the surface features were blotted out. Then, however, the dust cleared away, and Mariner IX could begin its photography.

This time the pictures were really detailed, and proved to be puzzling. Huge volcanic craters were shown; there could be no doubt of their origin, and some of them were remarkably similar in form to volcanoes of Hawaiian type. Such was the Nix Olympica, which is visible from Earth as a tiny patch; it is in fact a lofty mountain, with a 300-mile base, topped by a magnificent volcanic crater. Many other similar volcanic structures were found, and there were also some fascinating systems of canyons, each member of which would outrank the Grand Canyon of the Colorado. There were, too, some features which looked superficially like riverbed systems, even though we may be confident that there is no liquid water on the Martian surface today.

Astronomers are still uncertain as to the past history of the Martian surface, except that vulcanism has played the major rôle in shaping the formations. Meanwhile, there is the intriguing possibility that volcanic activity still continues there. Some of the Mariner results indicate clouds over the tops of some of the major calderas; these could be condensation clouds, but might also be volcanic ejecta. If this is the case, Mars will be a geologist's paradise. In any case it has proved to be far more interesting than appeared likely before the journey of Mariner IX.

The Russians, too, have paid attention to Mars. In 1971 they launched two probes which, like Mariner IX, went into closed paths round the planet. The second of the Soviet probes dropped a "landing craft" on to the surface, which came down safely, but which transmitted for only twenty seconds before signals from it ceased permanently. Further Russian probes also failed.

As for the canals, it seems that they are due either to chains of roughly-aligned craters or to lofty ridges. They are not pure illusions, as has been suggested in the past, but they are certainly natural features. Intelligent life on Mars cannot exist.

181. What does Mars look like through a telescope? With the type of telescope best suited for the observation of Mars, the planet appears as a tiny reddish disk. This disk seems to be boiling, seething and turbulent because of the motion and disturbance in the atmosphere of the Earth, through which the observer must look. For very brief intervals, widely separated, this turbulence will cease as the varying currents in the atmosphere quiet down, and the veil will be withdrawn from the surface of Mars so that some of the detail may be seen. These brief periods of clearing are called "good seeing" and may last, sometimes, for as long as ten seconds. The obscuring waves of disturbance will cease and Mars will appear as a clean-cut disk, generally a dull brick red, with distinct greenish markings and a startling white cap at one end. Other vague markings, just at the lower edge of visibility, will come and go, now standing out, now fading into the general background. Then, just as the observer begins to believe that he can see definite markings, the boiling starts again and the mysterious planet fades back into confusion. It is during these brief, tantalizing seconds that the specialists, who have been glued to the eyepieces of the telescopes for hours, perhaps, see and note down these elusive features.

182. What are the dark areas on Mars? Nobody is yet sure. It used to be thought that the dark patches, which sometimes look greenish but are more usually seen as grey, must be due to low-type vegetation, no more advanced than our lichens or mosses; but the new information sent back by the Mariner space-probes makes this seem rather less likely. Also, it has been found by several methods of investigations that the dark areas are not all low-lying, as used to be thought, so that they cannot be the beds of long-dried seas. For instance the Syrtis Major, the most prominent dark patch on Mars, is an elevated plateau. Most astronomers now incline to the view that the dark regions are due either to shifting dust-layers or else to a real difference of the colour of the surface. However, there is no general agreement, and we cannot yet rule out the chance that Mars may be able to support primitive organic matter. It is also notable that the dark areas seem to show seasonal changes which are bound up with the growth and shrinkage of the polar caps.

183. Does Mars have seasons? Yes. The axis of Mars is inclined at an angle of 23°.58 from the plane of its orbit—just a little more than the inclination of the Earth. Mars has seasons which resemble those of the Earth except that they are longer, because the Martian year is longer. Mars is also like the Earth in that it is at perihelion during its southern summer, which makes that season shorter and warmer than the northern summer.

184. Does Mars have any satellites? Two. They were discovered by Asaph Hall in 1877. Hall was an American astronomer then working at the Naval Observatory in Washington, D.C. The inner satellite, Phobos, is an extraordinary body. It moves round Mars at a distance of only about 4,000 miles from the surface, and completes one revolution in 7 hours 39 minutes—so that it goes round the planet three times every Martian day. It is the only natural satellite in the Solar System to have a revolution period shorter than the rotation period of its primary. Close-range photographs of it were obtained from the Mariner IX probe in 1971; Phobos is irregular in shape—about 16 miles long, 14 broad—and is pitted with craters. Deimos, moving at 12,000 miles from Mars and with a period of 30 hours, was also photographed from Mariner IX; it is rather smaller than Phobos, and it also is irregular, with a crater-pitted surface.

Phobos and Deimos are quite different in nature from our massive Moon. Their small size and irregular shape makes it seem likely that they are captured asteroids rather than genuine Martian satellites. Neither will be of much use as a source of illumination during the Martian night!

185. How long does it take us to see all of Mars's surface from the Earth? Because of the closely similar period of rotation of Mars and the Earth, there is only a very small difference in the region which Mars presents to us night after night. For that reason, it would take one observer who stayed in one place 40 days to see all of Mars. A number of observers, strategically placed about the Earth, however, could see all of Mars in 24 hours.

186. Why does Mars look red? Mars looks red because the material of which its surface is made reflects red light better than it does light of any other colour. Most of the surface of Mars reflects light very much in the same way that sandstone or very dark rock does.

187. Does Mars always look the same to us? No. When there is a favourable opposition, Mars appears to be about three times as bright as Sirius, the brightest star. It has an apparent diameter in a telescope at such times of about 23″ of arc. At its least favourable opposition, it is about as bright as Canopus, and its telescopic diameter is reduced to 3.5″, a difference of about 1 in 7 parts. At other times Mars may be invisible because it is in conjunction, or reduced to relative insignificance by distance.

188. When can we see Mars best? If there is an opposition with Mars at the time when Mars is at perihelion, it will be nearer to the Earth than at any other time. Mars reaches the perihelion point of its orbit during our August, so that if there is an opposition in August, Mars could be as near as about 34 million miles. A favourable opposition takes place every 15 or 17 years. There is no record of there ever having been a perfect opposition of Mars, that is, one that occurred at the exact time when Mars reached its perihelion point. There were favourable oppositions in September 1956 and in August 1971; on each occasion Mars was in the southern hemisphere of the sky, and very low down as seen from Europe.

When Mars is near opposition, details on it may be seen with very modest telescopes, but it is clear that our main information today is drawn from space-probes. At present the Americans are planning their Viking programme, which will result in an automatic probe being soft-landed on Mars in 1976; then we should find out, once and for all, whether there is any life there. The Russians also have plans for exploring the planet. Hostile though it may be, Mars is much less unfriendly than Venus—and we cannot exclude the possibility of underground water supplies.

It has been suggested that the first men may go there before the end of the century. We must wait and see; but if all goes well, there is every expectation that before A.D. 2070 there will be extensive, permanent colonies on Mars.

IV. JUPITER, SATURN, URANUS, NEPTUNE AND PLUTO

189. Which is the largest of the planets? Jupiter, with an equatorial diameter of 89,329 miles, more than 11 times that of the Earth, is the largest of the planets. Jupiter's surface area is 120 times that of the Earth and its volume is 1,318 times the Earth's volume. The mass of Jupiter is larger than the masses of all the other planets put together. It is 318 times the mass of the Earth and $1/_{1047}$ the mass of the Sun. Jupiter is considerably flattened at its poles because of its extremely rapid rotation. Its oblateness is large enough to cause a difference of 5,475 miles between its equatorial diameter and its polar diameter, which measures 83,854 miles.

190. How far is Jupiter from the Sun? Jupiter's mean distance from the Sun is 483,900,000 miles. Its orbital eccentricity, however, is fairly small, only $1/_{20}$; but even at that, Jupiter can be as far from the Sun as 599,623,980 miles, or as near to the Sun as 366,609,000 miles.

191. How fast does Jupiter travel around the Sun? Jupiter travels at an average rate of 8 miles per second.

192. How long does it take Jupiter to go once around the Sun? Nearly 12 of our years. The precise figure is 11.80 years.

193. How long is a day on Jupiter? Jupiter's day is the shortest of any of the planets' days. Jupiter rotates upon its axis in 9 hours and 55 minutes. The surface of the planet that we see is covered with dense clouds and these do not rotate evenly, as a solid surface would. For that reason the rotation of Jupiter appears to be more rapid at its equator than near its poles.

194. How cold is Jupiter? The surface temperature of Jupiter is about —150° F. This low temperature is the reason that some authorities suspected a deep layer of frozen menthane and ammonia is below the atmosphere, which is probably hydrogen.

195. How bright is Jupiter? When Jupiter is in opposition and most favourably situated for observation from the Earth, it is about 3 times as bright as Sirius, the brightest star. At its faintest, it is slightly fainter then Sirius. Among the planets only Venus and Mars can be brighter than Jupiter at the times of its greatest brilliance.

196. Has Jupiter an atmosphere? Yes, in fact, most of the volume of the visible planet is atmosphere. Rupert Wildt, a modern astronomer and an authority on the planets, has constructed theoretical models of both Jupiter and Saturn. His model of Jupiter shows an atmosphere of hydrogen, methane and ammonia which is about 8,000 miles deep. Beneath this, there is a layer of ice about 10,000 miles deep over the central core of Jupiter. These dimensions leave about 38,000 miles for the diameter of Jupiter itself. An alternative theory holds that Jupiter is made up mainly of hydrogen, so compressed near the centre that it starts to behave like a metal.

197. When can we see Jupiter? Jupiter is visible at some time each clear night for much of the year. The Earth, because of its much more rapid motion about the Sun, overtakes and passes Jupiter every 399 days. Therefore Jupiter comes to opposition about five weeks later each year than it did the previous year. At such times it rises as the Sun sets and is in the sky almost all night. This is the best time to see Jupiter.

198. Does Jupiter have any satellites? Jupiter has 12 satellites, more than any of the other planets. Four of these satellites are very bright and easy to see, and may be watched through ordinary field glasses. They were discovered by Galileo in 1610. Three of these four are larger than the Moon; and two of them, Ganymede and Callisto, fourth and fifth in order of distance from Jupiter, are each larger than Mercury. They move very rapidly around Jupiter in periods ranging from less than two days to a little more than two weeks. Their changes in position from night to night are extremely interesting to watch.

There is one tiny satellite closer to Jupiter than these four large satellites, and seven small satellites travelling in orbits far outside the orbits of the larger satellites. None of the 8 smaller satellites have names, but are designated by Roman numerals in order of their discovery. The four large ones, in order of their distance from Jupiter, are Io, Europa, Ganymede and Callisto.

199. Does Jupiter have any visible surface features? The atmospheric envelope of Jupiter is always streaked with belts which lie nearly parallel to Jupiter's equator. These bands are due to dense clouds, and their even, parallel distribution is the result of the rapid rotation of Jupiter. There are also irregularities in the clouds from time to time, in the form of large oval spots of various colours. One of the most famous of these is the Great Red Spot, which was first seen in 1831 and which in 1878 developed to become a deep red oval area about 30,000 miles long and 8,000 miles wide. Its colour deepened and diminished at various times in the years that followed, but the spot is still there, and was very prominent in the years following 1964. Excellent photographs of it were obtained from the probe Pioneer 10, which by-passed Jupiter in December 1973. It was still prominent in 1974.

200. What are Jupiter's satellites made of? Two of the four larger satellites of Jupiter are made of rock and are probably very much like our Moon. These two are Io and Europa. Their density is very much like that of the Moon. Callisto has a density which is very low —the satellite would float in water, and it may be made of ice. The density of Ganymede indicates that it is probably a combination of ice and rock. Nothing definite is known of the composition of the 8 smaller satellites, but they are considered to be rock.

201. Can we always see all four of Jupiter's large satellites? No. Frequently one or more of them may be either passing beyond the body of the planet as we see it from the Earth, or passing between the Earth and Jupiter. In either case, we cannot see the satellites. The positions of the four large satellites with reference to Jupiter are given in astronomical publications, and it is a most interesting sight to see one or more of them as they disappear behind the great planet, or to watch them emerging after they have been hidden.

202. Do all of Jupiter's satellites revolve in the same direction? No. The four outer satellites are moving around Jupiter in the opposite direction to Jupiter's rotation and counter to the way in which the other 8 satellites are travelling. Most of the satellites of the solar system revolve about their primary planets in the same direction that the planets rotate. One theory to account for this "retrograde" motion is that the outer satellites of Jupiter were once asteroids or minor planets and were "captured" by the strong gravitational field of Jupiter. The belt in which the asteroids travel about the Sun lies between the orbits of Mars and Jupiter, and the gravitational field of Jupiter is known to affect the revolutions of several of the asteroids. (See question 251.)

203. What great astronomical discovery was made through the observation of the satellites of Jupiter? Before 1675, the speed of light was considered to be infinite. Galileo had undertaken some experiments to support a suspicion he held that light had a definite and measureable velocity, but his efforts came to nothing. In 1675, a Danish astronomer, Olaus Roemer, was timing the occultations of the satellites of Jupiter. He found that there was a strange discrepancy between the recurrence of the hiding of the satellites behind the great planet when it was in various parts of its orbit. When Jupiter was in opposition—on the opposite side of the Earth from the Sun—the occultations were later than when Jupiter was near conjunction—on the same side of the Sun as the Earth. The difference was about 16½ minutes. Roemer reasoned that this difference could be accounted for by the time it took light to travel the greater distance—about the diameter of the Earth's orbit—186 million miles. The figure that Roemer arrived at was close to the figure for the velocity of light which more modern methods have found.

204. How big is Saturn? Saturn is second in size of all the planets, exceeded only by Jupiter. Its equatorial diameter is 75,021 miles and its polar diameter is 67,805 miles. This oblateness, or flattening at the poles, is greater than that of any of the other planets and is caused by Saturn's rapid rotation. Although Saturn's diameter is 9 times the diameter of the Earth, its mass is only 95.3 times the mass of the Earth. This means that Saturn has a low density and a small specific gravity. It would float in water—if enough water could be assembled in any place to accommodate it.

205. How far from the Sun is Saturn? The mean distance of Saturn from the Sun is 887,200,000 miles. When Saturn is at aphelion—farthest from the Sun—its distance is 937 million miles, and at perihelion, 839 million miles.

206. How long is Saturn's year? Saturn takes 29½ of our years to make one revolution about the Sun.

207. How long is Saturn's day? Saturn rotates on its axis in 10 hours, 38 minutes at the equator. The polar period is somewhat longer.

208. Does Saturn have an atmosphere? Yes. Saturn's atmosphere is very much like that of Jupiter, with certain changes due to the lower temperature of Saturn. There is much hydrogen and methane in the atmosphere of Saturn. Such ammonia as may be there must be frozen. Rupert Wildt considers that the atmosphere of Saturn must be even deeper than that of Jupiter. Wildt suggests that Saturn has an outer atmosphere of clouds of methane and of ammonia crystals which is about 12,000 miles deep. Beneath that is a layer of ice about 14,500 miles deep with a small planet inside this shell—a planet about 18,500 miles in diameter. Some astronomers have even suggested that both Saturn and Jupiter may lack a solid centre entirely, and be spheres composed completely of gases at low temperatures and with pressures rising at their centres to many millions of times the pressure of the Earth's atmosphere.

209. What is the origin of Saturn's rings? Saturn's rings may possibly be made up of the debris of what was once a tenth satellite of Saturn. Any satellite which comes within a certain distance of its primary—a distance known as Roche's limit—will theoretically be shattered into fragments by the tidal forces raised in its substance by the gravitational power of its primary. Roche's limit is 2.44 times the radius of the primary. The rings of Saturn lie entirely within Roche's limit, while Janus, the first satellite of Saturn beyond the rings, lies at a distance of three times the radius of Saturn and is therefore outside the theoretical danger zone.

A satellite which for some reason swung too close to its primary would, therefore, not be pulled to the primary intact, but would be torn to pieces. Its fragments would immediately assume orbits about the primary, although it is possible that some of them would be drawn to the primary's surface and strike it with considerable impact.

210. How big are Saturn's rings? The diameter of the entire ring system, including, of course, Saturn itself, is 171,000 miles. The outer ring is a little more than 10,000 miles wide. Then comes Cassini's division, a dark band in which there is little or none of the material which composes the rings. Cassini's division is difficult to measure, but it is probably about 3,000 miles wide. Inside Cassini's division is the second ring, which is slightly less than 16,000 miles wide. Inside this is the innermost ring, known as the crêpe ring because its appearance resembles the dress material known as crêpe. The crêpe ring is about 11,500 miles wide. The total width of the rings alone is 41,500 miles and the inner edge of the rings is 7,000 miles above the surface of Saturn. The crêpe ring has fewer particles per unit of space within it than do the other rings, which is why it has the almost transparent appearance of the material for which it is named. The depth of the ring system cannot be more than about 10 miles, which makes it one of the flattest things in nature, relatively speaking. This strange shallowness is the reason that the ring system almost disappears when the motion of Saturn brings that planet at such an angle to the Earth that the rings are presented edge on to our view. At such times, a powerful telescope is needed to show the rings at all.

211. Why is there an open space in Saturn's rings? The space between the components of Saturn's ring system, known as the Cassini division, is caused by the gravitational forces of Saturn's nearer satellites, which set up perturbations in the orbits about Saturn of the particles that make up its rings. A satellite will "sweep out" a region at a distance from the satellite corresponding to a simple fraction of the period of the satellite—the time it takes to make one revolution about the planet. The Cassini division is at a distance from Saturn, corresponding to half the period of Mimas, a third of the period of Enceladus, and a quarter of the period of Tethys. Other less noticeable divisions in the ring system correspond to other fractions of the periods of various satellites, and are, for that reason, kept gravitationally free of the particles that make up the ring system. The Cassini division, a definite dark line dividing the rings a little less than halfway out in the rings from Saturn, is named for Jean Dominique Cassini (1625–1712), who discovered the aperture.

212. Can Saturn's rings be seen through a small telescope at all times? No. The rings are so thin that when Saturn's inclination on its axis presents them to us edge on, they cannot be seen with a small telescope. Twice during Saturn's journey around the Sun the rings almost vanish, to appear again after a brief interval when the motion of Saturn carries them beyond the critical point. The rings are best seen when Saturn's inclination is such that they are presented at the greatest angle to our line of sight—at 15-year intervals opposite to those times when the rings are invisible.

213. How many satellites does Saturn have? Saturn has 10 satellites which have been given the names of Greek deities. They are, in order of distance from Saturn, Janus, Mimas, Enceladus, Tethys, Dione, Rhea, Titan, Hyperion, Iapetus and Phoebe. Titan is one of the largest satellites of the solar system, and has an atmosphere denser than that of Mars. It was the first of Saturn's family to be discovered, by Christian Huygens in 1655. Phoebe, the outermost of Saturn's satellites, is in retrograde motion about Saturn, moving in the opposite direction to the rotation of Saturn. The four outermost satellites of Jupiter are also revolving against the rotation of Jupiter. Like them, Phoebe may be a captured asteroid instead of a normal satellite. Its diameter is less than 200 miles.

214. How big is Uranus? Uranus has a diameter of 29,300 miles, a little less than 4 times the diameter of the Earth. Its oblateness is $1/14$, slightly more than that of Jupiter and less than that of Saturn. The mass of Uranus is 14.68 times the mass of the Earth.

215. How far from the Sun is Uranus? The mean distance of Uranus from the Sun is 1,784,800,000 miles. Uranus has a rather eccentric orbit, however, so that at perihelion it is 167,770,000 miles nearer to the Sun than at aphelion.

216. When and by whom was Uranus discovered? Uranus was discovered in 1781 by Sir William Herschel. Herschel's own account of the discovery declares that he found the planet during a systematic examination of the heavens in which he looked at every star. The visible planetary disk of Uranus was discernible at once. A few nights after he had first seen it, he verified the nature of the object he had seen by its change of position among the stars, though at first he believed it be a comet instead of a planet.

217. Was Herschel the first to see Uranus? No. Flamsteed, the first Astronomer Royal of England, had recorded it in his notes five times, from 1690 on, without realizing the true nature of what he had seen. Another astronomer had records of having seen it at least a dozen times. Altogether, it had probably been seen twenty times before Herschel found it and announced its true nature.

218. How fast does Uranus travel around the Sun? Uranus revolves at 4.22 miles per second.

219. How long is the year on Uranus? Uranus takes 84 years and 4 days to complete one revolution about the Sun.

220. How long is the day on Uranus? If by *day* is meant the alternation of light and dark, Uranus presents a complicated situation. Uranus is inclined at an angle of 98° from the plane of its orbit, so that it alternately presents first one pole and then the other to the Sun as it revolves about the Sun. For part of its year, one pole is lighted by the Sun and the planet's rotation has no effect upon the light and darkness of the planet's surface. For another season, the light of the Sun will be moving, as seen from Uranus, from one pole to the other, and the rotation of the planet will produce day and night until the original situation is reversed and the Sun is standing almost over the other pole. After this period, the second phase will occur again in reverse, with the Sun apparently moving in the opposite direction, from pole to pole, and Uranus having day and night as measured by light and dark, as before. The yearly cycle is complete when Uranus turns again to the Sun the same pole which faced the Sun at the beginning of its year. Uranus rotates upon its axis in 10 hours, 48 minutes.

221. Was Uranus always known as Uranus? No. Herschel himself called it *Georgium Sidus,* "George's Star," after George III of England. Other astronomers called it "Herschel." The name "Uranus," suggested by Johann Elert Bode as being more in keeping with the mythological names of the other planets, was the one that was finally adopted.

222. Does Uranus have seasons? Because of its inclination of 98° to the plane of its orbit, Uranus presents first one of its poles and then the other to the Sun, slowly circling on its side. It takes Uranus about 42 years to complete one half-circle during which time the planet will have presented one pole and all of one side to the Sun. The other pole and the other side would be toward the Sun during the other 42 years. Because of the rapid rotation of Uranus, however, the whole surface of the planet would be exposed to the Sun at some time, except during those seasons when the poles of Uranus were most directly pointed at the Sun.

223. Does Uranus have an atmosphere? Uranus, like Jupiter and Saturn, has a deep atmosphere made up mainly of hydrogen plus certain other elements, such as methane, which are able to retain their gaseous properties at a temperature of 305° F. below zero. Uranus also resembles the two giant planets in its internal composition.

224. Does Uranus have satellites? Uranus has 5 known satellites. Herschel discovered two of them; Lassell found two more in 1851 and Gerard Kuiper found the fifth in 1948. The sizes of these satellites are not known with sufficient exactness for figures to be given. They all move about Uranus in the plane of the planet's equator, and because of the strange position of Uranus, the satellite system of Uranus presents itself edge on to our view every 42 years. Although all of the satellites of Uranus are revolving around Uranus in the direction of the planet's rotation, the angle of Uranus' inclination to the plane of its orbit makes them technically retrograde as compared with the absolute directions of most of the other satellites of the solar system.

The satellites of Uranus have been named from Shakespeare and

Alexander Pope. "Titania" and "Oberon" are from *A Midsummer Night's Dream;* "Ariel" from *The Tempest;* "Umbriel" from *The Rape of the Lock* and "Miranda" from *The Tempest.*

225. What does Uranus look like? Uranus is too faint to be well seen without a telescope. Through a small telescope, Uranus presents a vague greenish disk. In a large telescope, some faint cloud bands may be seen, but these are much less distinct than those visible on either Jupiter or Saturn.

226. Where is Uranus now (1975)? Uranus takes about 84 years to make one circuit of the Sun. Hence, it spends about 7 years before each of the 12 zodiacal constellations. At this time, it is before the stars of Virgo, the Virgin, where it will remain for some years.

227. Who discovered Neptune? Neptune was the first planet to be discovered mathematically. Perturbations in the orbit of Uranus caused astronomers to believe that some outside force was influencing the regular progress of the planet. By the year 1800, Uranus had deviated so far from its calculated path that some reason had to be found to account for its wanderings. Several theories were advanced: a collision with some other body in space, possibly a comet; a cloud of cosmic dust or gas was holding it back from its scheduled journey; or a satellite which had not been discovered was tugging at it, gravitationally. There were those who held that the laws of Newton and Kepler, under which the path of Uranus had been calculated, were at fault.

Two young students set to work on the problem independently, without any knowledge of the efforts of the other. One was an Englishman, John Couch Adams; the other, Urbain J. J. Leverrier, was a Frenchman. Adams worked on the assumption that there was an eighth planet beyond Uranus which was the cause of the perturbations in the orbit of Uranus. He worked for 4 years on his problem and at last calculated the position in which the theoretical planet should be found. He sent his findings to the Astronomer Royal of England, George Airy. Airy held to the theory that the irregularities of Uranus could be accounted for by a failure in the law of gravitation, and gave Adams' papers only superficial examination. Leverrier, in France, was of the same opinion as Adams about the cause of

Uranus' wanderings. After much of the same sort of research, he, too, calculated the possible position of a planet that could be held responsible. Leverrier sent his work to Galle, a great German astronomer. Galle had recently acquired a new and excellent chart of the part of the sky in which Leverrier had predicted that the new planet might be found. He also had good weather and the time to make careful observations. On the very night that he received Leverrier's figures, he searched and found Neptune. This was on September 23, 1846.

Meanwhile, English astronomers had belatedly become aware of the importance of Adams' papers. They had been searching, too, but in a much more desultory and laborious fashion. Their work was finally successful and they too found Neptune. A controversy raged for some time over the question of who should receive credit for the discovery. Leverrier's work, while it had been completed at a later date than had that of Adams, had resulted in the actual discovery of Neptune. Both astronomers are now given equal credit for the discovery.

228. How big is Neptune? Neptune has a diameter of 31,200 miles, or about 4 times the diameter of the Earth. Its mass is 17.3 times that of the Earth. The oblateness of Neptune is only $1/45$, smaller than that of any of the other giant planets.

229. How far is Neptune from the Sun? The mean distance of Neptune from the Sun is 2,796,700,000 miles. The orbit of Neptune has a very small eccentricity, .009, but even this small amount represents a difference of 48 million miles between perihelion and aphelion.

230. How fast does Neptune travel around the Sun? The orbital velocity of Neptune is 3⅓ miles per second.

231. How long are the day and the year on Neptune? Like the other giant planets, Neptune rotates at high speed. Its day is about 14 hours. Neptune takes 164.79 of our years to make one revolution about the Sun.

232. Does Neptune have seasons? Yes, in that it is inclined 29.6° from the plane of its orbit. At Neptune's distance of almost 3 thousand million miles from the Sun, the difference between summer and winter is purely academic.

233. Does Neptune have an atmosphere? Yes. Neptune is so far away that no markings have been seen upon the visible surface of the planet, but spectroscopic analysis of the light it reflects indicates that its atmosphere is very much like that of the other major planets, with allowance made for Neptune's lower temperature at its greater distance from the Sun. It is probably very similar to Jupiter, Saturn and Uranus. The surface temperature on Neptune has been estimated at about —325° F.

234. Does Neptune have satellites? Neptune has two satellites. One of them may be the largest satellite in the solar system, although its exact size has not yet been determined. This is Triton, discovered by Lassell in 1846, shortly after the discovery of Neptune itself. Triton's diameter has been estimated at over 3,000 miles to 5,800 miles. The second satellite, Nereid, was found in 1949 by Gerard Kuiper. Its diameter is not known. The motion of Triton is retrograde—opposite to the direction of Neptune's rotation. The motion of Nereid is direct. Triton is about 219,000 miles from Neptune; Nereid has a very eccentric orbit round Neptune.

235. Where is Neptune now (1975)? Neptune is now to be seen in Ophiuchus (the Serpent bearer) which is not one of the official Zodiacal constellations, but does enter the Zodiac between Scorpio and Sagittarius.

236. What does Neptune look like? Neptune presents a faint, slightly bluish disk to the telescope observer. It is about 1½ magnitudes fainter than the limit of naked-eye visibility. Telescopes do not show any definite markings on the disk of Neptune.

237. How far is Pluto from the Sun? Pluto is almost 40 times as far from the Sun as the Earth is. Its mean distance from the Sun is 3,675,000,000 miles. Its eccentricity, however, is the largest of that of any of the planets, so at perihelion it is 2,761,000,000 miles from the Sun, but at aphelion, it moves out to a distance of 4,589,000,000 miles. Its perihelion distance is 49 million miles less than is Neptune's aphelion distance, so that, at certain times. Pluto's distance from the sun is less than that of Neptune.

238. How fast does Pluto travel around the Sun? Pluto travels at a speed of 2.94 miles per second.

239. How long is the day on Pluto? The period of Pluto's rotation is very difficult to determine, as are most of the items of physical information about the planet. In 1956, an estimate of 6½ Earth days was given for Pluto's rotation period.

240. How was Pluto discovered? Pluto was discovered mathematically, as was Neptune. There were irregularities in the motion of Uranus and Neptune which early led to the belief that another planet in an orbit outside the orbit of Neptune was affecting their even progress around the Sun. Percival Lowell and William H. Pickering were the first to attempt to calculate the position of this undiscovered planet, but many years passed before it was actually found. Research toward locating the planet was carried on sporadically for many years at the Lowell Observatory, Flagstaff, Arizona. Lowell died in 1916, and years later, a young astronomer, Clyde Tombaugh, was assigned to pursue the search. He worked steadily, photographing the suspected part of the sky through a telescope specially built for the purpose. His search was complicated by the fact that the new planet was supposed to lie in a part of the sky that was exceptionally rich in stars, in the constellation of Gemini, The Twins. Literally thousands of photographs were studied before a faint disk was seen to have moved slightly among the stars in plates taken at some time apart. The announcement of the discovery was made in 1930.

241. How big is Pluto? The size of Pluto is one of the many mysteries of astronomy. Pluto was discovered as a result of perturbations in the orbit of Uranus and Neptune. These irregularities called for an object which would have, at least, the mass of the Earth. The best estimates of the size of Pluto, drawn from observations made since its discovery, all seem to indicate that it is much smaller than the Earth. The best available value for the diameter is 3,700 miles. If this is correct, then Pluto is the smallest of the principal planets, apart from Mercury, and could not have produced measurable irregularities in the movements of Uranus or Neptune—so that its discovery was purely fortuitous.

242. How long is the year on Pluto? Pluto takes 247.7 Earth years to complete one revolution about the Sun.

243. Does Pluto have seasons? Even with the largest telescopes, Pluto reveals only a vague disk without definite outlines or any surface markings. There is no knowledge yet of Pluto's inclination to the plane of its orbit.

244. Does Pluto have an atmosphere? There is no evidence of any atmosphere on Pluto. If Pluto is the size that has been estimated—from 4,000 to 10,000 miles in diameter—the presence of any atmosphere is unlikely. If, however, its mass is sufficient to produce the observed perturbations in the orbit of Neptune, that mass might be sufficient to generate enough gravitational force to hold an atmosphere upon Pluto's surface. If this is so, any atmosphere there would be completely and solidly frozen, for at Pluto's distance from the Sun, its surface temperature must be extremely low.

245. Does Pluto have any satellites? No satellites of Pluto have been discovered. There is no gravitational evidence, such as irregularities in Pluto's motion, that would reveal the presence of satellites. If there should be any satellites of Pluto, they would undoubtedly be so faint that even photographs of long exposure through large telescopes would hardly be able to reveal them.

246. What does Pluto look like? Even through the largest telescopes, Pluto presents nothing but a faint, indefinite disk. This disk may not represent the complete diameter of Pluto, but may be a sort of high light covering a region about the centre of the side of the planet which is facing the observer. A telescope of considerable power is needed to show Pluto.

247. Could Pluto collide with Neptune? No. Even though there are times during the passage of the two planets about the Sun when Pluto is 49 million miles nearer to the Sun than is Neptune, the orbit of Pluto is inclined by more than 17° to the plane of the ecliptic. Even though they should reach the corresponding parts of their orbits at the same time, Pluto would never be in the same plane as Neptune. The nearest approach that it is possible for the two planets to make to each other is about 240 million miles.

248. Where is Pluto now (1975)? Pluto is near the boundary of Virgo (the Virgin) and Coma Berenices (Berenice's Hair). It is the only principal planet which can move outside the twelve constellations of the conventional Zodiac.

249. Was Pluto once a satellite of another planet? Possibly. Almost from the time of its discovery, many astronomers voiced their suspicions that Pluto may at one time have been a satellite of the planet Neptune. The size of Pluto did not indicate that its mass would be sufficient to account for the perturbations observed in the orbit of Neptune and blamed on the new planet. The orbit of Pluto had such a great eccentricity that at one point it lay closer to the Sun than did the orbit of Neptune. The first of these factors made Pluto seem rather more satellite-size than planet-size; while the second indicated that it had, at one time, been perhaps associated with Neptune. In 1955, Dr. Gerard Kuiper made public his own belief that Pluto had been once a satellite of Neptune and declared that the rotational period of Pluto—about 6½ days—confirmed this theory. This rotational period was so much slower than that of any of the other outer planets—slower indeed than any of the planets except Mercury and Venus—that Pluto must have spun in this fashion as a satellite. How it became detached from Neptune is not known, but since it is now revolving about the Sun, it is a planet in good standing, whatever its start in life may have been.

250. Are there supposed to be any planets beyond Pluto? The presence of additional planets beyond the orbit of Pluto is speculative. It will take about 100 years before enough of Pluto's orbit has been travelled by that planet to reveal any possible perturbations which could disclose the presence of a disturbing object beyond it. If Pluto turns out to be sufficiently massive to account for the perturbations in the orbit of Neptune, and if its own course is even and true, there may be no further planets. If the contrary condition transpires, there may be more members of the Sun's family, but their discovery is likely to be more a matter of chance than of plan, since they would be so faint that they could be seen only in long-exposure photographs taken through the largest telescopes. Since such photographs would probably be taken for purposes other than the location of a hypothetical trans-Plutonian planet, it would be purely by chance that such a planet would be found.

V. METEORS AND COMETS

251. What are asteroids? Asteroids—sometimes known as planetoids—are minor planets of the Sun. These bodies revolve about the Sun in a belt which lies between the orbits of Mars and Jupiter. The first asteroid was discovered on January 1, 1801, by the Italian astronomer Giuseppe Piazzi (1756–1826). In the next six years, three more were found, but the fifth was not found until 1845. Since that time many thousand have been seen or photographed and precise orbits calculated for about 2,000.

Ceres, the first asteroid discovered, is still the largest that has ever been found. It is about 700 miles across and has a mass which is about $\frac{1}{8,000}$ of the Earth's mass. From this size, the asteroids range down to small, irregular pieces of rock sometimes less than a mile in diameter. Most of them have been found photographically, for an asteroid, because of its swift relative motion, will make a short line across a photographic plate which has been moved to follow the apparent motion of the stars. While most of the asteroids are in the space between Mars and Jupiter, some few stragglers may be found at times coming inside the orbit of Mercury or outside the orbit of Jupiter. No one knows how many asteroids there are, and estimates of their number run from 30,000 to 50,000. Theories of their origin are divided between their being (a) the debris of an ancient planet, or (b) the material of which a planet might have been made, but was not, because of the gravitational influence of Jupiter, with the preponderance of opinion favouring the latter theory.

252. What is the name of the first asteroid to be discovered? Piazzi called the first asteroid Ceres. Ceres was the goddess of the harvests and the special deity of Sicily, where Piazzi was born.

253. How big is Ceres? Ceres measures about 700 miles in diameter. Its mass has been estimated as about $\frac{1}{8,000}$ the mass of the Earth.

254. How rapidly were other asteroids discovered? Pallas was found in 1802, Juno in 1804 and Vesta in 1807. After that, 38 years passed before the next one was found. This was Astraea. The asteroids then began to turn up more rapidly; and now many thousands have been photographed and about 2,000 have had their orbits plotted and determined, which is the equivalent of official acceptance.

255. How many asteroids have been discovered? About 2,000 asteroids have been studied sufficiently so that their orbits are known. Thousands more have been seen on photographic plates taken of the sky at one of the great observatories. The 2,000 or so that have been officially accepted because of the knowledge of their orbits are actually a small fraction of the number that have been observed.

256. Are all the asteroids in orbits between Mars and Jupiter? No. About 85% of them are at a mean distance of about 200 million miles from the Sun. Mars is 141 million miles from the Sun and Jupiter about 480 million miles. A few of the asteroids travel occasionally much nearer to the Sun than does Mars and still fewer paths lie farther from the Sun than is Jupiter.

257. Which of the asteroids has the largest orbit? Hidalgo, whose orbit has a mean diameter of 531 million miles.

258. Which of the asteroids has the smallest orbit? Icarus. The orbit of Icarus is such that the asteroid, at aphelion, is farther from the Sun than is the Earth. At perihelion, however, Icarus passes closer to the Sun than Mercury does.

259. Which is the smallest of the asteroids? Adonis is probably the best known of the small asteroids. It is about half a mile in diameter. There are, undoubtedly, many other asteroids which are smaller, but which have not yet been thoroughly investigated.

260. What are the Trojan asteroids? The Trojans are asteroids whose orbits are obviously influenced by the gravitational force of Jupiter. The periods of the Trojans—the length of time they require to make one complete journey around the Sun—correspond to the period of Jupiter, or to ¾ or ⅔ of Jupiter's period. The positions of the Trojans in their orbits always make an equilateral triangle which has the Sun and Jupiter at the other two corners of the triangle.

261. How are the asteroids named? The discoverer of an asteroid has the privilege of naming it. With the exception of the first few, asteroids must be named according to certain rules which were established as more and more of them were found. Most of the asteroids, which move in orbits normally located somewhere between Mars and Jupiter, must be given names which end in the Latin feminine ending *-a*. Those whose orbits bring them to within a smaller distance from the Sun than one astronomical unit are given Latin masculine names. Asteroids of the Trojan group are named for various heroes from the legend of the Trojan War. Most of the asteroids have feminine names, because most of them are within the normal asteroids limit. For example, "Ricea" is one named for a modern American astronomer, Dr. Hugh S. Rice. When the orbit of an asteroid has been determined the asteroid receives a permanent number in addition to its name. These numbers indicate roughly the order of discovery, and usually follow the names of the asteroids. Hidalgo, for example, is "Hidalgo (944)."

262. Can asteroids be seen without a telescope? One asteroid, Vesta, is technically within naked-eye visibility, but it is an extremely difficult object to see, even if its location is known precisely. Vesta is the brightest, but not the largest of the asteroids.

263. What is a comet? A comet is a lump of frozen gases of various kinds in which are buried a number of small stony or metallic particles.

264. How are comets discovered? Comets are sometimes found on photographs which are taken nightly by patrol cameras at the great observatories. Occasionally they are found by professional astronomers who chance upon a comet while looking at or for something else. Comets are often discovered by amateurs who study the skies through smaller telescopes as a hobby. One amateur, Leslie Peltier, of Delphos, Ohio, has a dozen comets to his credit.

265. Why are comets so named? Ages ago, when comets were not visible until they were large and bright and startling, with, as a rule, long tails, men called them "hairy stars," because a comet's tail bore some superficial resemblance to the customarily long hair of women. The word *comet* is from the Latin word *coma,* which means "hair."

266. What are the different parts of a comet? The nucleus of a comet is the brightest part, and is usually made up of most of the solid fragments in the comet. It is generally visible as a bright spot. The coma is a cloud of gases surrounding and including the nucleus. The tail which fans out from the head of a comet is variously shaped and, indeed, may change from time to time in any one comet. The coma and nucleus together form the head of the comet.

267. How many comets are there? No one knows. More than a thousand comets have been seen and recorded, but many of these have since disappeared and new comets are constantly being discovered.

268. Where do comets come from? The origin of comets is no better known than is the origin of the solar system. A Dutch astronomer, Jan Oort, has presented the theory that the solar system is located in a sort of cometary marshalling yard which contains billions of comets. Occasionally the gravitational force of the Sun, or sometimes of one of the major planets, draws a comet from this celestial storehouse and brings it into an orbit such that it comes close enough to the Sun and Earth to be observed.

269. How do comets travel in the solar system? Comets travel about the Sun in orbits which are either elliptical, parabolic, or hyperbolic. (See question 987.) A comet whose orbit is elliptical will continue to revolve about the Sun and may be seen periodically from the Earth. If the orbit of a comet is a parabola or a hyperbola it will not return but will plunge through the solar system and disappear.

270. How does a comet change as it approaches the Sun? When a comet has come near enough to the Sun to be affected by the Sun's radiation, the gases of which it is composed are vaporized. The comet grows tremendously in size without gaining mass, and is a vapour instead of a solid. The tiny particles which were embedded in the frozen gases are now loose and travelling along as they previously were, but no longer held tightly.

271. Why does a comet's tail usually point away from the Sun? A comet's tail is formed when the Sun's radiation and emitted particles or "solar wind" drive the very tenuous gases and tiny particles that form

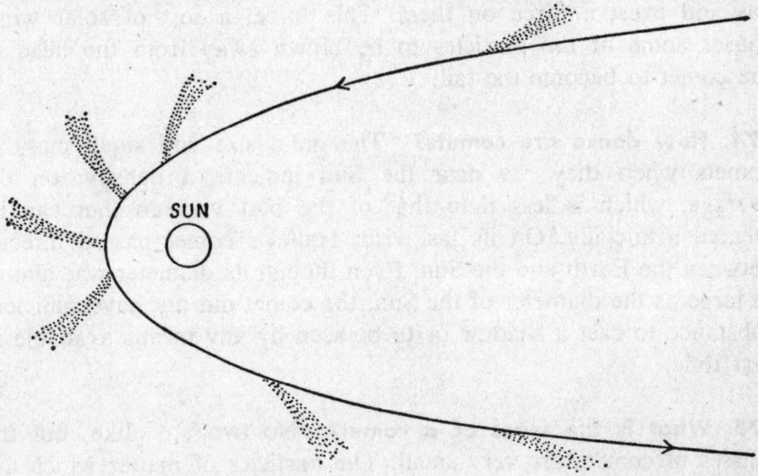

Forces from the Sun drive tenuous gases away from the head of a comet to make the comet's tail. Hence, the tail of a comet generally points away from the Sun, no matter in what direction the comet is travelling.

the head of a comet away from the head. Since the force that produces the tail of a comet comes from the Sun, the tail is driven out from the side of the comet's head which is away from the Sun, and would thus point away from the Sun.

272. How long is the tail of a comet? There have been comets that developed little or no tail as they passed near the Sun. As a rule, however, comets' tails are extremely long. The tail of Halley's comet, at the time of its last appearance, was about 100 million miles long, longer than the distance between the Earth and the Sun. The record appears to be held by the Great Comet of 1843, which had a tail over 300 million miles in length.

273. What produces the tail of a comet? The radiation of the Sun. Out in space, a comet is a cold, frozen body—a snowball containing dust and chunks of rock and metal. As the comet approaches the Sun, it is warmed and the frozen gases vaporize. These vapours are then excited by ultra-violet light from the Sun and begin to glow. The light rays from the Sun and the tiny particles of matter expelled from the Sun at times of solar activity strike the particles of glowing gas and exert a force on them. This force, a sort of solar wind, causes some of the particles to be blown away from the head of the comet to become the tail.

274. How dense are comets? The great size and small mass of comets when they are near the Sun indicate a density, on the average, which is less than that of the best vacuum that can be created artificially. On its last visit, Halley's comet passed directly between the Earth and the Sun. Even though its diameter was almost as large as the diameter of the Sun, the comet did not have sufficient substance to cast a shadow or to be seen by any means available at that time.

275. What is the mass of a comet? No two are alike, but the masses of comets are very small. The particles of matter which are held in the gases of a comet are relatively few and, when a comet is near the Sun, are widely separated. The orbits of many known comets have been carefully studied, for a knowledge of the motion of an object in its orbit will give a clue to its mass. Cometary orbits indicate that their masses are on the order of about one millionth of the mass of the Earth. One astronomer has called comets "bags full of nothing."

276. Is it true that the nucleus of a comet is made of ice? Yes, in a sense. When a comet is far enough from the Sun so that the radiation of the Sun does not affect it, the nucleus of the comet, which is composed of gas and tiny particles of matter, is frozen. This is ice, but it is gas-ice and there is probably none of the molecular combination that we call water, as such, in the nucleus of a comet.

277. How fast do comets travel? The speed at which comets travel varies according to the shape of their orbits. By the laws of Kepler, which explain the velocities of objects travelling in elliptical orbits, we know that a comet is travelling fastest when it is nearest the Sun. A comet will gradually slow down as it leaves the Sun. The deceleration of a comet leaving the Sun is about the same as its acceleration was as it approached the Sun. At the distance of the Earth from the Sun—93 million miles—the speed of an average comet would be 26 miles a second. If a comet were to strike the Sun, its speed at the moment of impact would be 386 miles per second. Comets have come sufficiently close to the Sun to reach speeds of more than 300 miles per second, but a speed so great is likely to throw the comet into a parabolic or hyperbolic orbit and expel it from the solar system.

278. How often do comets appear? There are about 70 comets that make more or less regular appearances in Earth skies at intervals of between three and nine years. There are about 40 more with periods ranging from 10 to 1,000 years. Still others are unpredictable. These will come in from outer space and be influenced by the Sun's gravitational forces to assume either an elliptical orbit about the Sun, or to be thrown into parabolic or hyperbolic orbits and leave the neighbourhood for good.

279. How many comets are seen each year? Five or six comets, on the average, appear in Earth skies each year. About a third of these are periodic comets describing regular orbits about the Sun. The rest are new comets which have never been seen before.

280. How many naked-eye comets are there? There is one naked-eye comet in an average period of a little more than a year. Even these are usually inconspicuous objects. Very few of them reach a brightness equal to that of the North Star.

281. How big are comets? When a comet is far from the Sun, it is a comparatively small ball of frozen gases and particles of matter. When it is near the Sun and the gases are vaporized by the Sun's radiation, the comet reaches enormous size. The average diameter of a comet when it is near enough to the Sun and the Earth to be visible by some means, is about ten times that of the Earth. There have been comets which became as big as the Sun.

282. What are the materials found in the gases of a comet? An analysis of the light of a comet shows that there are carbon, hydrocarbons, cyanogen, hydrides of nitrogen, ionized carbon monoxide, ionized nitrogen and ionized hydroxyl. The basic compounds of these materials are ammonia, water, hydrocarbons and carbon oxides.

283. Does a comet undergo any changes when it passes near the Sun? Yes. The Sun's gravitational power attracts the loosely associated particles that make up the nucleus of the comet. Much of this material is dragged away from the comet and scattered along its orbit. The gases and finer particles that are forced out from the head of the comet to make its tail are lost to the comet. Some comets do not survive the ordeal of even a single passage near the Sun, and most comets suffer considerable loss of the material of their nuclei each trip. It takes a robust comet to make as many trips as Halley's comet is known to have done.

284. Have comets been seen to change during one appearance? Yes. The most evident change is in the form of the comet's tail, which is far from stable. In most cases, the tails of comets vary in shape and in intensity within a very short time, sometimes in a few hours. Biela's comet, which was discovered in 1772, was found to have split in two at its appearance in 1846. Both comets appeared again in 1852, but neither has been seen since.

285. How are comets named? A comet is usually given the name of the individual who discovered it. Sometimes a comet is given two names, separated by a hyphen, as the Pons-Winnecke comet. This indicates that one of the men discovered it and the second was the first man to see it on its return passage. This two-name combination could also mean that the two men, working as a team, discovered the comet, as in the case of the Arend-Roland comet.

A temporary and more businesslike designation is given to each comet as it is first seen. This designation is the number of the year followed by a small letter whose alphabetical order tells the numerical order of the discovery for that year. Comet 1968b, for example, was the second comet seen in 1968. Comets cannot be recognized by any particular physical feature. They can be identified only by the orbits they follow. When sufficient observations have been made of a comet, its orbit may indicate that it is one that had been seen many years before. In that case, it will already have a designation. If the comet is a new one, it will be given a permanent designation as soon as all the data upon it have been collected. This permanent designation will again be the numeral of the year of discovery, followed this time by a Roman numeral which will indicate the order of its perihelion passage among all the comets for that year. This is not necessarily the same order as the order of discovery. Comet 1974b might become known as Comet 1974 II, provided it were the second comet to pass closest to the Sun in 1974.

286. Which is the most famous of all comets? The best known comet is undoubtedly Halley's comet, which last appeared in 1910 and which is due to appear again sometime in 1985 or 1986. Halley's comet has been seen at intervals of about 76 years ever since 240 B.C.

287. If Halley did not discover Halley's comet, why is it named for him? Three hundred years ago, comets were, and had always been, considered chance visitors from space or exhalations from the Earth's atmosphere. In 1682, Halley seized upon the idea that comets might be members of the solar system, making regular journeys around the Sun. Halley listed all the facts known about comets. Up to 1610, when the telescope was first used in astronomy, the only known comets were those which were bright enough to be seen with the naked eye. In Halley's list there was at least one regular series of appearances which stood out. To this series belonged the bright comet Halley had seen in 1682 and which had apparently been seen in 1607 and before that in 1531. Halley took his problem to Isaac Newton, who made the intricate mathematical calculations which seemed to justify Halley's theory that comets travelled in more or less regular but enormous orbits about the Sun. Newton's figures so convinced Halley that he made the prediction that the bright comet which was seen in 1682 would return in 1758. Halley also asked that, if the comet did return, credit for its discovery be given to an Englishman. Halley died in 1742. The comet did appear in 1758, and was hailed by the civilized world as a monument to the great astronomer for whom it was immediately named.

288. The Earth was said to pass through the tail of Halley's comet

in 1910. Did that happen? Yes. Considerable consternation developed when the announcement of this possibility was made, particularly since it was announced that the gases of the tail contained a lethal substance, cyanogen. The much greater density of the Earth's atmosphere, however, protected the Earth's inhabitants from any harm, and no noticeable effect was produced by this adventure.

289. Do the planets have any effect upon comets? Yes. The larger planets, Jupiter, Saturn, Neptune, and Uranus, are massive enough to influence the orbits of comets, helping to throw them into elliptical paths about the Sun and limiting the dimensions of the ellipses in which the comets travel. Jupiter is credited with having a "family" of more than 30 short-period comets. Saturn is supposed to have a family of two comets and Uranus is also credited with two. Neptune is given six, including Halley's comet, but there is still much to be learned about the continuing influence of these last planets upon the comets they are supposed to control.

290. Do the comets belonging to a planet's "family" always circle about that planet as well as about the Sun? No, but the gravitational influence of the major planets limits the point of the aphelion passage of their comets to about the same distance from the Sun at which the planet travels.

291. Has a comet ever hit the Earth? No record exists of a comet striking the Earth in historical times. It is possible that such a collision took place in the past, for there are scars on the surface of the Earth which look as though they had been inflicted by a group of solid objects from space, such as might make up the head of a comet. If such a thing were to happen, there would be a glorious display of meteors with the possible danger of larger chunks of metal or stone striking the surface of the Earth and causing a catastrophe. Most of the particles, however, would be destroyed by the friction of their swift passage through the Earth's atmosphere.

292. When was the most recent naked-eye comet? This answer has been written several times. When the first answer was complete, the first really bright comet in nearly 50 years appeared in the heavens. The original answer was discarded and a second was written to keep the text as much up to date as possible. Shortly after the second answer was complete, a second naked-eye comet suddenly appeared and the job had to be done over again.

The second comet appeared on August 2, 1957, and was discovered by A. Mrkos, a Czech astronomer with the Skalnaté-Pleso Observatory. It is known as Mrkos' comet and has been given the temporary designation of 1957d. When it was first seen, it had just passed its perihelion point and was as bright as first magnitude for a few days. It moved away from the Sun and from the Earth, very low in the northwestern sky, growing rapidly fainter until, after two weeks, it was too faint to be seen without optical aid.

Before that, in April of 1957, the Arend-Roland comet appeared. This comet was discovered by Arend and Roland, two Belgian astronomers at the Uccle Observatory in Belgium in November, 1956. It was approaching the Sun at that time and was found telescopically in the early morning sky. It came in, rounded the Sun and disappeared for a few months. It was predicted that this comet would reach naked-eye visibility the following spring and some predictions went so far as to say that it would rival Halley's comet in brightness. It did appear and was the brightest comet since Halley's, but it was a long way from being as bright. At its best, it was about half as bright as the North Star. It had a tail which could be seen without telescopes for a distance of about 5° from the head of the comet.

The last bright naked-eye comet was Bennett's Comet, discovered by a South African amateur astronomer. It became really spectacular, with a long tail, and it remained on view for several weeks. It cannot be compared with the great comets of the 19th century, but it was certainly the best for many decades. Kohoutek's Comet was expected to become bright in late 1973, but failed to do so.

293. For how long can any one comet be seen in the sky? A comet will remain visible only for a few weeks at the most. Halley's comet was seen telescopically in April, 1910, early in the month. It became visible to the naked eye toward the end of April. By the middle of June, it had gone too far from the Earth to be seen any longer. The Arend-Roland comet, in April, 1957, was visible without a telescope for about two weeks, as was the Mrkos comet in August, 1957, and Bennett's Comet in early 1970.

294. How can a comet be distinguished from a meteor? Meteors can be seen for a few seconds at the most. Sometimes the trail—not the tail—left by a meteor as it flashes through the sky may remain visible for as long as a minute. Comets do not appear to move any more than does the Moon, for example. They are in swift motion, but their distance from the Earth is measured by many millions of miles so that their motion is not apparent. They are usually much larger, visibly, than are meteors and seldom as bright.

295. What is the difference between a meteorite, a meteoroid and a meteor? How do these differ from an asteroid? A meteoroid is a fragment of stony or metallic material in space. Most meteroids are tiny—microscopic or dust size—but a few of them are large and massive. It is estimated that between 80 million and 100 million meteoroids strike the Earth's atmosphere every 24 hours. When one of sufficient size strikes the atmosphere of the Earth, it is usually moving at a speed which ranges from 10 to 40 miles a second. At this speed, the friction of the meteoroid's passage into and through the atmosphere heats the meteoroid to the vaporization point and it is destroyed. If the encounter occurs at night, the meteoroid, in its burning, and the glow of the intensely heated air marking its path, make a streak of light in the sky. This streak of light is a meteor. The word *meteor* means "something in the air." People speak of meteors as "shooting stars." When a meteoroid is big enough to make a considerable display in the sky, it is sometimes called a fireball. There is no rigid standard as to how big or bright a meteor must be to be called a fireball. One astronomer said that a meteor bright enough to be reported to the newspapers is a fireball. Under certain circumstances, a large meteoroid may explode under the stress of expansion caused by its rapid passage from the cold of space to the intense heat of friction. An exploding meteoroid is a "bolide." (See question 981.)

An asteroid—or planetoid—is a minor planet. Between the orbits of Mars and Jupiter there is a wide belt of space in which most of the asteroids revolve about the Sun, obeying the same laws that govern the motions of the major planets. The first asteroid, Ceres, was discovered by an Italian astronomer, Piazzi, in 1801. Many thousands have been found since, most of them photographically, and the orbits of about 2,000 have been calculated and verified, which constitutes confirmation of discovery. Ceres was also the largest ever seen—700 miles in diameter—and asteroids of only a mile or two in diameter have been found on photographic plates taken with large telescopes.

It would be possible for an asteroid to be a meteor, provided it came into the Earth's atmosphere. If this happened, the result would undoubtedly be a meteorite, for a meteorite is a meteoroid which has survived the ordeal of its swift passage through the atmosphere and reached the surface of the Earth. Bulk is the main factor in this survival, and there have been many meteorites discovered upon the Earth's surface. One tremendous shower of meteorites was seen to fall in Siberia in June, 1908. The largest meteorite on display is in the American Museum–Hayden Planetarium, resting upon a Toledo scale which shows its weight to be 34 tons, 85 pounds.

296. What is a meteor? The word *meteor* is from a Greek word which means "something in the air." It is the name for a brief, almost instantaneous streak of light seen in the dark sky—a shooting star.

297. Are meteors really stars? No. A meteor is the visible manifestation of the vaporization of a particle of metal or stone which has been heated to the point of destruction by the friction of its passage through the Earth's atmosphere at very high speed.

298. How far away from the Earth are meteors? The bright streak of a meteor is usually about 60 miles above the surface of the Earth, although large meteors may appear at a greater distance and approach more closely before being destroyed.

299. How fast do the particles that become meteors travel? From 10 to 45 miles per second, relative to the Earth.

300. What is the name of the particles that become meteors? The stony or metallic particle that causes a meteor is called a meteoroid.

301. Where do meteoroids come from? No one knows. Until we solve the problem of the origin of the solar system itself, we are unlikely to find the answer.

302. What are meteorites? Meteorites are meteoroids which have landed upon the Earth.

303. When is the best time to see meteors? Meteors may be seen on any clear dark night from anywhere on Earth. They are best seen after midnight because there are more of them between midnight and dawn than before midnight. (See question 304.) There are certain nights in the year, however, when the chances of seeing meteors are much better than average.

304. Why are there more meteors seen after midnight than before? After midnight, the Earth's rotation places us on the forward side of the Earth in relation to its revolution about the Sun. Since the motion of the Earth has a great deal to do with the velocity of meteoroids, more of them are picked up in front of the Earth as it goes around the Sun, than would overtake the Earth from behind it. Those seen before midnight are the few whose speed is sufficient to enable them to catch up with the Earth. After midnight, the Earth itself is bumping into them.

305. What is a meteor shower? A meteor shower is the appearance of meteors with much more than average frequency. A meteor shower marks the Earth's passage through a swarm of meteoroids. Sometimes meteors appear, during a shower, at the rate of one or two per minute and upon a few very special occasions, many thousands have been seen during one night.

306. Where do the meteoroids that make meteor showers come from? The origin of meteoroids is uncertain; possibly they are "debris" left over when the planets were formed. This also applies to the comets, with which meteoroids are associated, and to the asteroids.

307. What is the "radiant" of a meteor shower? The radiant of a meteor shower is the point in the sky from which the meteors seem to radiate. The location of that point before a given constellation usually gives the name to the shower.

308. How is the radiant of a meteor shower determined? The paths of the observed meteors are plotted on a chart of the heavens.

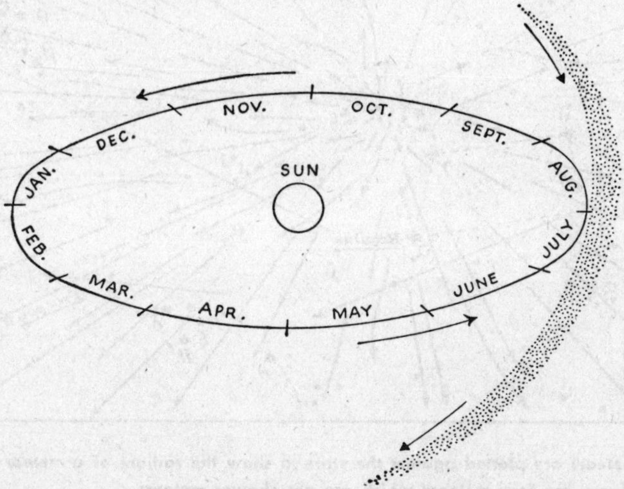

Meteoroids left in space by a passing comet are encountered annually by the Earth. These meteoroids cause meteor showers.

It will be found that most of the meteor paths, if projected, will cross at a given point. That point is the radiant.

309. How many meteor showers are there in a year? A considerable number, about a dozen of which may be regarded as major.

310. What are the major meteor showers called and when do they occur?

Shower Name	*Date*	*Hourly Rate for One Observer*
Quadrantids	January 3	35
Aurigids	February 9	12
Lyrids	April 21	12
Eta Aquarids	May 4	12

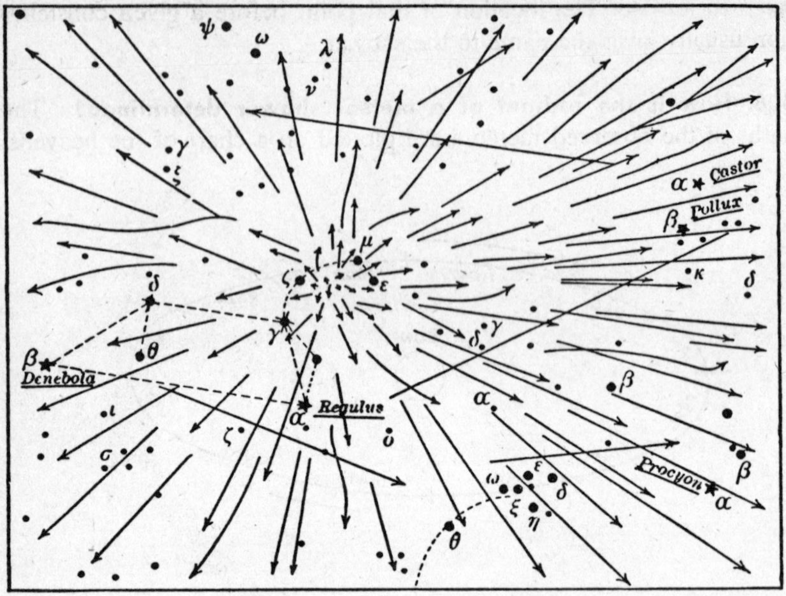

Paths of meteors are plotted against the stars to show the radiant of a meteor shower. Note that there are four meteors which are not shower meteors.

Draconids	June 28	12
Delta Aquarids	July 29	20
Perseids	August 12	50
Orionids	October 20	20
Taurids	October 31	12
Arietids	November 12	12
Leonids	November 16	20
Geminids	December 13	40
Ursids	December 22	15

311. What do the names of the meteor showers mean? The names of meteor showers give the name of the constellation and sometimes the designation of an individual star in the constellation before which the central point or radiant of a meteor shower is found. The Aurigids, for example, seem to radiate from a point in the constellation of Auriga, the Charioteer; the Draconids, from Draco, the Dragon; the Eta Aquarids, from near the star Eta (η) Aquarii; and so on.

312. Do the meteors really have any association with the constellations after which the showers are named? No. The meteors are really light years this side of the stars, but as the meteors are seen from the Earth, they flare up against the background of the vastly more distant stars.

313. Which is the best of the meteor showers? The Perseid shower is consistently the best and the most dependable of the meteor showers. The forerunners of the Perseid shower begin to appear a week or so before the climax of the shower, and the stragglers continue for a week or two afterward. The peak of the show is on August 11–12, from midnight on. The Orionids, in October, are good and so are the Geminids, in December.

314. Are the meteor showers really showers? The number of meteors usually seen by one observer in a given time will answer that. Twelve meteors an hour is one every five minutes. In the Perseid Shower, the writer logged 62 an hour on one memorable night and saw absolutely none on an equally memorable occasion. The Perseid Shower usually furnishes a good show and the average number of meteors seen, from personal experience, is about 30 an hour, or one every two or three minutes. The meteors do not appear with clocklike regularity. One crowded minute may produce five or six, and may be followed by a long, dull wait until the next swift streak cuts through the dark sky.

315. Has there ever been a real shower of meteors? Yes. In November 1799, and again in November 1833 and 1866, the Leonid meteors were magnificent for a few hours, and the sky seemed "ablaze". Throughout history the Leonids had given displays of this kind every 33.3 years, but the showers of 1899 and 1933 did not materialize, as the orbit of the swarm had been affected by planetary perturbations. In November 1966, however, the Leonids were back again, and the New World had a display fully equal to those of the past. Unfortunately, European observers missed it, since it occurred during European daylight. Whether a comparable display will occur in 1999 remains to be seen. The Draconids were spectacular in 1933 and 1946, but did not equal the Leonids.

316. What are Bielids? Bielids are meteors which were associated with a meteor shower which has not been observed for a number of years. The meteors which composed the Bielid meteor shower were believed to be the debris of a comet known as Biela's comet. This comet was observed to split in two in 1846; this has not been seen, as a comet, since 1852. The tiny stony and metallic fragments of which the comet's head was composed were strewn along the comet's orbit and were encountered by the Earth as a meteor shower each year for a number of years, whenever the orbit of the Earth crossed the path in which the Bielids travelled. The meteor stream, however, probably because of the gravitational influence of one of the larger planets, has been scattered or diverted, and the Earth no longer has a noticeable Bielid meteor shower.

317. What is a fireball? A fireball has been defined as a meteor bright enough to cast a shadow.

318. Where can a bolide be seen? A bolide is the generally accepted name for a meteoroid which explodes in the air, usually with a loud noise. The word itself is taken from a Greek word which means "to project" or "to throw." Bolides are the result of a rare combination of circumstances. On any clear night, one observer at one place may see several meteors. It has been calculated, however, that millions of meteors brighter than tenth magnitude strike the Earth's atmosphere daily. Most of these are sub-microscopic and their aggregate mass is only about one ton. At great average intervals, however, large bodies penetrate the atmosphere. Some of them are big enough to withstand complete vaporization by the frictional heat of their passage through the air and eventually fall to the surface of the Earth. Such missiles are called meteorites. If they become sufficiently bright during their journey to cast a shadow, they are called fireballs. When a meteoroid enters the atmosphere, the friction of air heats its surface, but the interior remains intensely cold. This range of temperatures sets up stresses within the body of the meteoroid which sometimes cause it to explode in the air with a sound that may be heard over an enormous area. This is a bolide. They may be seen anywhere, but their appearance is rare.

319. What does an observer look for in watching meteors? The brightness of the meteor, in terms of stellar brightness; the point at which the meteor was first seen; the duration in time of the flight of the meteor; the direction the meteor travelled and the part of the sky traversed by it; any trail or other unusual appearance; the colour; an estimate of the speed of the meteor, and, of course, the number of meteors in a given time.

320. What is the best way to observe meteors? An official meteor watch consists of at least eight observers who lie on couches which are facing outward from a central point like the spokes of a wheel. Each observer has a clerk who is to note what the observer sees and keep time. The observer calls out the location of the meteor, its brightness, its direction and signals when it vanishes. The clerk notes the time of appearance and disappearance and takes down the rest of the data from the observer.

321. What benefit is gained from meteor observation? Meteor observation can give us a knowledge of the amount of meteoric material in space. We may also gain some knowledge of the comet from which a meteor shower originated. The velocity of the meteors may tell whether they originated from the solar system or came from outer space. Visual observation of meteors is now giving way to photographic recording, and radar studies.

322. How can a picture be taken of a meteor? Pictures have often been taken of meteors, accidentally, when a meteor has blundered into the field of a camera which was taking a picture of something else entirely. Now, meteor traps are set up at many observing stations. These meteor traps are made of pairs of wide-angle cameras aimed at a point about 50 miles above the surface of the Earth. Both cameras photograph the same meteor simultaneously. A device inside the cameras interrupts the streak of the meteor 20 times a second. The compared angle of the two pictures will give the distance of the meteor from the cameras and the breaks in the meteor trail will give its speed and duration.

323. Can a meteorite be distinguished from an ordinary piece of stone or metal? Microscopic examination of the crystalline structure usually reveals the terrestrial origin of a suspected meteorite.

Chemical tests, X-ray examination and the use of the spectroscope will also identify a meteorite.

324. How big a particle of matter does it take to make a meteor? A bit of meteoroid no bigger than the head of a pin will make a startling streak of light across the night sky. Most of the light thus seen will come from the atmosphere through which the meteoroid is speeding. The friction of the swift passage of the meteoroid through the gases of the atmosphere will heat them to the point where they will become ionized and will glow. Much the same effect is caused by lightning, which is the passage of invisible electric current through the atmosphere.

325. What are meteorites made of? Meteorites are most often composed of nickel-iron, sometimes of stone and sometimes of a substance that resembles glass more than anything else. These last are called tektites, and their presence, in numbers considerably smaller than the nickel-iron or stone variety, has given rise to some interesting speculation as to the origin of meteors. The meteoritic nature of tektites is, however, questioned by some authorities.

326. What are the Widmanstätten figures? If a meteorite is cut so that there is a smooth surface, and the surface is then polished and treated with a dilute acid, an intricate pattern of straight lines which intersect will usually be produced. This is the Widmanstätten figure. It cannot be duplicated in terrestrial or manufactured metals and appears in meteorites which have a normal or better nickel content. Meteors which are poor in nickel may not show the figure, but only a meteorite does show it.

327. Has there ever been a meteoroid explosion as powerful as an atom bomb? This can be answered only theoretically. Meteoroids are caused to explode by two sets of circumstances. First, if a large meteoroid enters the Earth's atmosphere, it is cold. Its surface is heated to the vaporizing point by the friction of its passage through the atmosphere, while its interior remains cold. This may cause the meteoroid to explode in the air. Such exploding meteoroids are called bolides and it is doubtful that the force of the explosion approaches the force of an atomic bomb explosion.

At rare intervals, a meteoroid will survive its passage through the Earth's atmosphere and strike the Earth's surface. The energy of the motion of a large mass at a tremendous velocity through space is changed to heat when the motion of the body is stopped, as it would be when it struck the Earth. This energy is released then as an explosion. There are several large craters in the surface of the Earth which are suspected of having been caused by such explosions. Some of the craters on the Moon are believed to be made by the same forces. It is not possible to measure exactly the size of the meteoroid in relation to the crater so produced, so it is not possible to estimate the amount of energy in relation to the size of the object.

We know that many of the forces of nature are more powerful than hydrogen bombs. The ordinary thunder storm has many times the potential of the ordinary hydrogen bomb. It is, therefore, probable that a sufficiently large meteoroid would cause more damage than the hydrogen bomb of today.

328. Has anyone ever been struck by a meteorite? Yes. The most recent incident was in September, 1954. At that time, a meteorite weighing about 10 pounds crashed through the roof of a house in Sylacauga, Alabama, and struck Mrs. Hewlett Hodges a glancing blow as she was napping on the couch in her living room. There is also a report of a Japanese girl's having been hit a glancing blow by a small meteorite. This was in the late 1930's.

As far as is known, these are the only instances of human beings having been struck, in spite of the countless meteors that have entered the Earth's atmosphere and the vast number of meteorites that have fallen upon the Earth in historic times.

A very interesting meteorite landed at Barwell, in Leicestershire (England) on December 24, 1965. Its weight before break-up during the final descent was about 200 pounds, and many fragments of it have been found. The last British meteorite is known as the Bovedy Meteorite, because fragments of it were found near that locality in Northern Ireland; the meteorite was seen by many people as it flashed across England and Wales, but the main mass fell in the sea, and only scattered fragments of it came down on dry land where they could be collected.

VI. MISCELLANEOUS TOPICS

329. What is astrology? Astrology was the root from which astronomy developed, just as alchemy was the foundation of chemistry. The Babylonians, about 6,000 years ago, became possessed with the idea that the heavenly bodies were deities and that each of them had some influence or control over man and his fate. They assigned the various hours of the day to each of various gods and demigods and, by implication, to the bright planets which were supposed to be or to represent those deities. They also began the practice of assigning certain sequences of events to certain constellations and of claiming to be able to foretell not only the fate but the character of an individual who happened to be born at a moment when a certain deity, in the person of a planet, was in a certain constellation.

Unfortunately, astrology, unlike alchemy, while thoroughly discredited by both science and religion, has not lost its hold upon gullible or weak individuals. Many clever and unscrupulous persons today make their livings by a pretence of being able to foretell the future or analyze the character of a person because he happened to be born when a planet, many millions of miles out in space, was standing before a certain group of stars, light years distant. Astrology is defined by most objective dictionaries as a pseudo science.

330. What is the Denderah Stone? Denderah is the name of a village on the Nile in northern Egypt. In a temple dedicated to the Egyptian god Hathor, in Denderah, there was a carved stone showing the sky as it was known to the Egyptians of about the beginning of the Christian Era. In this carving, the constellations of the Zodiac are recognizable, but many of the other constellation figures differ from the modern picture. Ursa Minor is a small dog, which may account for the name of Cynosura. (See question 580.) Sirius is shown as a star between the horns of a cow which is riding in a boat. The Denderah Stone is now in the Louvre, in Paris.

331. What is a deferent? In the Ptolemaic system, which considered the Earth as the motionless centre of the universe, it was difficult to account for the motions of the other planets. An attempt to explain their varying motions held that the planets revolved in small circles whose centres moved along the circumferences of large circles

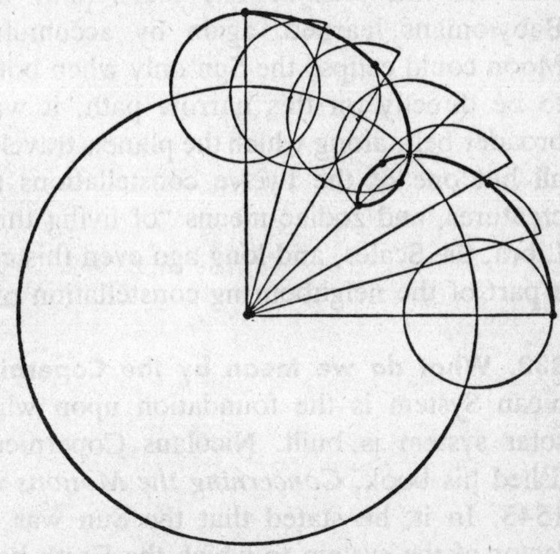

The large circle is the deferent. Around its circumference travelled the centres of the small circles, which were, according to the Ptolemaic System, the orbits of the planets. The moving points which were the centres of the small circles were epicycles.

about the Earth. These large circles were called deferents, and the small circles were epicycles.

332. If the ancients did not understand our solar system, how did they discover the ecliptic and the zodiac? (See questions 335 and 336.) The general belief of the world before the acceptance of the Copernican theory was that the Sun and the planets moved about a motionless Earth. In many respects, a moving Sun would produce the same effect as a moving Earth. The Sun does appear to travel about the Earth in a day, and around the sky in a path among the stars in a year. During years of observation, by handing down bits of information from one generation to another, the ancients learned that during one month, certain stars rose just before the Sun and during the next month another set of stars took the place of the first set. This happened year after year at the same time and it was not long before they were able to trace the apparent path of the Sun around the year through one after another of the Zodiacal constellations.

The strange, wandering stars which were, to the ancients, different from other stars only in their ability to move through the skies, also marched from one to another of these same constellations in various periods of time. The twelve star groups which formed a background for the movement of the Sun and the planets were especially significant for this reason. The Sun's path was narrow and when the Babylonians learned, again by accumulated observation, that the Moon could eclipse the Sun only when both Sun and Moon happened to be directly on this narrow path, it was called the Ecliptic. The broader belt, along which the planets travelled, was the zodiac because all but one of the twelve constellations that marked it were living creatures, and zodiac means "of living things." The one exception is Libra, the Scales, and long ago even this constellation was considered a part of the neighbouring constellation of Scorpius, the Scorpion.

333. What do we mean by the Copernican System? The Copernican System is the foundation upon which our conception of the solar system is built. Nicolaus Copernicus, a Polish scholar, published his book, *Concerning the Motions of the Heavenly Bodies,* in 1543. In it, he stated that the Sun was the centre and controlling factor of the system to which the Earth belonged, and that the Earth was not the centre of the system but a planet that moved, like the other planets, around the Sun, which stood still in the centre. This theory contradicted the Ptolemaic System which had been accepted for over 1,000 years and which said that all heavenly bodies moved about a stationary Earth.

Copernicus' system was not accepted at once. Only after years of conflict and struggle did man finally consent to relinquish his egotistical belief that the Earth upon which he lived had to be the centre of the universe. The work of men like Giordano Bruno, Kepler and Galileo, beside many others, was needed to convince man of the essential truth of Copernicus' theories.

334. What symbols are used to indicate the various members of the solar system and how did these symbols originate?

Name	*Symbol*
Sun	☉
Moon	☾
Mercury	☿
Venus	♀
Earth	⊕

Name	Symbol
Mars	♂
Jupiter	♃
Saturn	♄
Uranus	♅
Neptune	♆
Pluto	♇

The symbols for the Sun, Moon and Earth are stylized, conventional drawings of those objects. The symbol for Mercury has always been the caduceus—a rod with two serpents twined about it—and the symbol for that planet is again a conventional picture of the caduceus. Venus, as the goddess of love, is represented by a mirror with a handle. Mars has a shield and a spear and Jupiter is a stroke of lightning. Saturn is a scythe. The symbol for Uranus represents the sceptre and orb as befits the ruler of the skies, which was the role of Uranus in mythology. The trident is the traditional symbol for Neptune, who was the god of the seas. The symbol for Pluto is made up of the first two letters of Pluto and, quite fittingly, of the initials of Percival Lowell, the astronomer who initiated the research that led to the planet's discovery.

335. Which constellations comprise the signs of the zodiac and what are their symbols?

Aries	♈	Horns of a ram
Taurus	♉	Horns of a bull
Gemini	♊	Twins
Cancer	♋	Claws of a crab
Leo	♌	Capital lambda; first letter of Leo
Virgo	♍	ΠAP, first three letters of Parthenos, the Virgin, conventionalized.
Libra	♎	A conventionalized balance
Scorpius	♏	A conventionalized scorpion
Sagittarius	♐	Arrow
Capricornus	♑	Conventionalized drawing of a half-fish, half-goat
Aquarius	♒	Egyptian symbol for water
Pisces	♓	Conventionalized drawing of two fishes joined

336. What are the seasons and the arbitrary symbols of the signs of the zodiac? Spring begins when the Sun's apparent motion in the heavens, really caused by the combination of the Earth's revolution about the Sun and the Earth's inclination upon its axis, brings the Sun over the equator of the Earth, while the Sun is within the boundary of the constellation of Pisces, the Fishes. This moment occurs about the 20th of March. When man first began to keep track of the position of the Sun at various seasons, the Sun lay before the constellation of Aries, the Ram, at this time, and the point in the sky at which the Sun crosses the equator on its way north to begin the Spring is still known as the first point of Aries, although its location is actually in Pisces. The slow circling eastward of the constellations is caused by the precession of the equinoxes. (See question 59.) The Sun's crossing of the boundaries of the zodiacal constellations still bears the original designations. The conventional entry of the Sun into each constellation takes place about the twentieth or twenty-first of each month. Although at such times the Sun may be only about one third of its way across an entirely different constellation, custom still decrees that the seasonal change be spoken of as the entrance of the Sun into the sign of the original constellation. On March 20, then, the Sun enters the Sign of Aries, although it is really only part of the way through the constellation of Pisces.

Near April 21, the Sun will actually cross the border of the constellation of Aries, but will be said to enter the Sign of Taurus, the Bull. On May 21, its position will be at the border of Taurus, and it will enter the Sign of Gemini, the Twins. Summer will begin when the Sun reaches that point in the heavens when it is farthest north as seen from the Earth. This happens on June 21. The Sun will then be 23½° north of the celestial equator and before the constellation of Gemini. It will be said, then, to enter the Sign of Cancer, the Crab. This is the longest day of the year. It marks the summer solstice and is the beginning of summer. From this time on, the progress of the Sun will be southward. On July 21, the Sun will enter the Sign of Leo, the Lion, but its position will be before the stars of Cancer. On August 21, its position will be before the stars of Leo, and it will enter the Sign of Virgo, the Virgin. On September 21, the Sun will again cross the equator, this time on its way south, and the autumnal equinox will take place. The Sun will be in Virgo then, but in the

Sign of Libra. October will see the Sun in Libra, and in the Sign of Scorpius. In November, the Sun is passing before Scorpius, the Scorpion, and is in the Sign of Sagittarius, the Archer. December 21 brings the winter solstice, when the Sun is overhead 23½° south of the equator, in the constellation of Sagittarius, but in the Sign of Capricornus, the Sea Goat. January moves the Sun into the constellation of Capricornus and into the Sign of Aquarius, the Water Bearer, and February sees the Sun before Aquarius and in the Sign of Pisces. The round of the year is completed in March when, on about the twenty-first, the Sun is again where we first found it, before the stars of Pisces and in the Sign of Aries.

For the symbols of the constellations of the zodiac and their probable significance, see question 335.

337. What was the Star of Bethlehem? Science does not know the answer to this question. Man's attempts to find the answer, however, are interesting.

The Star is mentioned in one passage in the Bible and nowhere else. Matthew 2, verses 1 through 12, quoted below with an omission here and there, is the only mention of the Star of Bethlehem.

Now when Jesus was born in Bethlehem of Judea in the days of Herod the king, behold, there came wise men from the east to Jerusalem, saying, Where is He that is born king of the Jews? for we have seen his star in the east, and are come to worship him. When Herod the king had heard these things, he was troubled, and all Jerusalem with him. And when he had gathered all the chief priests and scribes of the people together, he demanded of them where Christ should be born. And they said unto him, In Bethlehem of Judea. . . . Then Herod, when he had privily called the wise men, enquired of them diligently what time the star appeared. And he sent them to Bethlehem and said, Go and search diligently for the young child and when ye have found him, bring me word again, that I may come and worship him also.

When they had heard the king, they departed; and, lo, the star, which they saw in the east, went before them, till it came and stood over where the young child was. When they saw the star, they rejoiced with exceeding great joy. And when they were come into the house, they saw the young child with Mary his mother, and fell down, and worshipped him: and when they had opened their treasures, they presented unto him gifts; gold, and frankincense, and myrrh. And being warned of God in a dream that they should not return to Herod, they departed into their own country another way.

There are several clues in this story that tell us a great deal if we take into account the status of astronomy of that time. Anything that was visible in the sky was called a star. A comet was a hairy star; meteors were falling stars; the planets were wandering stars. When the Biblical account speaks of a "star," it does not necessarily mean a star under the rather narrow definition we use today. It means "something in the sky." It could have been a single heavenly body, or a combination of objects.

It is also apparent from the story that very few individuals saw the star. The Wise Men saw it. Herod, obviously, did not see it, nor did his advisers. The Wise Men were scholars, men of special training. They practised astrology, which bears the same relation to astronomy that alchemy does to chemistry. The sight of an object or group of objects in the sky would have a special meaning for them, but it might have no significance whatever to people whose culture and beliefs held astrology in abhorrence, as did that of the Hebrews.

The Wise Men claimed to have seen "his star in the east." This may be a misinterpretation of the language of the original account. The Wise Men came from what is now either Iran or India. In order to see something in the sky over Bethlehem and to have it appear east of them, they would have to be somewhere in the Mediterranean. If, however, we read "We, in the east, have seen his star," the account begins to make sense. Now the story might go, "When we were home, far to the east of here, we saw something in the sky that makes us believe that a great king is to be born here in Jerusalem. Can you help us find him?" Herod was jealous of his kingship and would not countenance a rival. He asked the Wise Men to let him know what they found. Because they *were* wise men in addition to being scholars, they decided not to return to him with their story. They continued their search and, ultimately, with the aid of their "star," reached their goal. The "star" could have been many things. It might have been a meteor, but this seems most unlikely because of the extremely short duration of visibility which meteors have. Even a fireball or a bolide could be bright for only a few seconds. The Wise Men were impelled to leave their own country and travel for months, apparently, with the "star" in view.

The "star" might have been a comet. Astronomical research has sought in vain for any record of a comet at or near the time of the birth of Christ. Several comets, including Halley's comet, made their appearances within a few years of this time, and it is not beyond the bounds of possibility that serious perturbations may have caused one of them to vary its scheduled visit by as much as several months. There is, however, no contemporary record of the appearance of a naked-eye comet upon which we can depend.

The "star" might have been a nova, or a supernova, that flared up to extraordinary brilliance and remained bright for weeks and possibly months before subsiding to its normal faintness. Records of novae are found in the annals of ancient times, but there is none to correspond with the proper date. There was nothing in the behaviour of what we call "stars" today which could be linked with that great event.

The date of the birth of Christ is universally celebrated on December 25. It is quite certain, however, that he was not born at that time of the year. In Luke's beautiful account of the Nativity, we are told that shepherds were watching their flocks by night in the fields. Shepherds did not do that except at one time of the year—in the spring when the lambing season was at hand. At other times, particularly in the winters, mild though they may have been in Judea, the sheep were shut in sheepfolds at night. Christ was probably born in the spring.

It is most unlikely that Christ was born in the year that we call A.D. 1. The practice of numbering the years from the birth of Christ did not begin for nearly 500 years after that event, and many changes and errors had taken place during that interval. We know that Herod was king at the time Christ was born. Between the birth of Christ and the death of Herod, about 2 years passed. Josephus, the great Hebrew historian, relates that Herod died in a year in which the Feast of the Passover was followed by an eclipse of the Moon. Astronomy finds that such a year is the one corresponding to 4 B.C. Thus, Christ was born in what was probably the spring of the year that we call 6 B.C.

There is one class of "stars" to which we can turn for a clue as to what the Wise Men saw. This class is made up of five visible wandering stars, the planets, whose motions through the other stars had always been a source of wonder and amazement to man. Mars, Jupiter, and Saturn, at the time of the probable birth of Christ, stood very close together before the dim stars of the constellation of Pisces. The association of these three planets caused them to form a small triangle against the sky. They were not so near to each other that their light blended into one shining object. In fact, they could not have been seen at all because they were too nearly in line with the Sun, rising and setting at almost the same time as the Sun. This unusual configuration, then, could only have been known to men who were familiar with the motions of these wandering stars, and who might know, from previous observations, made months before, exactly where these "stars" would be at this time. To the Wise Men, this strange little invisible triangle, which could have remained in place for only a short time because of the rapid motion of Mars, may have been the "star" which drove them out on their quest.

Johannes Kepler, whose contributions to astronomy are gigantic, but who cast horoscopes for a living, saw this same configuration in the year 1604. Its occurrence excited him, for he was probably the one man then alive whose knowledge of planetary motions and orbits was sufficient to permit him to calculate the recurrence of this juxtaposition of the three planets. Kepler spent weeks in calculating the cycles of Mars, Jupiter and Saturn in order to find out how often they might be in the same position in the sky, relative to the other stars, as they were in 1604. His calculations gave him an interval of 805 years between one such triple association and the next. If Kepler saw this planetary triangle in 1604, it must have happened before that in A.D. 799 and again in 6 B.C.

That is all that science knows of the Star of Bethlehem. If the Star was none of these things, it could only have been a miracle.

338. Where did we get our calendar? Our calendar came to us through man's efforts to keep track of the obvious intervals of time. The observed apparent motions of objects in the heavens give us these intervals of time. First and most apparent is the interval of light and darkness caused by the rotation of the Earth, which exposes the Earth's surface alternately to sunlight and to shadow. It is this rotation which makes the Sun appear to rise and set and which gives us day and night.

The next most obvious time division is the interval taken by the Moon in its passage about the Earth, during which the Moon is first seen as a slender crescent in the west just after sunset and thereafter appears to increase regularly in size to full Moon and then to diminish to vanishing. This period of time was the original and natural month —indeed, we get the word *month* from *moon,* and the lunar month is still used in some calendars.

The next division is the interval taken for the Earth to move about the Sun or, as was considered to be the case in less enlightened times, the Sun to move about the Earth, until it reached the place whence it started. This period of time was first reckoned by the change in seasons and then by the varying positions of the Sun at the horizons as it rose and set day after day.

None of these natural time intervals is a multiple of the others. There is not an even number of days in the natural month or in the year, and there is not an even number of natural months in the year. Man learned this to his cost after some thousands of years of trying to keep time upon the supposition that the various time intervals did come out even. A number of compromises had to be used. The ancient Egyptians had 12 months of 30 days each, varying the true lunar month by half a day. They also added five days to the end of their 360-day year to make things come out as near even as possible. The primitive Romans began their year in March and reckoned it by lunar months. At the end of 12 lunar months, they still had 11 days to go, and added another month every 2 years. Confusion grew under this system, but it did furnish us with the names for many of our months.

The year began with March, named for Mars, who at that period had general charge of the year's growth and harvest before he was given supervision over war and rapine. April was named Aprilis from the Latin word *aperire,* "to open," because the plants began their growth—opened their leaves. Maia was a minor Roman goddess in charge of growth, too, and gave her name to May. Junius, youth in general, from the Latin *juvenis* gave June to the year. The rest of the months, numbered from five to ten inclusive, were named for their numerical order and were called Quintilis, Sextilis, Septembris, Octobris, Novembris and Decembris. Between Decembris and Martis, no time was counted at first. This period was considered the dead part of the year. That thought remains with us when we speak of "the dead of winter." The primitive ancestors of the Romans so regarded this interval in their original home in northern Europe, but their descendants, after several generations in the milder climate of Italy to which they had migrated, gave up the idea of leaving that period anonymous and filled the space with January and February. January is named for Janus, the two-faced god of entrances and doors. Februila was the deity to whom atonement was made annually for the misdemeanours of the preceding year.

The combination of a lunar-solar calendar with the clumsy system of piecing out every second year with an intercalary month soon threw the calendar and the seasons hopelessly out of step. Then Julius Caesar, with the dubious help of an astronomer named Sosigenes, attempted to set things straight again. Caesar decreed that the year numbered 708 A.U.C. (*ab urbe condita*—"from the founding of the city") in the Roman calendar was to have 60 extra days. This was the year 46 B.C. in our calendar. That adjustment again put the seasons in the part of the calendar where they belonged and, to keep them that way, Caesar also declared that the year was one quarter of a day longer than 365 days, so that an extra day should be added every fourth year. Thus he invented the leap year. To do himself honour, he also renamed Quintilis, the fifth month of the year, Julius, or July, and said that January, March, May, July, October and December were to have 31 days and February 29, or, in a leap year, 30. This worked out very well until Julius Caesar's nephew, Augustus Caesar, decided that he was entitled to equal honours with his uncle. He took the month of Sextilis, named it August after himself and, in order that his new month should not be inferior to that of his illustrious uncle, took another day from poor February and added it to the 30 days that August had. August now has 31 days and February either 28 or 29, according to whether or not there is a leap year.

The Julian Calendar, as Caesar's new reckoning was called, was an improvement, but it was still out of true by 11 minutes and 14 seconds a year, which adds up to an entire day every 128 years. Again the calendar and the seasons began their slow syncopation, but in the opposite direction. After about 1,000 years, various wise men suggested another reform. In 1582, Pope Gregory revised the calendar by dropping the accumulated error of ten days, and established the rule that only the century years that could be evenly divided by 400 should be leap years. This reform was not generally adopted at once. In England, its acceptance was postponed until 1752. By that time, it was necessary to drop 11 days, to the riotous objection of a large segment of the population who demanded the return of their lost days. In 1918, the Russian government adopted the Gregorian calendar, but the Russian Orthodox Church still goes by the old style.

Even this last change does not bring the year and the days into complete agreement. There are 28 seconds to be accounted for each year and further changes will have to be made in another 1,500 years, when that discrepancy will have amounted to a full day. Perhaps it will be necessary to make the millennial years ordinary years unless they can be evenly divided by 4,000.

339. How many different calendars are there? At present, the civilized world uses essentially two calendars. The conventional calendar of 365 days, with a leap year every fourth year, is used everywhere except in Mohammedan countries. The Mohammedan calendar uses the lunar month of 29½ days. For this reason, the Mohammedan months gain on the seasons and their calendar has 33 years to our 32. The lunar month is also used in the official calendar of the Hebrew religion. The Russian Orthodox Church calendar is the Julian calendar as it was before the Gregorian reform of 1582, so that Russian religious festivals occur ten days or two weeks later than the corresponding feasts of the rest of the world.

The astronomical calendar is one which avoids the use of months and years. It reckons dates from an arbitrary starting point, January 1, 4713 B.C. This system was adopted upon the suggestion of Joseph Justus Scaliger in 1582. Scaliger named his calendar in honour of his father, Julius, and the dates under Scaliger's reckoning are called Julian dates, but they should not be confused with the Julian calendar. (See question 338.) According to Julian dating, the year 1957 began at noon on January 1 and the date was Julian Day number 2,435,840. The year ended at noon on January 1, 1958, which day was number 2,436,205. Hours are disregarded in Julian Days, and the day is divided decimally instead, with 144 minutes making one tenth of a day.

340. Why was the ancient day, from sunrise to sunset, divided into twelve rather than ten parts? This is a relic of the sexigesimal system of the Babylonians. Since twelve is a multiple of six and a division of sixty, it lent itself readily to their calculations and was considered as useful a number as ten is today. It was very probable that the division of the day into twelve parts was as natural to them as our decimal system is to us.

Before man developed his present means of measuring time accurately, the hour was flexible. Each day—from sunrise to sunset—consisted of 12 hours. The hours of summer were longer than the hours of winter because the light endured longer on summer days than on those of winter.

341. Why does the hour have sixty minutes instead of a decimal number? The Babylonians of 3,000 years ago used the sexigesimal system of numbering, which was based upon a multiplication of 6, instead of the decimal system which we now use. They divided the circle, for example, into 60 × 6 parts—the 360 degrees. Each degree, in turn, was divided into 60 parts, and each of these smaller parts was again divided into 60. Claudius Ptolemy took this method of division from the Babylonians and called the first division of the degree the *pars minutae primae,* or the first small part. The division of this first small part he called the *pars minutae secundae,* or the second small part. Ptolemy's names became known in English as "minutes" and "seconds."

When man began to divide the hours, the round face of the clock led him to use again the divisions of a circle, and the first small parts of the hour were again called minutes and the second divisions called seconds. Recently, particularly in science and in the timing of sporting events, there is a tendency to use seconds only, without reference to minutes, and to divide the seconds into tenths, where necessary. A decimal system of time is thus beginning to come into use.

342. What is the meridian and what is the zenith? The meridian is an imaginary line that begins at the north point of the horizon, goes completely across the sky directly over your head and meets the horizon at the south point. The point on the meridian that is exactly overhead is the zenith. As a matter of record, the point opposite the zenith is the nadir. The nadir is directly beneath your feet.

343. What is an astronomical unit? An astronomical unit is the mean distance between the Earth and the Sun. Because the Earth's orbit about the Sun is not a circle, but an ellipse, the Earth is considerably nearer to the Sun at one time during the year than at any other time, and it also reaches a point at which it is farthest from the Sun. (See question 982.) The actual difference between aphelion and perihelion is about three million miles. Half the difference between aphelion and perihelion is the mean distance, which is 93,000,000 miles. This is an astronomical unit.

344. How is an astronomical unit measured? One method of determining the length of the astronomical unit is to obtain what astronomers call the Sun's mean equatorial horizontal parallax. If an observer could stand upon the Sun and see the Earth, he would find that the Earth would cover a very tiny amount of the circumference of an imaginary circle drawn completely around the Sun so that it crossed the Earth. The Earth's diameter would thus cover a certain number of seconds of arc as seen from the Sun. The measure of the radius of the Earth—half of its diameter—seen in this manner would be the Sun's equatorial horizontal parallax.

It is, of course, impossible to get the measurement in this way, but it can be obtained by taking a sight of the Sun from two points upon the Earth separated by the equatorial diameter of the Earth, or about 7,927 miles. The angles at which the Sun is viewed are very accurately determined from each point. These angles are extremely small because the Sun is far away, and the nearer an object is, the larger these angles would be and thus easier to measure. With the Moon, for example, the measurement of these angles is fairly simple. The Moon's parallax is almost one degree—57′2″.7. The Sun's parallax is less than 9″—8″.790, with a probable error of 1/1,000 of a second of arc. Refinements of this measurement are made from time to time by means of radar observations, principally of the planet Venus. By the latest and best calculations, the astronomical unit is 92,957,000 miles.

345. What is meant by the "first point of Aries"? The first point of Aries is the vernal equinox, the point at which the Sun's path, the ecliptic (see question 57), crosses the equator from south to north. It marks the Sun's position in the sky at the instant of the beginning of spring. When man first began to gain a knowledge of the apparent motions of the Sun, this important crossing was located before the constellation of Aries, the Ram. In the more than 5,000 years that have passed since that time, the precession of the equinoxes (see question 59), has moved the location of this point westward into the constellation of Pisces, the Fishes, but it is still called the first point of Aries.

346. Is there any place where something weighs nothing? The condition of weightlessness may be induced when one body moves at a speed sufficient to counteract the gravitational attraction of any other body upon it. An astronaut moving beyond the atmosphere in a spacecraft is weightless unless he brings his rocket motors into play. There is still some doubt as to the long-term effects of weightlessness; the Russian astronauts of Soyuz XI (Dobrovolsky, Volkov and Patsayev) died after being weightless for just over three weeks, though this tragedy may have been due to a fault in their capsule. More information will be drawn from the American Skylab programme, which involves setting up a full-scale artificial station in orbit round the Earth. If weightlessness does prove to be a serious hazard, it is inevitable that manned space research will be very much delayed. Few scientists believe that this will happen; however, the possibility cannot be discounted. Unfortunately, conditions of weightlessness are very difficult to simulate on the Earth by artificial means.

If a body could be placed at the precise centre of the Earth, the gravitational forces acting upon it would also be at balance, and a body in this position would also be weightless.

347. Why is the sky blue? The sky is blue because of the effect of our atmosphere upon the light of the Sun. The atmosphere itself, together with dust particles and tiny droplets of water vapour, scatters, diffracts and reflects the Sun's light, turning it into a blue dazzle that we call the sky, but which really hides the sky from us. The less dust, water vapour and so on there is in the air, the darker blue the sky becomes. Above the atmosphere, the sky is black and the stars are visible all the time.

348. What is albedo? Albedo is the measure of the ability of a body to reflect light from its surface, and is usually expressed in terms of percentage. The albedo of the Moon, for example, is 7%, which means that the Moon reflects only 7% of the light of the Sun that falls upon its surface. The albedo of the planet Venus is much higher, because of its highly reflective atmosphere.

349. What is zodiacal light? Zodiacal light is a faint, wedge-shaped glow that may sometimes be seen in the evening to the west, just after sunset, or in the morning before sunrise to the east. It extends upward from the horizon about 40°, and is about as bright as the Milky Way. It is believed to be the reflection of sunlight from countless small particles of matter, about dust-grain size, which form a vast cloud whose centre is the Sun. This cloud extends outward from the Sun, approximately in the plane of the Earth's orbit, to a distance somewhat beyond the Earth. The zodiacal light is not easy to see from our latitudes, but the best chance of viewing it is to the west in March or to the east in September.

350. Is twilight longer in the temperate or torrid zones? Twilight officially lasts until the Sun has reached a point 18° below its setting horizon at night. Morning twilight begins when the Sun reaches a point 18° below its rising horizon. The angle at which the Sun approaches the horizon in the torrid zone is always much nearer to 90° than it is in the temperate zones. Hence, its motion to and from 18° below its horizon is more direct and shorter. For that reason, twilight is always shorter in the torrid zone than in the temperate zones.

351. Who was Bode and what was his law? Johann Elert Bode was born in Hamburg, Germany, in 1717, and published his first astronomical paper in 1766 on the eclipse of the Sun that took place in August of that year. This paper was followed, two years later, by an elementary text on astronomy which led to his appointment to a board for improving the yearly tables of positions of stars and planets. He founded the *Astronomisches Jahrbuch,* the widely known German astronomical publication, and edited it for 51 years until his retirement in 1825. He was made Director of the Berlin Observatory in 1786 and died in 1826.

In 1772, Bode brought to general notice a law which might account for the distance relationships of the planets then known and the Sun. This law, which is generally known as Bode's Law, may have been discovered by J. D. Titius, of Wittenberg, several years earlier.

At that time, the three telescopic planets, Uranus, Neptune and Pluto, had not been discovered, nor were the asteroids known. Indeed, it was Bode's Law which was, in a measure, responsible for the discovery of the first asteroid. (See question 251.)

Bode—or Titius—took a series of numbers, beginning with zero and increasing geometrically in steps of 3: 0, 3, 6, 12, 24, 48, 96, 192, 384, 768. To each of these numbers he added 4, and then divided the result by 10. This gave him a series which matched rather closely the distances from the Sun of the planets which had always been known to man, when these distances were given in astronomical units. (See question 343.)

Planets	*Bode's Number*	*Add 4*	*Divide by 10*	*Distances in Astronomical Units*
Mercury	0	4	0.4	0.39
Venus	3	7	0.7	0.72
Earth	6	10	1.0	1.00
Mars	12	16	1.6	1.52
Asteroids	24	28	2.8	2.65 (average)
Jupiter	48	52	5.2	5.20
Saturn	96	100	10.0	9.54
Uranus	192	196	19.6	19.19
Neptune	384	388	38.8	30.07
Pluto	768	772	77.2	39.52

A close agreement with actual distance is shown here in the case of the nearer planets, particularly the asteroids, whose existence was not known until 1801. The legend is that Piazzi discovered Ceres, the first asteroid, when he was sweeping the skies at the proper distance between Mars and Jupiter, though actually he made the discovery by sheer chance. Uranus falls beautifully into place, and it was not found until 1781. Neptune is very much out of line, but Pluto fits much better into the Neptune slot than it does into its own place. Nothing is known to exist at the distance which was assigned to Pluto in this strange "law." It is possible that Bode's law was nothing more than an exercise involving the distances of the planets then known, and that the agreement of Uranus and the asteroids is pure accident. Some recent theories of the formation of the solar system and of the distribution of its members, however, have tried to make use of Bode's figures.

352. Who was Roche and what is Roche's limit? Edouard Roche was a French astronomer who did considerable research into the nature of the Earth and of the other planets, particularly as regards their construction. He proved, in 1850, that the liquid satellite of any planet, if at a distance from the planet greater than a certain definite limit, would be distorted by the tide-raising forces resulting from the planet's gravitational attraction. If, however, the satellite were nearer to the planet than this critical distance, the tide-raising forces of the primary would overcome the mutual gravitation of the satellite's parts and would tear the satellite to pieces. If the satellite and the planet are of the same density, the critical distance, known as Roche's limit, is 2.44 times the radius of the planet.

If the satellite, or the material of which it might be composed, is solid rather than liquid, the forces of the planet would prevent the coalescence of this material into a satellite and force it to remain in fragments which would assume orbits about the planet according to Kepler's laws relating to the mass of each fragment.

353. What is meant by an inferior planet? An inferior planet is one which is nearer to the Sun than the Earth is. Mercury and Venus are the inferior planets.

354. What is meant by a superior planet? Those planets which are farther from the Sun than is the Earth are the superior planets. They are Mars, Jupiter, Saturn, Uranus, Neptune and Pluto.

355. What is a terrestrial planet? The planets in the solar system which are most like the Earth, particularly in size, are the terrestrial planets. They are Mercury, Venus, Earth, Mars and Pluto.

356. What do we mean by the eccentricity of the orbit of a planet? The shape of the orbit of a planet is an ellipse. (See question 987.) The Sun is always at one focus of the orbit of a planet. A line drawn through the two foci of an ellipse terminating in the border of the ellipse is the major axis of the ellipse. A line bisecting the major axis and at right angles to it is the minor axis, and the point where the two lines cross is the centre of the ellipse. The eccentricity of an ellipse is the ratio between half of the major axis and the distance from one focus to the centre of the ellipse. If the ellipse were a circle, both foci would be at the centre and the eccentricity would be zero. The eccentricity of an ellipse is represented by a figure which is always a fraction between zero and 1, for if the figure were 1, both foci would be at the circumference of the ellipse and it would be a straight line. The nearer to zero this fraction is, the nearer to being a circle is the ellipse, and the smaller will be its eccentricity.

The planet whose orbit has the smallest eccentricity and which is, therefore, nearest to being circular, is Venus. Pluto has the largest eccentricity of orbit. Here is a table showing the eccentricities of the orbits of the various planets.

Mercury	0.2056244
Venus	0.0067968
Earth	0.0167301
Mars	0.0933589
Jupiter	0.0484190
Saturn	0.0557164
Uranus	0.0471842
Neptune	0.0085682
Pluto	0.2486438

357. What are the aspects of a planet? The aspects of a planet are its appearance, particularly with regard to its phasing at its various configurations, its motion through the sky and its position at various points in its orbit. Part of the aspects of the inferior planets—Mercury and Venus—are their phasing and an apparent oscillation from side to side. The aspects of the superior planets—those farther from the Sun than is the Earth—are like the aspects of the Moon in that they appear to move completely around the sky, but, with the exception of Mars, which shows a slightly gibbous phase at times, none of the superior planets exhibits phasing.

358. What is meant by conjunction? When a planet or the Moon is on the same side of the Earth as the Sun, it is in conjunction. At such times, the planets cannot be seen from the Earth because they are in the sky at the same time as is the Sun. Two planets, or a planet and the Moon are in conjunction with each other when they are in the same part of the sky; that is, when they have the same right ascension or celestial longitude. They are then said to be in conjunction, even though they may not be so precisely in a line as to eclipse or occult one another.

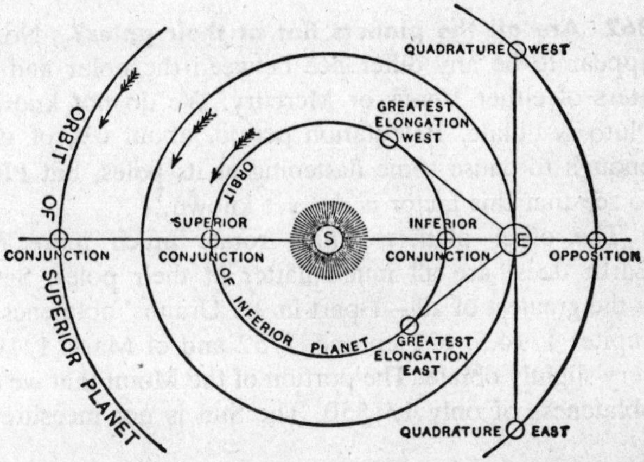

Various planetary positions

359. What is meant by superior conjunction? A planet is in superior conjunction when it is on the same side of the Earth that the Sun is, but beyond the Sun. The phrase is generally used only in speaking of Mercury or Venus, since these two planets may also be in inferior conjunction—that is, on the same side of the Earth that the Sun is, but between the Sun and the Earth. The planets more distant from the Sun than is the Earth can only be in superior conjunction, so that, as a rule, the word *superior* is omitted as being understood when speaking of them.

360. What is meant by opposition? When a planet or the Moon is on the opposite side of the Earth from the Sun, it is in opposition, and will rise in the east as the Sun sets in the west. Only the Moon and those planets which are farther from the Sun than is the Earth can be in opposition.

361. What is meant by quadrature? When a planet's position, as seen from the Earth, is 90° from the Sun, it is in quadrature. When the planet is east of the Sun, it is in eastern quadrature, and it is in western quadrature when it is west of the Sun. Mercury and Venus, because they are nearer to the Sun than is the Earth, can never be in quadrature. Mercury can never be seen more than 28° from the Sun, and Venus never more than 48°. When a planet is in quadrature, it is on the meridian at sunrise or sunset.

362. Are all the planets flat at their poles? No. There does not appear to be any difference between the polar and equatorial diameters of either Venus or Mercury. We do not know whether or not Pluto is oblate. Its rotation period, about 6⅓ of our days, is rapid enough to cause some flattening at its poles, but Pluto is so difficult to see that this factor is not yet known.

The other planets, which rotate much more rapidly than the Earth does, are all much flatter at their poles. Saturn's oblateness is the greatest of all—1 part in 9.5 Uranus' oblateness is 1/14, that of Jupiter 1/14.5, of Neptune, 1/52 and of Mars, 1/192. The Moon is very slightly oblate. The portion of the Moon that we can see shows an oblateness of only 1/1550. The Sun is not measureably oblate.

363. Has a planet or a planet-satellite combination ever been verified outside of the solar system? Yes, several nearby stars show motions indicating that they are associated with less massive bodies which may well be planets. The most famous case is that of Barnard's Star, 6 light-years away from us.

364. Was there ever a planet in our solar system called Vulcan? No. There is an irregularity in the orbit of Mercury, however, which was contrary to any motion which could be explained by the law of gravity. The long axis of Mercury's orbit—the line of apsides (see question 153)—was revolving more rapidly than predicted. The French astronomer Leverrier, who was one of the discoverers of Neptune, theorized that this departure from gravitational lawfulness could be caused by the influence of a planet between Mercury and the Sun. This hypothetical planet received the name of Vulcan. No such planet was ever found, however, and Mercury's eccentricity went unexplained until the time of Albert Einstein. In his general theory of relativity, Einstein declared that, at high relative velocities, orbital motions will deviate from the Newtonian standard. The deviation in the case of Mercury fitted the amount of excess velocity within a fraction of a second of arc. Vulcan was thus relegated to limbo.

365. What is meant by escape velocity? What are the escape velocities of the Earth, Sun, Moon, and Jupiter? Escape velocity is the velocity necessary to overcome the gravitational pull of one body upon another. The force of gravity is directly proportional to the mass of a body, hence a greater speed is required to escape frcm a more massive body. The velocity of escape from the Earth is 7 miles per second; from the Sun, 383 miles per second; from the Moon, 1.5 miles per second and from Jupiter, 37 miles per second.

VII. MOTIONS AND ECLIPSES

366. What causes an eclipse of the Sun? An eclipse of the Sun is caused by the Moon's passage between the Sun and the Earth in such a way that the body of the Moon blots out our view of the Sun.

367. How many eclipses of the Sun can there be in one year? There can be as many as five eclipses of the Sun in one year. There must be at least two and the usual number is four.

368. Why isn't there an eclipse of the Sun every month? The orbit of the Moon about the Earth is tilted at an angle of almost 6° from the plane of the Earth's orbit about the Sun. For this reason, most of the passages of the Moon between the Sun and the Earth are either above or below the direct line between the Earth and Sun, and do not cause eclipses.

369. When do we have solar eclipses? When the Moon is near one of its nodes—those places where the plane of the Moon's orbit intercepts the plane of the Earth's orbit—at the same time that the Sun is in line with the Moon's node, we may have an eclipse.

370. What is the phase of the Moon during a solar eclipse? Eclipses of the Sun always occur at new Moon, since that is the phase of the Moon when it is on the same side of the Earth as is the Sun.

371. What is an annular eclipse? When the Moon, at the time of an eclipse, is at apogee—the point in its orbit when it is farthest from the Earth—its apparent diameter is smaller than the apparent diameter of the Sun. At such times, the Moon will not entirely cover the Sun but will leave a narrow ring of Sun still visible around the Moon. Such eclipses are called "annular" from the Latin word *annulus,* which means "a ring."

372. What are the mechanics of an eclipse of the Sun? The Moon, like all opaque objects, casts a shadow. When the Moon passes directly between the Sun and the Earth at the proper place in its orbit, the shadow of the Moon touches the surface of the Earth. The Moon's shadow is a cone which is cast out into space by the Moon from the side away from the Sun. The length of the Moon's shadow at the time of new Moon may be as short as 228,000 miles or as

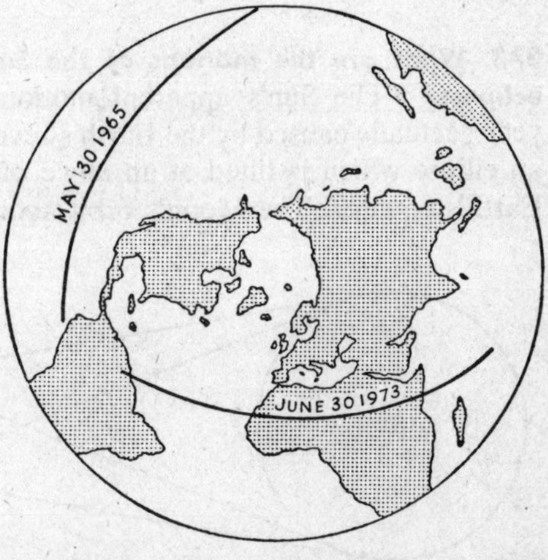

Typical paths of solar eclipses. The path of the Moon's shadow as it passes over the Earth's surface.

long as 236,000 miles. The distance of the Moon from the Earth varies from 222,000 to 253,000 miles, so that the Earth, under ideal conditions, could never be more than 14,000 miles inside the tip of the Moon's shadow. Thus, if the Moon's shadow is produced under conditions favourable for an eclipse, it will make a dark spot on the surface of the Earth—a spot whose greatest diameter will be 167 miles, if it strikes the Earth directly. If it strikes the Earth obliquely, it may be an ellipse with its longest diameter 167 miles. To anyone within this shadow spot, the Sun will be totally eclipsed, and the period of totality will be longer for those who are nearer the centre of the shadow spot. For about 2,000 miles to the north or south of the Moon's shadow, the eclipse will be partial.

The motion of the Moon makes the Moon's shadow move over the surface of the Earth at an average speed of about 2,100 miles per hour. The Earth is also rotating in the same direction in which the shadow is moving, from west to east. This reduces the average speed of the shadow at any one point on the surface of the Earth to about 1,040 miles per hour or about 1,600 feet per second. If, at the time of the eclipse, the Moon is at or near apogee, its cone of shadow will not reach the Earth, and there cannot be a total eclipse. The cross section of the Moon's shadow, extended, will strike the Earth, however, and to anyone standing within the area covered by this cross section, there will be an annular eclipse. Annular eclipses have a possible maximum length of about 12 minutes.

373. What are the motions of the Sun and the Moon that cause eclipses? The Sun's apparent motion about the Earth during the year (actually caused by the Earth's revolution about the Sun) follows an ellipse which is tilted at an angle of 23½° from the plane of the Earth's equator. The Moon's orbit about the Earth is inclined to the

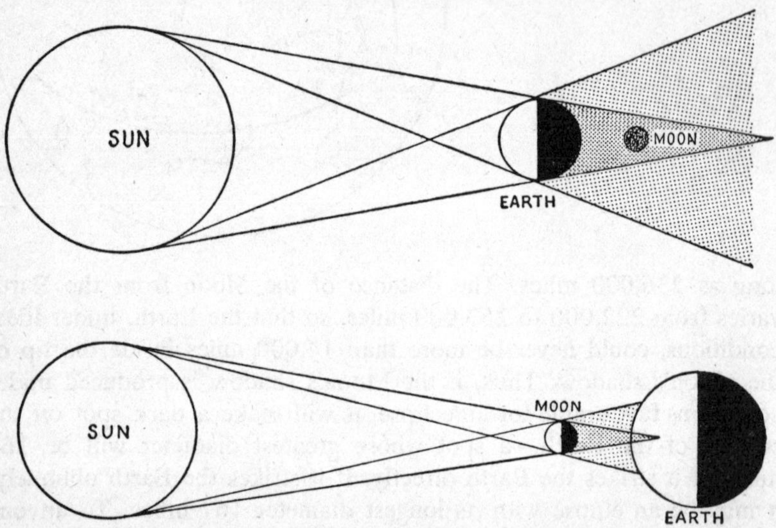

Upper diagram shows the Moon eclipsed by passing through the shadow of the Earth. Lower diagram shows a solar eclipse, which can be seen only by those upon that part of the Earth touched by the shadow of the Moon.

plane of the Earth's orbit about the Sun by an angle of 5.8°. Each time the Moon makes one revolution about the Earth, it crosses the plane of the Earth's orbit about the Sun twice, once from north to south and once from south to north. These crossings are the nodes of the Moon's orbit. The crossing from south to north is the ascending node and that from north to south is the descending node. The nodes of the Moon's orbit move slowly westward, taking nearly 19 years to go completely around the Moon's orbit. Since the Moon moves in the opposite direction to this regression of its nodes, it meets the same node in slightly less time than it takes to make a complete revolution about the Earth, with respect to the Sun's position. This nodical month is 27.21 days.

The apparent motion of the Sun, far outside of the Moon's orbit, is also from east to west, in the opposite direction to the regression of the nodes of the Moon's orbit. Hence the Sun would pass directly beyond the same node in slightly less than a calendar year. This period of time is an eclipse year and contains 346.62 days. When the Sun is in line with the descending node of the Moon at the same time that the Moon is at or near that node, there must be an eclipse of the Sun. If the Sun or the Moon is within 15°31″ of the node, there will be an eclipse. Because of the converging outlines of the Earth's shadow, through which the Moon passes to cause a lunar eclipse, there are about 5,700 miles of the Moon's orbit in which a lunar eclipse is possible as against about 10,000 miles of the Moon's orbit in which it is possible for a total eclipse of the Sun to occur. For this reason, there may be only two lunar eclipses in an eclipse year while there may be as many as five solar eclipses in an eclipse year, although this does not happen very often. There were seven eclipses in the calendar years of 1805 and 1935, and there will be seven in 2160, two of the Moon and five of the Sun.

374. What are nodes? Nodes are the points at which the planes of the orbits of two celestial bodies intersect. A node most frequently

Nodes are the points where two orbits intersect. ☋ marks the descending node and ☊ marks the ascending node.

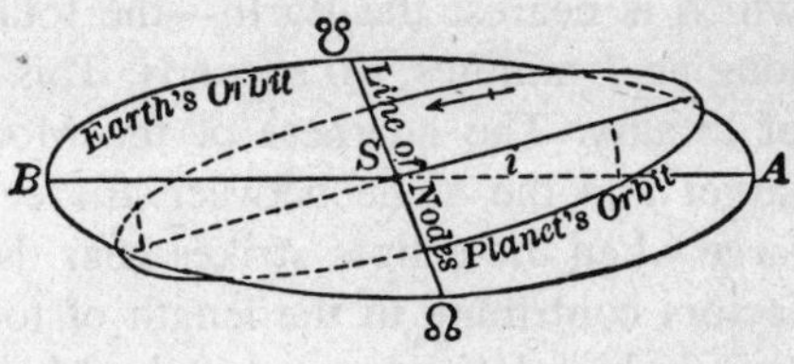

refers to the point at which a planet's orbit intersects the plane of the ecliptic—the path which the Sun apparently describes about the Earth. When the planet crosses the plane of the ecliptic from south to north, the node is called the ascending node, and it is the descending node when the passage of the planet is from north to south. The points at which the Moon's path crosses the Sun's path, as seen from the Earth, are nodes. It is only when the Moon reaches one of its nodes at a time when the Sun is also at or near that point that an eclipse is possible.

375. Is it possible to have more than seven eclipses in one year? It is possible to have more than seven eclipses in one calendar year, but it has never happened. If a solar and a lunar eclipse occurred in January, and there was the usual complement of eclipses during the eclipse year which began after January, it would be possible to have three lunar and five solar eclipses during one calendar year.

376. What is an eclipse month? Eclipses will occur at two opposite seasons of any given year, and the months in which eclipses are due to occur are called eclipse months. In 1957, for example, there were two eclipses, one in April and one in October. April and October were the eclipse months of 1957.

377. In a total eclipse of the Sun, which side of the Sun's disk is first covered by the Moon? The west side—or the right-hand side for observers in the northern hemisphere; the left-hand side for observers in the southern hemisphere. The Moon's actual motion is from west to east, and it overtakes the Sun.

378. How long does a total eclipse last? That depends on many factors. If an eclipse occurs so that it may be seen from near the Earth's equator and the Moon is at perigee—the point in its orbit which is nearest the Earth—the total part of an eclipse may last as long as 7 minutes, 40 seconds. This is the greatest possible duration of totality. The nearness of the Moon makes its apparent diameter larger and the angle between the eclipse and the Earth's surface is large when the eclipse strikes near the Earth's equator. Both of these factors contribute to the length of totality. If the eclipse can be seen only in high latitudes and if the Moon is not at perigee, totality may be as brief as a few seconds.

379. What is the first visible manifestation of a total eclipse of the Sun? The appearance of the leading edge of the Moon in front of the Sun, creating a tiny black arc cut out of the edge of the Sun's disk. This is called "first contact."

380. What are some of the phenomena seen between first contact and second contact? The increasing coverage of the disk of the Sun by the Moon, in the form of an ever-growing black crescent; the slight cooling of the atmosphere; the growing darkening of the sky; the change in colour of the entire landscape; the retirement of birds and small animals which behave as if night were coming; the passage of swift rippling shadow bands over every light-coloured surface —this is the effect of the Earth's atmosphere upon the edge of the solid shadow of the Moon; sometimes the actual approach of the shadow itself, sweeping over the world like a thundercloud. Then, for an instant, just before totality, the appearance of a row of tiny, bright points of light, irregularly spaced around the black edge of the Moon, will flash on and off. These are Baily's beads. (See question 387.)

381. When is second contact during a total solar eclipse? Second contact is the instant when the Moon completely covers the disk of the Sun.

382. What may be seen between second contact and third contact? At second contact, the chromosphere, the prominences and the corona spring into visibility. The brighter stars may be seen, and any bright planets that are in the vicinity are visible. The sudden darkness of an eclipse is so completely different from the gradual darkening of twilight that it is startling. There is still considerable light from the corona and the prominences and from the particles of the Earth's atmosphere outside the cone of the Moon's shadow. It is in this period between second and third contact that astronomers do the important part of their work on solar eclipses. Photographs are taken of each phenomenon with the greatest possible speed. Spectra of the corona and the prominences and of the chromosphere itself are photographed. An eclipse team will drill and practise every move it is to make for weeks before an eclipse occurs in order to wrest the greatest possible amount of information out of the few minutes available.

383. When is third contact in a total eclipse of the Sun? Third contact is the instant when the Moon's motion permits the edge of the bright disk of the Sun to reappear.

384. What happens at third contact? Just before third contact, Baily's beads are again visible for an instant. Then the bright edge of the photosphere appears, instantly dispelling the darkness. The prominences, the chromosphere and the corona vanish in a flash. The disk of the Sun grows as the Moon moves on. The stars disappear, the air grows warmer and the puzzled birds and animals emerge from their shelter.

385. What is fourth contact? Fourth contact marks the instant, in a total eclipse of the Sun, when the last edge of the Moon disappears from before the Sun. This is the end of the eclipse.

386. How long does the entire eclipse last, from first contact to fourth contact? Under favourable conditions, the length of the entire eclipse could be about four hours.

387. What are Baily's beads? Baily's beads are a number of bright points of light which can be seen for a few seconds at the instant when an eclipse of the Sun becomes total. A total eclipse of the Sun is caused by the passage of the Moon before the face of the Sun in such a way as to hide the Sun completely from a small part of the Earth's surface. The gradual encroachment of the Moon, visible only as a black "something" over the face of the Sun, is accompanied by many thrilling phenomena which are described in the questions on eclipses. Baily's beads flash on for an instant just before the Sun's light is completely cut off by the black bulk of the Moon, and they flash on again on the lower side of the Sun just as the light of the first limb of the Sun is uncovered by the Moon's motion.

Francis Baily (1774–1844), an English astronomer, described these flashing points of light so vividly as they appeared to him during the eclipse of May 15, 1836, that they were named after him. Baily's beads are caused by the last rays of sunlight shining through inequalities on the surface of the Moon—valleys and great gorges—or, in the opposite sense, by the first rays of sunlight shining through inequalities on the other limb of the Moon as the eclipse enters its final stage.

388. Do astronomers take pictures during an eclipse? So much so that many astronomers never actually get around to seeing an eclipse. They are too busy with their cameras. A total eclipse provides the only opportunity for taking pictures of the Sun's corona and, for many astronomers whose observatories are not equipped with spectroheliographs or coronagraphs, the chromosphere and prominences. An eclipse also provides the only opportunity for obtaining a spectroscopic analysis of the light from the corona. Pictures are also taken of the region of the sky near the blacked-out Sun in order to determine whether there may be any planets nearer to the Sun than is Mercury. On at least one occasion, a comet was discovered near the Sun during an eclipse. No planets, however, have been found inside the orbit of Mercury.

389. Can astronomical observations be made during a solar eclipse? Yes. The relative positions of the Sun and Moon can be checked. This can be done by timing the four contacts, by determining the exact boundaries of the shadow of the Moon upon the Earth and by taking photographs, exactly timed, of the progress of the passage of the Moon before the Sun. Photographs are taken of the chromosphere, the prominences and the corona. At the instant of totality, the usually dark lines in the solar spectrum are reversed and stand out brightly. This is the flash spectrum and can be seen at no other time. The positions of stars which are nearly in line with the Sun and which can be seen only when the Sun is obscured by the Moon are very carefully checked.

390. Are there any other bits of information that can be gained during a total eclipse? The bolometric measurement of the radiation from the corona, which can be isolated only at such times; the timing and measurement of the shadow bands; the drop in the temperature of the air and the effects of the eclipse upon electric means of communication, such as radio, teletype, etc.

391. What can we find out during an eclipse about the theory of relativity? One part of Einstein's theory said that light would be affected by gravity. One way of determining this effect would be to pass a beam of light through a strong gravitational field to see if and by how much it was deviated from its normal course. The Sun provides a gravitational field strong enough to produce the effect and the light of stars lying beyond the Sun, but closely in our line of

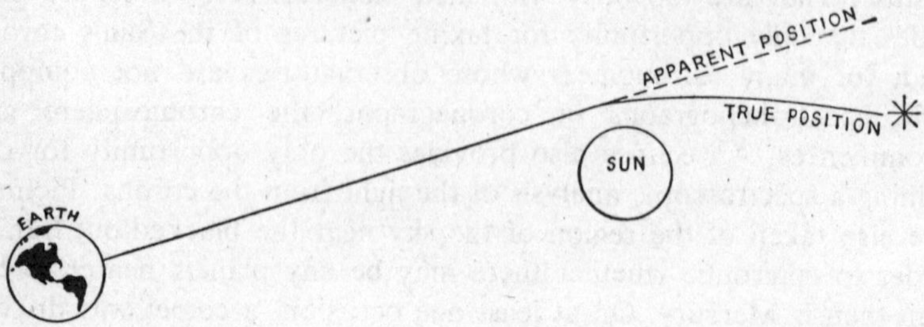

The gravitational field of the Sun attracts the light of a distant star, drawing it away from its normal path and displacing it from its true position, as seen from the Earth.

sight, provides the light for the experiment. The position of the star to be used is carefully checked at some time when it is not in the same part of the sky as is the Sun. Then, during the eclipse, the star is observed and its position rechecked. If gravity affects light, the light of the star will be deflected toward the Sun in passing, and the position of the star will be apparently displaced away from the Sun. This experiment has been performed and the results confirm Einstein's theory.

392. If a person could travel in the Moon's shadow during an eclipse, when would his journey start and finish? The Moon's shadow strikes the Earth somewhere at sunrise and the shadow travels until the sun sets, somewhere. On June 30, 1954, for example, a total eclipse began at sunrise in Nebraska, U.S.A. The shadow path travelled northeast across Wisconsin, Lake Superior, Canada, touched the tip of Greenland, crossed the Atlantic, the Scandinavian Peninsula, Central Europe, Asia Minor and wound up in Jodhpur, India, at sunset. Sunrise in Nebraska and sunset in India are about 3½ hours apart.

393. What is meant by an annular eclipse? An eclipse of the Sun is caused by the passage of the Moon between the Sun and the Earth so that the Moon blocks out the face of the Sun from a small part of the Earth's surface. Because of the much smaller distance of the Moon from the Earth as compared with the distance of the Sun from the Earth, the shadow of the Moon which strikes the Earth at such times has a diameter of only a little more than 100 miles.

The Sun and the Moon are the same apparent size in space solely because of a fortuitous relationship of size and distance. The Sun is about 400 times the diameter of the Moon and it is also about 400 times as far away from the Earth as is the Moon. There are times, however, because of the Moon's rather eccentric motion about the Earth, when the Moon may be farther than usual from the Earth and will consequently have a smaller apparent diameter. At such times, if an eclipse occurs, the Moon will not entirely cover the disk of the Sun, but will allow a narrow ring of Sun to be seen around its edge at the moment when the centres of the two bodies coincide. The word *annulus* is Latin for "ring" and describes such an eclipse. Because of the interference of the light from this ring of Sun, observers are not able to see the usual phenomena of corona, prominences and stars nearby. For that reason, an annular eclipse, while it is a magnificent spectacle, does not have the scientific value of a total solar eclipse.

394. What is the cause of a partial eclipse of the Sun? If the position of the Moon is not precisely at the point where it will completely cover the Sun, as seen from the Earth, its passage between the Earth and the Sun will produce a partial coverage, from an Earth viewpoint. Earth observers can see then the blacking out of a part of the disk of the Sun by a part of the Moon. At such times, the region in which a total eclipse would be visible will lie out in space a short distance from the Earth.

When some parts of the Earth are able to witness a total eclipse, there will be a wide path on either side of the narrow path of totality, in which a partial eclipse will be visible. The angle at which the Moon and Sun are seen, from this bordering path, is too small to make the Sun and the Moon coincide, and only part of the disk of the Sun will be covered.

395. How often is there an eclipse that can be seen from the same place from the Earth? About once every 360 years, on the average, an eclipse will take place in almost the same part of the Earth.

396. When will the next eclipse of the Sun be seen from Britain? There will be a total eclipse of the Sun which can be seen from Great Britain in 1999. The line of totality will cross Cornwall but will miss London.

397. What are the dates of some recent and forthcoming total eclipses?

Date	*Eclipse Path*
July 10, 1972.	Japan, North Canada, part of the Atlantic.
June 30, 1973.	North Africa, from Mauritania to the east coast. Totality lasts for over 7 minutes—almost a record—from some sites.
June 20, 1974.	South Indian Ocean. Short totality.
October 23, 1976.	South Indian Ocean again—this time totality lasts for 5 minutes.
October 12, 1977.	Parts of the Pacific area.
February 26, 1979.	Hudson's Bay area of Canada.
February 16, 1980.	Parts of East Africa.
August 10, 1980.	West Pacific.

398. What are the dates of recent and forthcoming annular eclipses?

Date	*Eclipse Path*
January 16, 1972.	South Atlantic.
January 4, 1973.	Antarctica.
December 24, 1973.	Mid-Atlantic area.
April 29, 1976.	Mediterranean area.
April 18, 1977.	South-West Africa.

399. Do eclipses occur in a regular series? Yes. The Babylonians discovered the sequence of eclipses. Eclipses take place near the same location at intervals of 18 years, 11½ days (10⅓ days if there happen to be 5 leap years during that time). This period is almost equal to 223 synodic or "Moon" months, or 19 eclipse years. During that time, the Moon and the Sun have gone through all their possible changes and are back where they started in relation to each other. The new cycle of eclipses, each having the same characteristics of duration and so on, will, however, strike the Earth about 8 hours westward of the original series. The Babylonian word *saros* was the name given to this eclipse cycle; it is said that the Babylonians were able to predict eclipses by the use of the saros, but this is questionable.

400. What is a saros? The saros is the interval of time that passes between the occurrence of eclipses at or near the same place upon the surface of the Earth. The saros is an interval of 223 lunar months and covers 18 years, 11⅓ days, with one day less or more, according to the number of leap years that may have occurred during that time. The number of days in the saros, 6,585.32, is just a few hours shorter than the number of days in 19 eclipse years, 6,585.78. The saros was discovered by the Babylonians.

401. What is the difference between an eclipse and an occultation? An eclipse occurs when, because of size or position, one body passes before another body in such a way as to hide it from view as seen from the Earth. The Sun is eclipsed by the Moon, and the Moon is said to be eclipsed by the Earth's shadow, although at such times, there is no solid body between the Earth and the Moon, and the Moon is not completely hidden by the shadow.

An occultation occurs when one body, because of size or position, hides from the Earth's view another body which is—or appears to be—much smaller than the concealing body. The Moon, for example, often occults stars.

402. What does an occultation look like? An occultation must be observed with a telescope; otherwise, the light of the Moon is usually enough to overwhelm the much fainter light of the star. Occultations can be accurately timed when the disappearance of the star—known as immersion—or its reappearance—known as emersion—can be clearly seen. This would make an occultation by the Moon at a phase other than full, so that one edge or the other of the Moon would be dark, more valuable. The Moon always occults a star from the west to the east—or from right to left as the observer in the northern hemisphere sees it. If the Moon is before full phase, the star will be shinning brightly and then, so suddenly that the effect is startling, it will disappear. In about an hour, the star will appear on the other side of the Moon. If the Moon is before full, the reappearance or emersion will not be as spectacular as or sudden as the immersion. The reverse will be true if the Moon is past full, for then the leading limb of the Moon will be illuminated and the following edge will be dark. Here the emersion will be sudden and startling.

403. Of what value to astronomy are occultations? The accurate timing of occultations helps to determine the location of the Moon in the sky with great precision. The observation of an occultation by a number of persons from widely separated points on the Earth's surface gives a lunar parallax (see question 437), which will help to refine the estimate of the Moon's distance from the Earth. The timing of such observations will also help to determine the exact longitude of the observing stations, since all of them will see the occultation at a different Earth time, depending upon their locations east or west of each other.

404. Is a transit the same as an eclipse? No. A transit is the passage of an apparently small body across the face of an apparently large body. Mercury and Venus, for example, transit the Sun when they pass so directly between the Earth and the Sun that we can see them as tiny black dots during their passage across the face of the Sun. If we were proportionately as close to Mercury as we are to the Moon, a transit of Mercury would then be an eclipse, for Mercury would appear to us as large as the Moon does now, and would completely hide from us the face of the Sun.

The word *transit* is also used to denote the passage of any heavenly body across the meridian.

405. What causes an eclipse of the Moon? The passage of the Moon through the shadow of the Earth is the cause of an eclipse of the Moon.

406. Does the Moon completely disappear during an eclipse? No. There is no solid body to come between the Earth and the Moon to cause a lunar eclipse. The Moon passes through the Earth's shadow, which is but a shadow, and cannot hide the Moon completely.

407. How big is the Earth's shadow? The shadow cast by the

Earth out into space away from the Sun is 859,000 miles long, on the average. This distance may vary by about 14,000 miles, depending upon whether the Earth is at perihelion or at aphelion. The diameter of the shadow tapers from the Earth's diameter of 7,927 miles to nothing.

408. What does an eclipse of the Moon look like? The Moon will be full. It must be, for if it is to pass through the Earth's shadow, it must be on the opposite side of the Earth from the light source which causes the shadow. This light source is the Sun. When the Moon is opposite the Sun, it is seen as full from the Earth. The full Moon will pass slowly through the shadow of the Earth from west to east. The Moon will change from its normal bright colour to a coppery red and its light will be greatly diminished. The red colour of the Moon will vary, depending on the weather. If the night is clear, the red will be lighter. The colour will be deeper and darker according to the amount of haze and cloudiness in the sky.

409. What is the cause of the red light that strikes the Moon during an eclipse? The thin shell of atmosphere that surrounds the Earth refracts the sunlight striking the daytime side of the Earth. This refraction excludes the short waves of light, that we see as blue, and permits the passage of the longer waves, which look red to us. The atmosphere also bends this red light around the Earth, filling the Earth's shadow cone with it. Hence there is some light, of a weird red colour, always within the Earth's shadow. We can see this light only when it is reflected from the surface of the Moon during an eclipse.

410. How wide is the Earth's shadow at the point where the Moon passes through it? That will depend upon the exact distance of the Moon from the Earth at the time of its passage. The Moon's distance from the Earth is, on the average, 239,000 miles. At that distance, the average diameter of the Earth's shadow cone is 5,700 miles. This distance may vary about 200 miles either way.

411. How long does the total phase of a lunar eclipse last? The diameter of the Moon is 2,160 miles. The Moon moves a little more than its own diameter through the sky in one hour. Since the diameter of the Earth's shadow is about 5,700 miles at the Moon's mean distance, the time taken for the passage of the Moon through that shadow is about one hour and forty minutes.

412. What is the penumbra of the Earth's shadow? For a wide region to either side of the central part of the Earth's shadow, some, but not all, of the sunlight is excluded. This part of the shadow is the penumbra, from the Latin word-combination meaning "almost shadow." By contrast, the dense portion of the shadow is the *umbra,* the Latin word for "shadow."

413. Can the penumbra be seen upon the Moon? Yes, but to a very slight degree. A skilled observer, under excellent conditions, would be able to mark it, but to the casual eye, its effect is not noticeable.

414. How long does the Moon take to pass through the penumbra? About an hour, all told. The penumbra is so faint and its edge so diffuse that it is difficult to time exactly. From its dimensions, however, it should require the Moon about half an hour to pass through the penumbra before entering the umbra, and about half an hour to pass through and emerge from the penumbra after crossing the umbra.

415. Is there any scientific knowledge to be gained from observing an eclipse of the Moon? Very little, as compared with the information that can be obtained from a total eclipse of the Sun. The edge of the Earth's shadow as seen upon the Moon is vague, rather than clear cut, because of the atmosphere of the Earth, so that accurate timing of a lunar eclipse is not possible. The reduced light of the Moon may give observers an opportunity to watch the Moon occult stars which are so faint that their light would otherwise be completely overwhelmed by the Moon's glow. A study of the drop in the surface temperature of the Moon during an eclipse gives some idea of the absorbing qualities of the surface of the Moon. An eclipse of the Moon can be used to determine longitude. A lunar eclipse is observed at the same absolute instant of time everywhere on the surface of the Earth from which it is visible at all. Hence, any variation in the time of the eclipse will give the observer his distance from the prime meridian, passing through Greenwich, England, from whose location all lunar eclipses are calculated.

416. Why is there not an eclipse of the Moon every month? The orbit of the Moon around the Earth is at an angle of 5.8° from the plane of the orbit of the Earth around the Sun. (See question 51.) For this reason, the Moon usually passes either over or under the Earth's cone of shadow when it is on the opposite side of the Earth, and there is no lunar eclipse.

417. Where on the Earth does an observer have to be in order to see a total eclipse of the Moon? A total eclipse of the Moon can be seen from anywhere on the night side of the Earth, under a clear sky. Lunar eclipses usually cover about half of the Earth's surface, by contrast with total eclipses of the Sun, which can be seen as total only from a very narrow path along the surface of the Earth. (See question 372.)

418. Can there be a partial eclipse of the Moon? Yes. The Moon may not pass through the centre of the Earth's shadow. It may ride either above or below the centre to an extent where only a part of the Moon will be touched by the shadow.

419. What are the dates of some recent and forthcoming total lunar eclipses?

Date	*Region Covered*
May 25, 1975.	Visible from the USA, but not from England.
November 18, 1975.	Visible from England. Partly visible from the USA.
March 24, 1978.	Not visible from England or the USA.
September 16, 1978.	Partly visible from England. Not visible from the USA.

420. Why do the Sun and the Moon appear to have the same diameter? The Sun is about 400 times the diameter of the Moon, and it is also about 400 times as far away from the Earth as is the

Moon. Also, both the Sun and the Moon are just about 108 times their own diameters away from the Earth. Any round or spherical body—a penny or a soup plate or a basketball—if it is held about 108 times its own diameter away from the eye will have the same apparent diameter as the Sun or the full Moon.

VIII. STARS

421. What is a star? A star is a gaseous self-luminous body in space.

422. Why do stars twinkle while the Moon does not? The stars are so far away that all we can see of even the nearest star—apart from the Sun—is a point of light which has only one dimension—length. The atmosphere of the Earth with its air currents, varying temperatures and turbulences, affects that point of light, sometimes dimming it, sometimes intensifying it, sometimes quenching it altogether. The result is a twinkle. In the cases of the Moon and the planets, we see not one single point of light but a perceptible disk made up of an infinite number of points of light. When any one of these infinite points of light is affected by the atmosphere, others are not, and we see, therefore, a steady image. Even the planets, however, when they are seen near the horizon through the densest portion of the atmosphere, will twinkle.

423. Why do the stars rise a little earlier each night? The Earth's revolution (its passage about the Sun that gives us the year) is from west to east, in the same direction as the Earth's rotation (its spinning upon its axis, which gives us day and night). Because both of these motions are in the same direction, the Earth gains almost one degree by revolution in the period of time required for it to complete one rotation. Hence, it will meet a given star slightly earlier—3 minutes, 56.56 seconds—than it met that star on the previous night.

424. Do all of the stars have names? No. Only about 200 of the brighter stars have been given individual names.

425. How did the stars get their names? About 200 of the brightest stars have individual names, in addition to the official designations by which they are known today. The naming of the stars began with man's first conscious knowledge of them. The Babylonians were probably the first historical people to name the stars. The Chinese named them, too, but since cultural contact with the Chinese was non-existent, we have little knowledge of Chinese star names. Most of the names which have come down to us are Arabic, with some Persian, Greek, Latin and Babylonian names included. Most of the Latin names are relatively modern. The preponderance of Arabic names is due to the fact that the Arabs carried on the science of astronomy during the Dark Ages, and many of their names seem to be translations of the Greek names that Ptolemy used in the *Almagest.* In that catalogue, Ptolemy's names were descriptive of the locations of the prominent stars in their various constellations. *Deneb,* for example, is an Arabic word meaning "tail," and is part of the original Arabic name for Alpha (α) Cygni, the brightest star in Cygnus, The Swan. Ptolemy called this star "The Star in the Swan's Tail." The Arabs translated this name literally as *Dhanab ad-Dajaja,* "The Hen's Tail," and it has come down to us in its present brief form. *Cor Caroli,* Alpha (α) Canum Venaticorum, was named by Sir Charles Scarborough in the seventeenth century. The Latin words mean "The Heart of Charles" and do honour to Charles I of England. *Sirius* is a corruption of a Greek word meaning "The Scorcher."

426. Who were the first people to name the stars? It was probably the first civilization that began to wonder about the stars who named them. The names that the stars bear have come to us, in the main, from about 2,000 years ago. Even then, they were drawn, in some instances, from more ancient times. The Babylonians, for example, had a system of star and constellation names which has been, for the most part, superseded by the star and constellation names of the Arabs and the Persians. Some of the star names, but only a few of them, come from Greek or Latin. Still fewer are modern—within the last 500 years.

427. Who started the Greek letter system of star designation? Johann Bayer (1572–1625), published a star chart and catalogue in 1603 that he called *Uranometria.* In it he added 12 new constellations of his own devising to the 48 that had been used until that time. To designate the stars, he used the letters of the Greek alphabet, followed by the genetive—or possessive—case of the constellation name. He usually lettered the stars in order of brightness within a constellation, but sometimes, probably because it was difficult to differentiate between stars of almost the same brightness, he lettered them in order of position, as in the case of the stars in the Great Bear or Plough in Ursa Major.

428. What happened when there were more visible stars in a constellation than there were letters in the Greek alphabet? There are 24 letters in the Greek alphabet. When Bayer ran out of Greek letters, he used Roman letters.

429. Are the Greek letters identifying the stars in a constellation assigned by order of the stars' brightness, or on the basis of their location in the constellation? Both. In most cases, the order of the Greek letters is according to the visual magnitude of the stars, but there are some instances in which the lettering is according to location. In the constellation of Ursa Major, for example, the stars are lettered in order of position, beginning with the outer edge of the quadrilateral of the Plough and going thence down around the bottom of the bowl, up the other side and on out to the end of the handle. In many cases, the order of the letters is not accurately given according to magnitude, and should not be depended upon too greatly in tabulating the stars in any given constellation.

430. What are the letters of the Greek alphabet?

Alpha	α	Iota	ι	Rho	ρ
Beta	β	Kappa	κ	Sigma	σ
Gamma	γ	Lambda	λ	Tau	τ
Delta	δ	Mu	μ	Upsilon	υ
Epsilon	ε	Nu	ν	Phi	φ
Zeta	ζ	Xi	ξ	Chi	χ
Eta	η	Omicron	ο	Psi	ψ
Theta	θ	Pi	π	Omega	ω

431. Who used Arabic numbers for star designations? John Flamsteed, the first Astronomer Royal of England, published a star catalogue called *Historia Coelestis*. In this, he numbered the stars in each constellation from east to west across the constellation. This system, of course, allowed for many more stars than the Greek-letter system.

432. What is meant when a star designation has a number following it, as "SS Cygni, 213843"? The number following the designations of such stars is the Harvard number. It gives the approximate location of the star in the celestial sphere. The six-digit number must be broken up into three sets of two digits each. The first two digits give the hour of right ascension—distance east of the vernal equinox—which in this case is 21 hours. The next two numbers give the minutes of right ascension following that hour, or 38 minutes. The last two digits give the distance north of the celestial equator, in this case, 43°. If a star is south of the equator, the whole number is either printed in italics or underlined. The designation SS is used for a variable star and is fully explained in question 535. Thus, this symbol means that the star is the variable star known as SS in the constellation of Cygnus, the Swan, and that it may be found at the co-ordinates 23 hours 38 minutes of right ascension and 43° north declination.

433. How far away are the stars? The nearest of the visible stars is 4.3 light years distant. Translated into miles, this distance is about 26 billion miles. This star is Alpha (α) Centauri, a bright star in the constellation of Centaurus, the Centaur, in the skies of the southern hemisphere. Alpha (α) Centauri is a multiple star. There are two bright components, one of which is of magnitude 0.01 while the other is slightly fainter—magnitude 1.40. It is this pair that we can see as one point of light and which makes up the nearest visible star. The third member of the multiple system is a very faint star of magnitude 11.3 which is really nearer to us than are the brighter stars. This is *Proxima* Centauri—*Proxima* is Latin for "nearest." This faint star is 4.25 light years distant.

434. What star is nearest the Earth? Which is second nearest? The Sun is the star nearest the Earth. The mean distance of the Sun from the Earth is 93,000,000 miles. The distance from the Earth to the second nearest star, whose designation is Proxima Centauri, is about 4.2 light years from the Sun. That distance translates into approximately twenty-five billion (25,000,000,000,000) miles.

435. What is the nearest and what is the farthest visible body, outside our solar system? There are two answers to this. First, without a telescope or any instrument to supplement the eye, the nearest visible star is Alpha (α) Centauri, which is 4⅓ light years distant. The most distant body that can be seen with the naked eye is the Great Spiral Galaxy in Andromeda, which is over 2,000,000 light years distant. With the 200″ reflector at Mount Palomar, provided the matter of location were discarded, Proxima Centauri, a companion to Alpha Centauri, at a fraction of a light year nearer to the Earth could be seen, while the most distant object visible through this great instrument would certainly be a galaxy or a quasar far out in space.

436. How are the distances to the nearer stars measured? The distances to the nearer stars are measured mathematically, by trigonometric parallax.

437. What is trigonometric parallax? *Parallax* is a word which is made up of two Greek words whose meaning is "apparent motion." To illustrate this apparent motion, try an experiment. Hold out your hand, with the fist clenched, at arm's length, with your thumb pointing upward. Now sight along your thumb at a point on the wall of your room. Close your right eye. Now, open your right eye, and at the same time, close your left eye. Your thumb will appear to make a tiny jump from right to left. Your thumb did not really move—it appeared to move. That is parallax. Instead of his thumb, an astronomer sights at one of the nearer stars. Instead of the wall of his room, he sees the star against a background of more-distant stars. Instead of closing one eye and opening the other, he sights the star in the spring, let us say, and again in the autumn, six months later. By doing this, he has made use of the diameter of the Earth's orbit—186 million miles—instead of the distance between his eyes. Thus the astronomer has constructed a triangle with a small base—the diameter of the Earth's orbit—and a tremendous altitude—the distance to the star. By the laws of trigonometry, if the base of a triangle is known and if the angles at each end of that base are known, the other dimensions of the triangle can be determined. The astronomer now knows the angle at the distant apex of his triangle.

To make use of this, he constructs another imaginary figure. He draws a circle with the star under investigation as the centre of the circle and the Earth at a point on the circumference of the circle. The radius of such a circle is the distance between the Earth and the star. In any circle, the distance measured along the circumference equal to the radius of the circle is called a radian, and is equal to 206,265 seconds of arc—57½°. The tiny angle that the astronomer found as

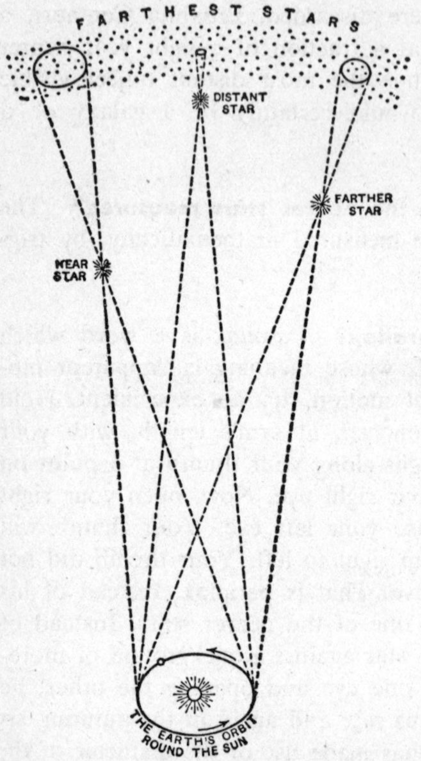

The apparent displacement of a star against the background of more distant stars is parallax. The more distant the star, the smaller its apparent displacement when viewed from two positions of the Earth in its orbit.

the parallax of the distant star has the same relation to the radian as the distance from the Earth to the star has to the distance of the Earth to the Sun. The parallax of the nearest visible star, Alpha (α) Centauri, is 0.751 seconds of arc—about $1/275,000$ of a radian. Therefore, Alpha (α) Centauri is 275,000 times as far from the Earth as the Earth is from the Sun. All we have to do now is to multiply the Earth's distance from the Sun—93 million miles—by 275,000. The answer, in round figures, is 26 billion miles.

The method of trigonometric parallax of star measurement decreases in accuracy as the distance increases. An angle of 1″ of arc, which is the angle of the parallax of a star at a distance of one parsec —3.26 light years—is about the angle subtended by a sixpence at a distance of three miles. This angle is larger than the parallax of any star. Angles of $\frac{1}{20}$ of a second of arc carry a probable error of about 10%. Thus, the direct method is not sufficiently accurate for stars whose distance is more than 50 parsecs, or about 160 light years.

438. How are distances greater than 160 light years measured? An analysis of the light of a star or other object at a greater distance will give a fairly accurate measurement. The various lines in the star's spectrum will disclose the elements which make up the star and tell approximately the temperatures at which they are reacting. From these clues, we can get a good idea of the real brightness of the star, disregarding distance. It is known that the brilliance of any object diminishes as the inverse square of the distance to the object—if a star is twice as far away, it is one quarter as bright. Hence, a comparison of the visual brightness of a star will tell its distance.

439. What is meant by "the modulus of the distance"? The modulus of the distance expresses the relation between absolute magnitude (M) and visual magnitude (m), and indicates the effect of distance upon the brilliance of a celestial object. It is known that light diminishes as the inverse square of the distance—that light is one quarter as bright at twice the distance—and that factor has produced a formula for finding the distance of a body whose absolute magnitude and whose visual magnitude are both known. This formula is log d (distance in parsecs) $= 0.2\ (m\text{-}M) + 1$, or log d (distance in light years) $= 0.2\ (m\text{-}M) + 1.513$. Thus, if any two of the factors are known—the star's distance and visual magnitude for example—the absolute magnitude of the star can be determined. If both absolute and visual magnitudes are known, its distance can be found.

440. What is a parsec? The word "parsec" is made up of the major parts of the words *parallax* and *second*. It means the distance from the Earth to a star whose parallax is one second of arc. There is no star as near as one parsec, for the parallax of the nearest star is 0.751″. A parsec is 3.26 light years.

441. How accurate are the measurements of distances in astronomy? Edwin Hubble, in his book *The Realm of the Nebulae,* says: "Beyond the nearer stars, distances cannot be determined with precision. Errors of 10% are considered to be very small and uncertainties of 25% represent reasonable accuracy. Under these conditions, distances are generally expressed in round numbers, using only one or two significant figures."

In determining the distance of stars by parallax, the angles are tiny. An angle of 0″.75, the parallax of Alpha (α) Centauri, the nearest star to the Sun, is about that subtended by a sixpence at a distance of three miles. The estimation of brilliance, upon which so many measurements of distance depend, is uncertain, since light is affected by so many factors. The tremendous speed and motion of the stars and their almost incredible brilliance are both rendered infinitesimal by distance and become increasingly difficult to measure with accuracy.

442. How many stars can we see on a clear night? On a perfect night, the human eye can see stars as faint as the sixth magnitude. (See question 459.) There are about 5,000 stars of that magnitude and brighter. If we eliminate all obstructions such as houses, trees and so on, a man would be able to see one half of the celestial sphere, or one half of all the stars bright enough to register upon his eyes. If he remained out all night, the original stars he saw would set—disappear beyond the western horizon as the Earth turned. As these disappeared, however, new stars would rise above the eastern horizon. If the time of the year he selected was at one of the equinoxes, he would have approximately twelve hours of darkness and during that time he would be able to see about half of the stars which were not included in his first sight of the sky. If he were at a latitude north or south of the equator, there would be a region of the sky far to the south, or north, which he would never be able to see without changing his location. Hence, on any one night, from one place, one man could see about three quarters of the naked-eye stars, or between 3,500 and 4,000.

443. Why are some stars brighter than others? In general, a star's brightness depends upon three factors: its distance from us, its temperature and its size. With stars of the same size and temperature, the nearer stars will appear to be the brighter. In stars at the same distance and of the same size, the hotter stars will be the brighter. With stars at the same distance and the same temperature, the larger stars will appear brighter.

444. How many stars are there that are too faint to be seen? In our Galaxy, there are about 100 billion stars. We can see about 4,000 of these, under ideal conditions, without optical aid. In all space, grouped in countless other galaxies, there are untold billions of stars. Sir James Jeans has said that there are as many stars in the universe as there are grains of sand on all the beaches of all the oceans of the world.

445. What does magnitude mean? Magnitude is the measure of the brightness of a star or of any luminous body in the heavens.

446. How does the magnitude of a star differ from its luminosity? When the word "magnitude" is used alone, it is considered to mean visual magnitude, or the brightness of a heavenly body as seen from the Earth. There is a term, "absolute magnitude," which refers to the brightness of a heavenly body as seen from the theoretical distance of 32.6 light years. Through the convention of absolute magnitude, all bodies are placed at the same distance and their various luminosities can be compared. Luminosity is the intrinsic brightness of a body, regardless of distance, and is generally referred to in terms of comparison with the Sun.

447. Are there different kinds of magnitudes? Yes. The word *magnitude* used alone without any modifying word is generally accepted to mean the brightness of a star as seen by the unaided eye from the Earth. This is also called "apparent magnitude," and is expressed by the small letter m. There are other magnitudes based on different methods of observation or on other astronomical conventions which are variously expressed. The symbol for absolute magnitude is the capital letter M. Visual magnitude is expressed by a small m with a subscript, as m_v; m_{pg} is photographic magnitude and m_{bol} is bolometric magnitude. (See questions 449, 450, 451.)

448. What is visual magnitude? Visual magnitude is the same as apparent magnitude or simply magnitude. It is the brightness of an object as seen with the unaided eye.

449. What is meant by absolute magnitude? Distance has a tremendous effect on the brightness of the stars as we see them. If two stars are of equal brilliance but at different distances from us, the nearer star will appear to be brighter. The Sun, for example, looks as bright as it does only because it is so near to us. If all the stars could be placed at the same distance, we would be able to get a much better idea of their actual brightness. Absolute magnitude is an attempt to eliminate the distance factor. Astronomy places all stars at the theoretical distance of 10 parsecs—32.6 light years—and declares that their calculated magnitudes at this distance shall be known as absolute magnitudes.

Most stars would appear much brighter if they were at this distance than they do as we see them because they are, for the most part, at much greater distances. There are a few which are nearer than 32.6 light years. For such stars, their visual magnitudes—most often called "magnitudes" without any modifying adjective—are brighter than their absolute magnitudes. The vast majority are more distant, and their absolute magnitudes are brighter than their visual magnitudes. There are a few, like Alpha Boötis (*Arcturus*), and Alpha Lyrae (*Vega*), which happen to be roughly 32 light years distant. In these special cases, absolute magnitude and visual magnitude are approximately the same. The symbol for absolute magnitude is the capital letter M.

450. What is photographic magnitude? Photographic magnitude is the brightness of a star as indicated by the diameter of its image on a photographic plate. This differs somewhat from visual magnitude because photographic plates are generally more sensitive to blue light. Thus the bluer stars will photograph brighter than they appear visually. Conversely, the red stars photograph less bright than they appear to be.

451. What is meant by bolometric magnitude? Bolometric magnitude is the measure of the total radiation of a star, not merely the small segment of radiation that we call light. Bolometric magnitudes are measured with a thermocouple or a bolometer instead of with the eye. Bolometric magnitude includes the light, heat and all forms of energy emitted by the star as radiated in waves shorter than the violet waves of light and longer than the red waves.

452. How are magnitudes determined? Magnitudes are determined usually by comparison of the object whose brightness is unknown with some star or object whose brightness is known. Polaris, the North Star, is used when possible as the standard. Polaris is of magnitude 2.0. Its light is brought, by mechanical means, into the same field with the star to be measured and the brightness of the two stars is compared either visually or by mechanical means.

453. How did the magnitude system start? The classification of stars according to their brightness was first formally used by Ptolemy, about 150 A.D. Ptolemy may have inherited the system from Babylonian times, since the division into six magnitudes seems to follow the Babylonian method of counting by sixes. Ptolemy divided all stars into six classes. Into the first class, he put about 20 of the brightest stars, and into the sixth class, all those stars which were just visible with the naked eye under good conditions.

454. What is the modern system of magnitudes? Ptolemy's original division of magnitudes into six groups has been retained, but the allocation of various stars into the different magnitudes has been greatly refined and the differences between magnitudes has been made more definite. It was found that each step in the magnitude scale was about 2½ times brighter than the next fainter step. About 100 years ago, Pogson suggested a logarithmic relation between magnitudes which has since been adopted and is now in use. There are five steps in the scale of six magnitudes which are visible to the naked eye, and one star of the first magnitude is 100 times as bright as a star of the sixth magnitude. The number which, multiplied by itself five times, will equal 100 is 2½. The logarithm of 100 to the base 10 is 2—in other words, 10 raised to the second power equals 100. One fifth of 2 is 0.4. The logarithm of 2.512 is also 0.4. Thus the modern interval between magnitudes of 2.512 lends itself readily to a logical mathematical process.

Once this definite relationship had been established, it was found that a few of the stars had to be moved from the magnitudes in which they had been originally placed to others which more accurately described their brightness. There were also some stars which proved to be brighter than first magnitude. To provide a classification for these stars, zero magnitude was established. Even then there are four stars which are brighter than zero magnitude. These are Alpha (α) Canis Majoris (Sirius); Alpha (α) Carinae (Canopus); Alpha (α) Centauri (Rigil Kentaurus); and Alpha (α) Boötis (Arcturus). For these, and for other objects which might fall into their class, minus magnitudes were provided. The magnitude scale, then, may extend as far to either side of zero as the need may require. The larger the magnitude number on the plus side, the fainter is the star. Minus magnitudes are brighter on an increasing scale of minus numbers. The Sun, for example, is magnitude −26.8; Sirius is −1.43; Polaris +1.99, and so on.

455. How bright is "first magnitude"? The flame of a standard plumber's candle seen from a distance of 1 mile is as bright as a star of first magnitude. The standard plumber's candle was defined in 1860 as the unit used in determining candle-power. It is officially a sperm candle weighing six to the pound which burned 120 grains an hour.

456. Are there any stars brighter than zero magnitude? Four. Alpha (α) Canis Majoris (Sirius) is of magnitude −1.43; Alpha (α) Carinae (Canopus) is of magnitude −0.73; Alpha (α) Centauri (Rigil Kentaurus) is of magnitude −0.27; and Alpha (α) Boötis (Arcturus) is of magnitude −0.06.

457. What are the gradations or intervals of the magnitude scale? Each magnitude is 2.512 times brighter than the next fainter magnitude.

458. Are there fractional intervals in the magnitude scale? Yes. Modern magnitudes are measured in tenths of a magnitude.

459. How does the magnitude scale work out? A star of the first magnitude is:

2.512 times as bright as a star of the		2nd	magnitude, and
$(2.512)^2$, or 6.30	"	3rd	"
$(2.512)^3$, or 15.84	"	4th	"
$(2.512)^4$, or 36.8	"	5th	"
$(2.512)^5$, or 100.0	"	6th	"
$(2.512)^{10}$, or 10,000	"	11th	"
$(2.512)^{15}$, or 1,000,000	"	16th	"
$(2.513)^{20}$, or 100,000,000	"	21st	"

460. How bright is the Sun? The visual magnitude of the Sun is −26.8. The absolute magnitude of the Sun is +4.86. It would therefore be visible without optical aid on a good, clear night.

461. How bright is the full Moon? The visual magnitude of the full Moon is —12.6. The visual magnitude of the Sun is —26.8, so that there are 14.2 steps in the magnitude scale between the Sun and the full Moon. This means that the Sun is about 400,000 times as bright as the full Moon.

462. Are there any other objects brighter than zero magnitude? Yes. The Sun and the Moon are, of course, brighter than zero magnitude. Some of the planets, when they are favourably situated, are also brighter than zero magnitude. Venus can be as bright as magnitude —4.4; Mars, as bright as —2.8; Jupiter, —2.5, and Saturn, —0.4.

463. How can an astronomer tell from a photograph of a star how bright it is? Astronomers work from photographic negatives in which the star images are black against a white background. The effect of light is to enlarge the image of a star so that the brighter the star is, the larger will be the black image of it against the white background. A careful control of exposure time, type of film and of all factors in taking the photograph will make it possible to obtain in this way a very accurate measure of the brightness of a star.

464. How much light do all the stars send to the Earth? All the stars in space combine to send to the Earth an amount of light equivalent to 1,092 stars of the first magnitude, or about 1/15 the light of the full moon.

465. What are the brightest stars, by absolute magnitude, in the Southern and in the Northern Hemispheres? As recently as 1956 the brightness of the stars was checked and rechecked by a very competent team of astronomers, Drs. G. H. Herbig and C. E. Worley. Their findings list two stars as equal in absolute magnitude, and one of these stars is in the Southern Hemisphere and one in the Northern. Both of them can be seen from latitudes north of the equator. The Northern star is Alpha (α) Cygni (Deneb) and the Southern star is Beta (β) Orionis (Rigel). The absolute magnitude of each is listed as —7.1. This gives each of them a luminosity of about 25,000 times the luminosity of the Sun.

466. Which is the brightest star by absolute magnitude? The brightest star that we can see, by absolute magnitude, is either of two stars. Alpha (α) Cygni (Deneb) is of absolute magnitude —7.1. Beta (β) Orionis (Rigel) is equally as bright. There is a star lying in the Larger Magellanic Cloud which has a greater absolute magnitude than either of these, but it is so far from the Earth that its visual magnitude is about 9—far too faint to be seen without a fairly large telescope. This star is S Doradus, a variable star, and its absolute magnitude may be on the order of —8 or —9.

467. Which are the 20 brightest stars? The most recent tabulation of the stars in order of brightness was made in 1956. Here they are with their visual magnitudes, their absolute magnitudes, their distances from us and their luminosities in terms of the Sun.

Star Name	*Designation*	*Visual Magnitude*	*Absolute Magnitude*	*Distance Light Years*	*Luminosity x Sun*
1. Sirius	α Canis Majoris	—1.43	1.43	8.7	23
2. Canopus	α Carinae	—0.73	—3.1	98	5,200
3. Rigil Kentaurus	α Centauri	—0.27	4.7	4.3	1
4. Arcturus	α Boötis	—0.06	—0.3	36	100
5. Vega	α Lyrae	0.04	0.5	26.5	75
6. Capella	α Aurigae	0.09	—0.6	45	120
7. Rigel	β Orionis	0.15	—7.1	900	25,000
8. Procyon	α Canis Minoris	0.37	2.7	11.3	5.8
9. Achernar	α Eridani	0.53	—2.3	118	800
10. Hadar	β Centauri	0.66	—5.2	490	12,000
11. Betelgeuse	α Orionis	0.7	—5.6	520	13,000

Star Name	Designation	Visual Magnitude	Absolute Magnitude	Distance Light Years	Luminosity x Sun
12. Altair	α Aquilae	0.8	2.2	16.5	8.3
13. Aldebaran	α Tauri	0.85	—0.7	68	120
14. Acrux	α Crucis	0.87	—3.6	370	5,900
15. Antares	α Scorpii	0.98	—5.1	520	12,000
16. Spica	α Virginis	1.00	—3.3	220	5,500
17. Fomalhaut	α Piscis Austrini	1.16	2.0	23	11
18. Pollux	β Geminorum	1.16	1.0	35	30
19. Deneb	α Cygni	1.26	—7.1	540	25,000
20. (Beta Crucis)	β Crucis	1.31	—4.6	490	6,000

468. Can we see all the stars in the vicinity of the Sun? There are about 30 stars within 15 light years of the Sun. Only seven of these are bright enough to be seen without a telescope.

469. Are any of the stars within 15 light years of the Sun brighter than the Sun? Only three of the neighbouring stars are brighter than the Sun. These stars are Alpha (α) Canis Majoris (Sirius), Alpha (α) Canis Minoris (Procyon), and Alpha (α) Centauri.

470. What are the stars within 15 light years of the Sun? The stars within 15 light years of the Sun are listed below with their names or designations, the constellations in which they can be found, their visual and absolute magnitudes and their luminosities in terms of the Sun's luminosity.

Name	Constellation	Vis. Mag.	Abs. Mag.	Distance Lt. Years	Luminosity x Sun
α Centauri	Centaurus	0.3	4.7	4.3	1.1
Barnard's Star	Ophiuchus	9.5	13.2	6.0	.0004
Wolf 359	Leo	13.5	16.6	7.7	.000017
Luyten 726–8	Cetus	12.5	15.6	7.9	.00004
Lalande 21185	Canes Venatici	7.5	10.5	8.2	.0048
Sirius	Canis Major	—1.43	1.45	8.7	23.0
Ross 154	Sagittarius	10.6	13.3	9.3	.00036
Ross 248	Andromeda	12.2	14.7	10.3	.00010
ε Eridani	Eridanus	3.8	6.2	10.8	.25
Ross 128	Hydra	11.1	13.5	10.9	.00030
61 Cygni	Cygnus	5.5	7.9	11.1	.052

Name	Constellation	Vis. Mag.	Abs. Mag.	Distance Lt. Years	Luminosity x Sun
Luyten 789–6	Aquarius	12.2	14.5	11.2	.00012
Procyon	Canis Minor	0.37	2.7	11.3	5.6
ε Indi	Indus	4.7	7.0	11.4	.12
2398	Draco	8.9	11.1	11.6	.0028
Groombridge 34	Andromeda	8.1	10.3	11.7	.0058
τ Ceti	Cetus	3.6	5.8	11.8	.36
Lacaille 9352	Pis. Aust.	7.2	9.4	11.9	.013
BD 5°1668	Canis Minor	10.1	12.2	12.4	.0010
Lacaille 8760	Microscopium	6.6	8.6	12.8	.028
Kapteyn's Star	Pictor	9.2	11.2	12.9	.0025
Kruger 60	Cepheus	9.9	11.9	13.1	.0013
Ross 614	Orion	10.9	12.9	13.1	.00052
BD 12°45238	Scorpius	10.0	11.9	13.4	.0013
Van Maanen's Star	Pisces	12.3	14.2	13.8	.00016
Wolf 424	Virgo	12.6	14.3	14.6	.00014
Groombridge 1618	Ursa Major	6.8	8.5	14.7	.030

471. What is meant by the "extinction" of a star? All stars look brighter when they are observed near the zenith—the overhead point of the sky—than they do when they are seen at the point of their rising or setting, near the horizon. The difference in brightness is caused by the greater density of the Earth's atmosphere near the Earth's surface. When a star is seen at such an angle that its light must pass through a great length of dense atmosphere, its brightness may be dimmed by as much as two magnitudes—a factor of more than six. This dimming is called extinction.

472. Do the stars have colour? Yes, very definitely. To anyone who is accustomed to looking at the stars, their different colours soon become apparent to the naked eye. When the stars are seen through a telescope, their colours are striking.

473. What are the common star colours? The colours most often seen in the stars are red, orange, yellow, white and blue. There are

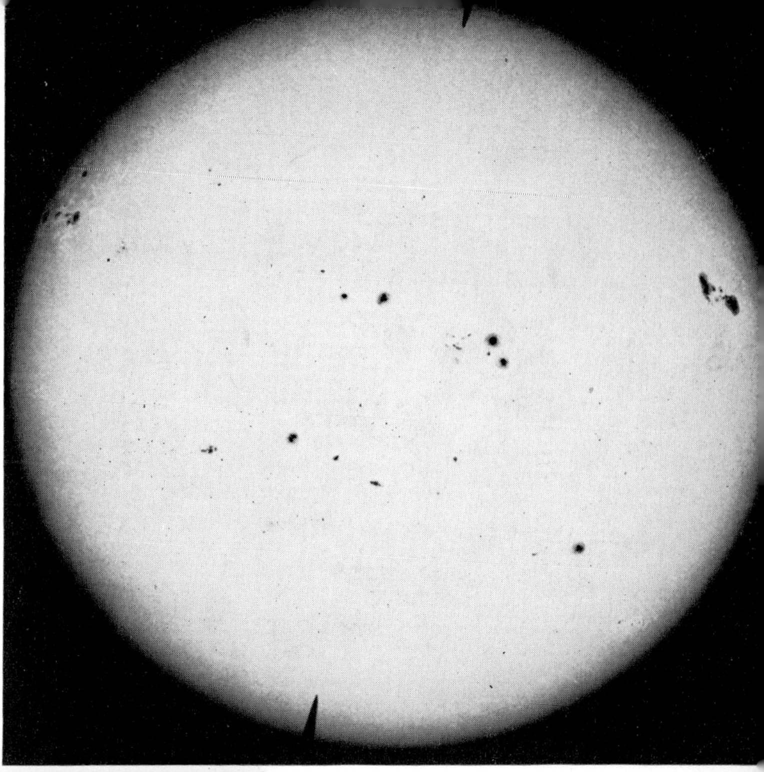

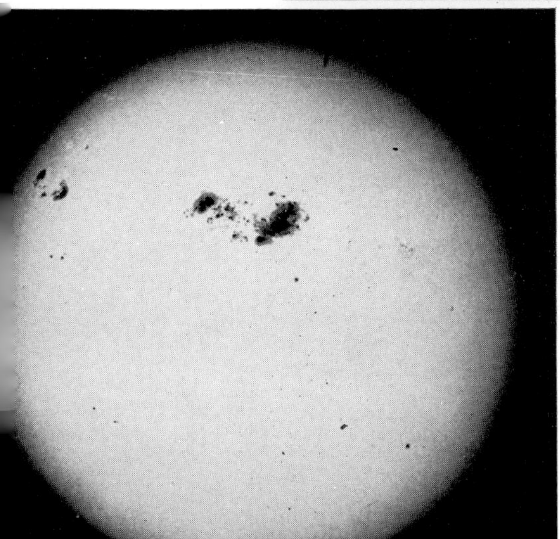

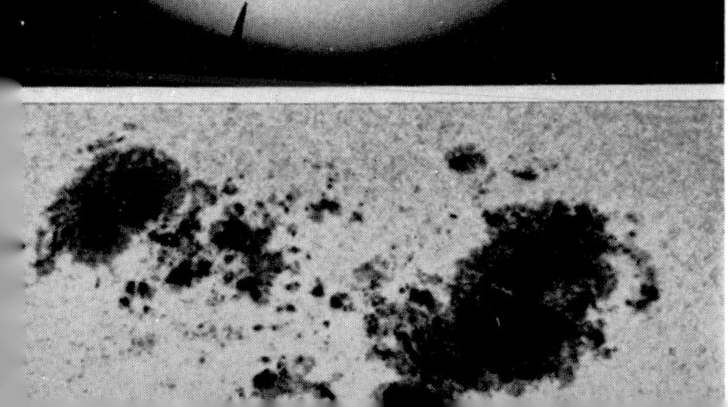

PLATE 1. (*Above*) The Sun at a time of maximum sunspot activity. There are 23 sunspot groups containing more than 90 individual sunspots. (Questions 28 to 37.)

(*Top left*) The sunspot group of April 7, 1947—one of the largest ever recorded. A close-up of the group is shown (*bottom left*) on which the spot on the right is about 96,000 miles across its greatest dimension.

PLATE 2. (*Above*) The corona taken during a total eclipse of June 8, 1918. Prominences are visible at various places around the border of the Moon. (Questions 366 *et seq.*)

(*Right*) The whole edge of the sun, taken in the light of the calcium K line of the spectrum. A variety of prominences can be seen. (Questions 24 to 39.)

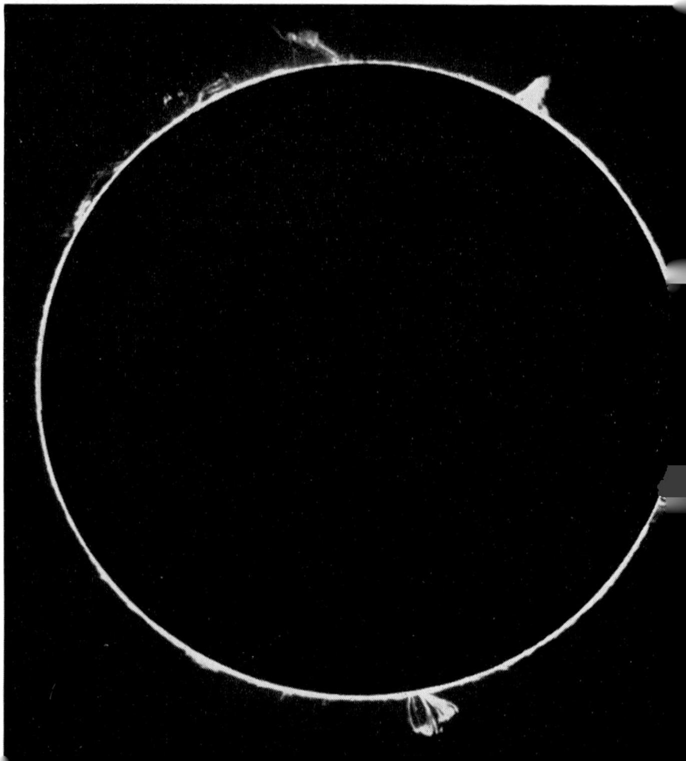

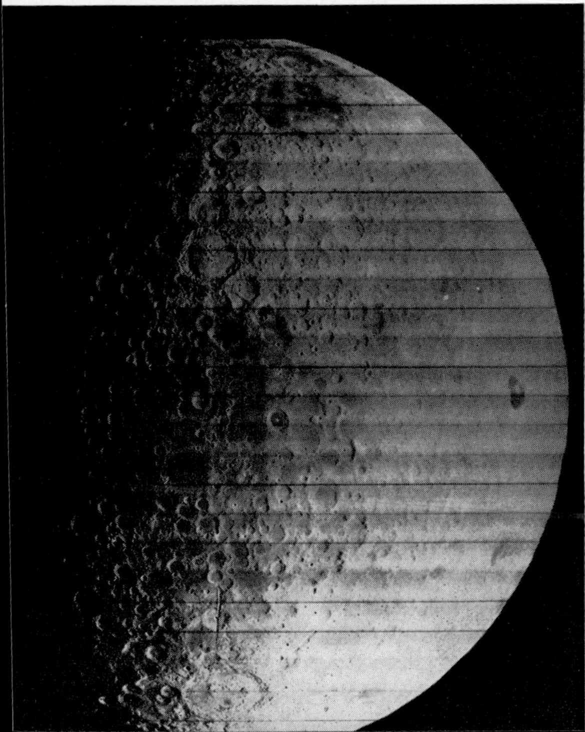

PLATE 3. (*Above*) The full Moon. (Questions 72 to 127.)

(*Left*) Photograph of the Moon, from the U.S. Orbiter series. This shows the side of the Moon turned away from the Earth. The cleft feature in the bottom half of the photograph is a vast crater valley, 240 kilometres long and in places 8 kilometres wide; the associated crater (below) is Hale, just visible as a very foreshortened object from Earth under suitable conditions of libration. (Questions 145, 910.)

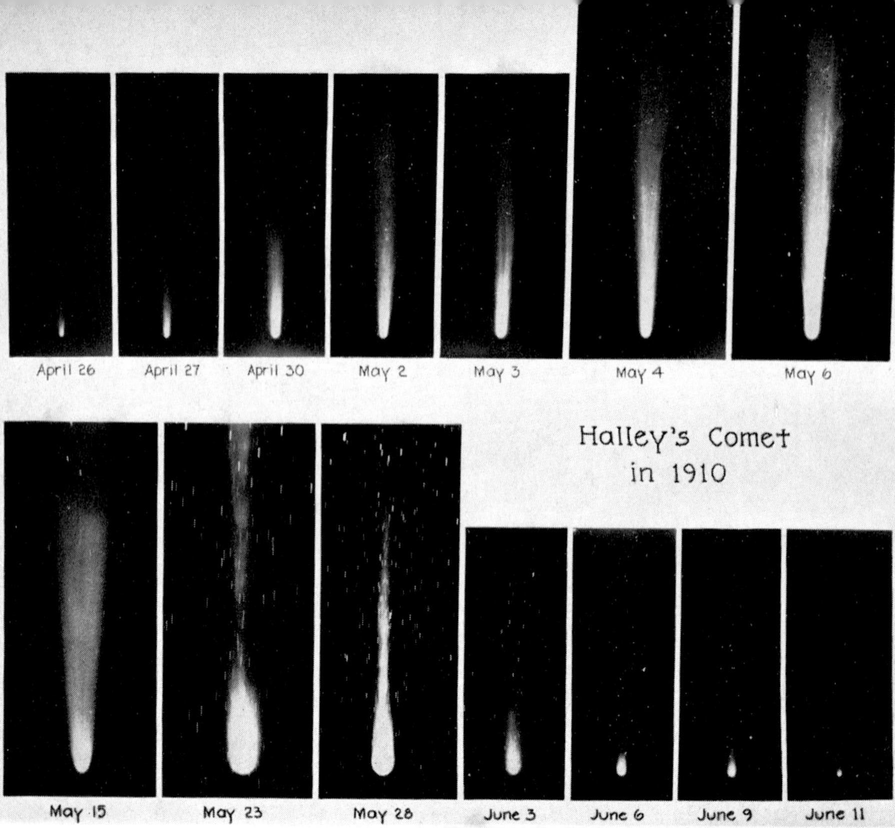

Halley's Comet in 1910

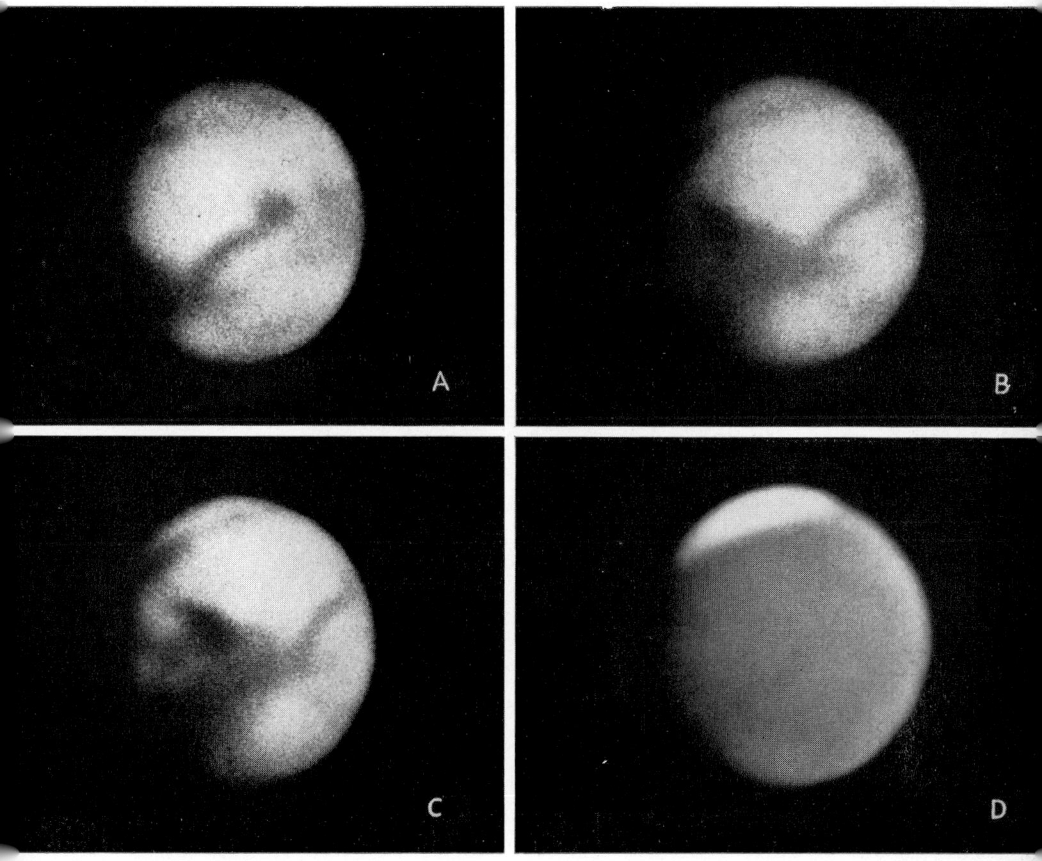

PLATE 4. (*Top left*) Fourteen views of Halley's Comet in 1910, showing its development and decay. (Questions 272 *et seq.*)

(*Bottom left*) The trail of a bright meteor crossing a nebula in Cassiopeia. (Questions 294 *et seq.*)

PLATE 5. Four views of Mars. A, B, and C, taken in red light, show the rotation of the planet. D, taken in blue light, reveals Martian atmosphere and shows clouds at the polar cap. (Question 172 *et seq.*)

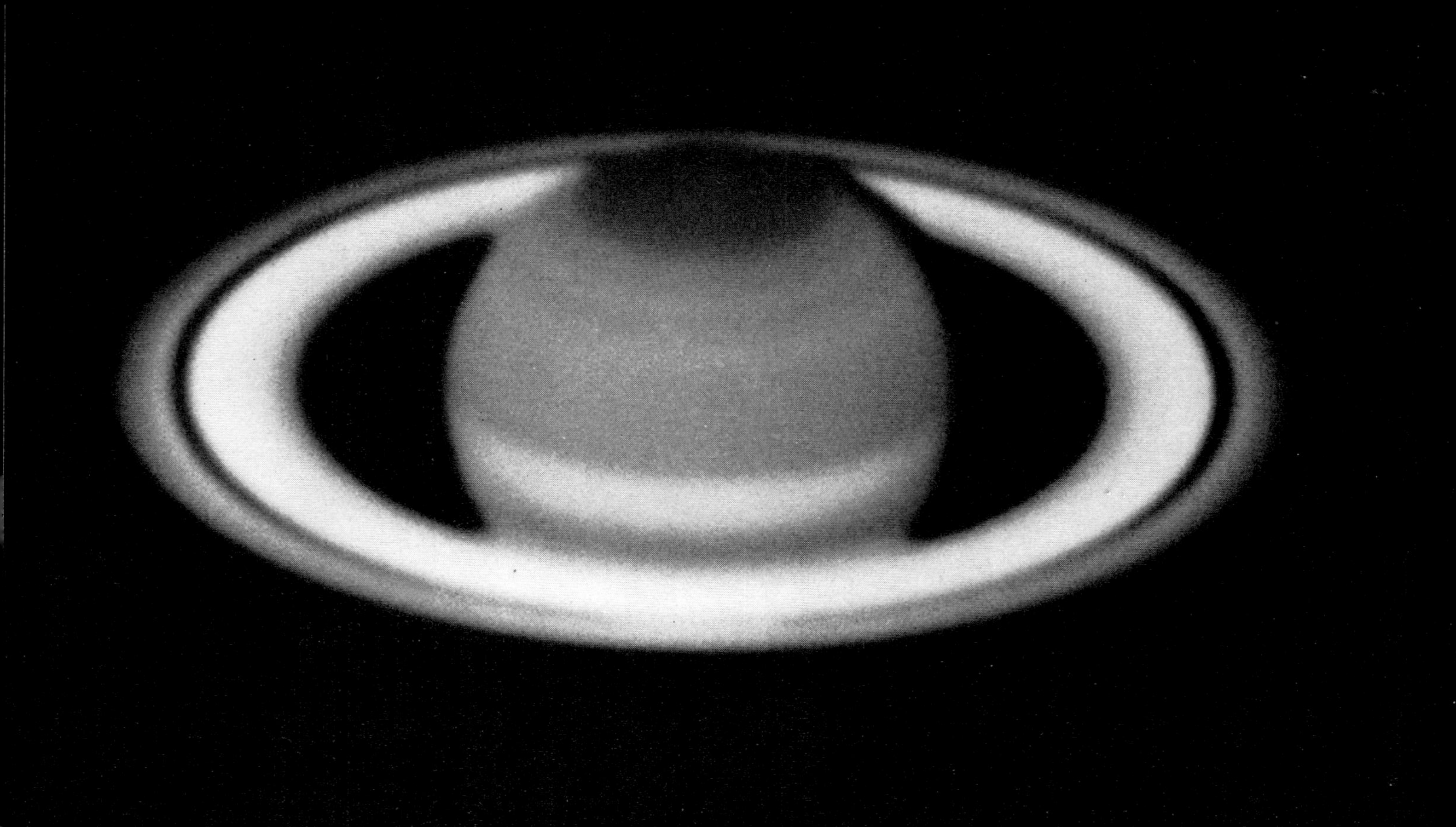

PLATE 6. Saturn and its rings—"one of the truly beautiful things in Nature." (Question 204 *et seq.*)

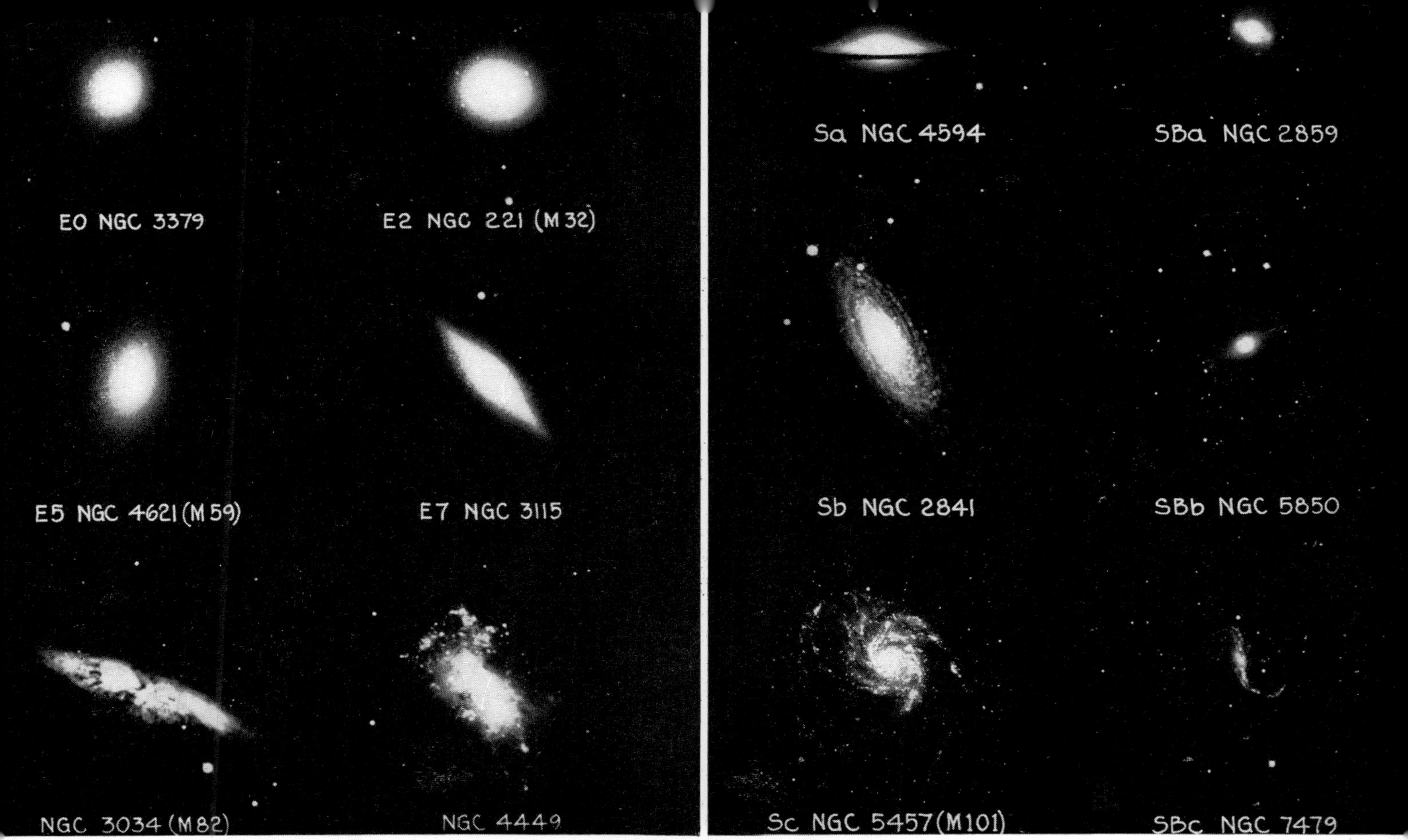

PLATE 7. Various types of galaxies. Stages of elliptical galaxies (EO-E7) and two irregular galaxies, on the left. To the right are typical spiral and barred spiral galaxies. (Question 820.)

PLATE 8. The Globular Cluster in Hercules. Messier 13—N.G.C. 6205. (Questions 532, 533, 591, 614, 685, 690.)

PLATE 9. The Great Spiral Galaxy in Andromeda. Messier 31—N.G.C. 224. Also shown are N.G.C. 205 and N.G.C. 221. (Question 584.)

PLATE 10. (*Above*) The Pleiades. (Questions 534, 563, 785.)
(*Left*) The Horse's Head Nebula in Orion. (Question 735.)

PLATE 11. (*Right top*) The Trifid Nebula in Sagittarius. Messier 20—N.G.C. 6524. (Question 768). (*Right below*) The Omega Nebula in Sagittarius. Messier 17—N.G.C. 6618 (Question 768.)

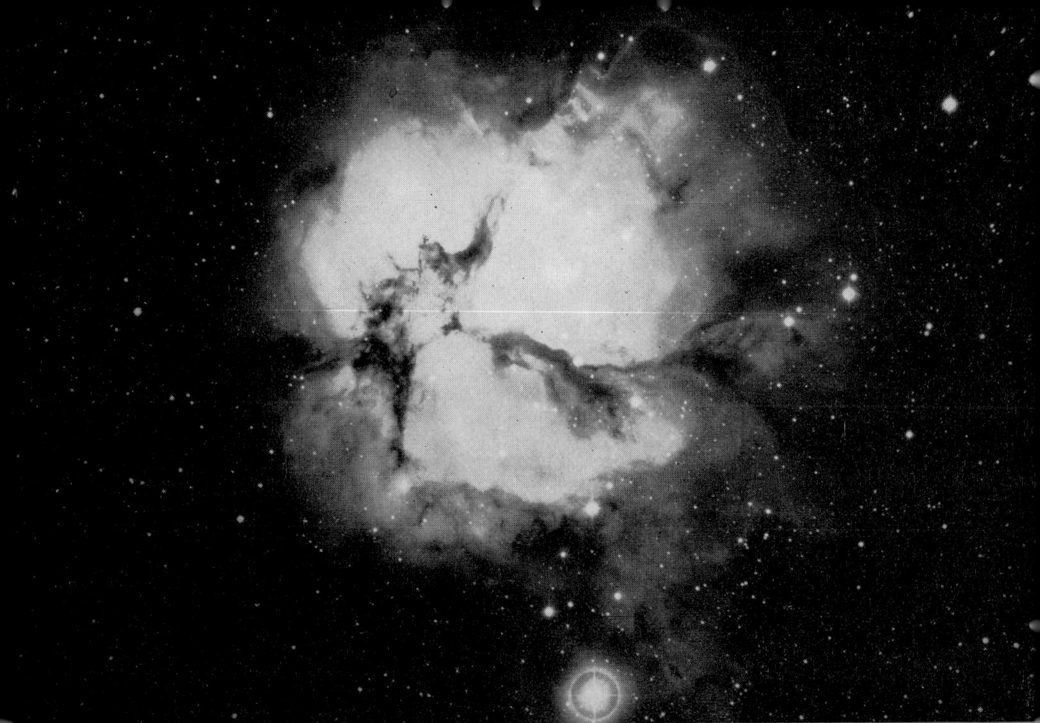

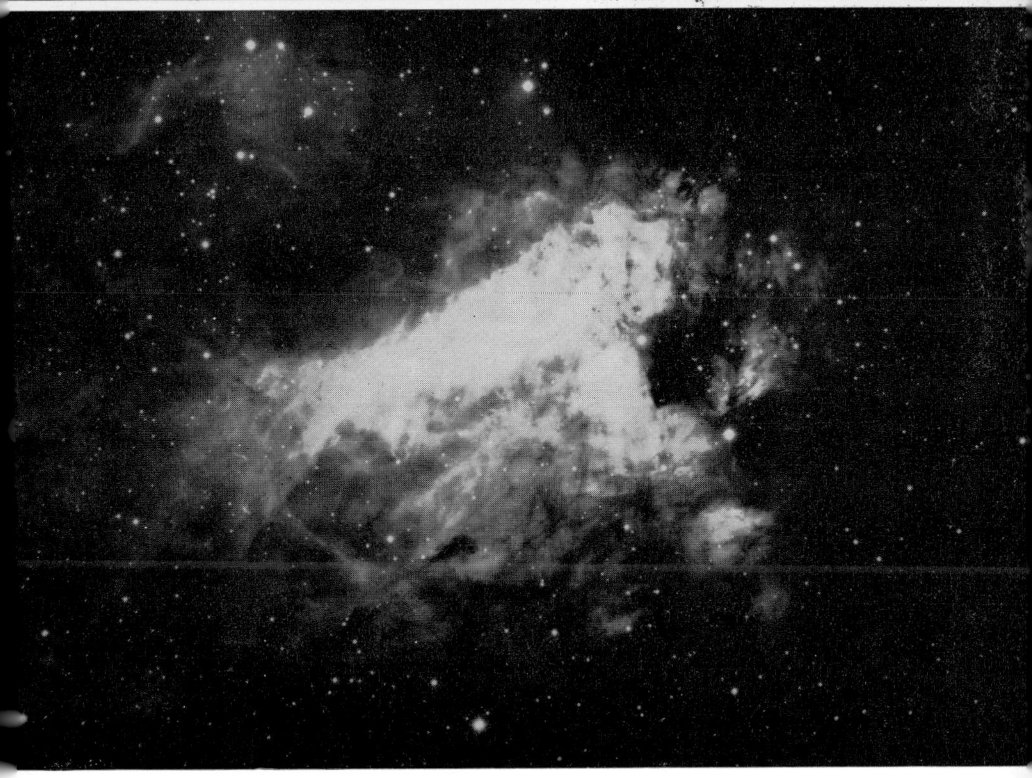

PLATE 12. Spiral galaxies in Ursa Major. Messier 81 and 82. Two other more distant galaxies are also shown. (Question 799.)

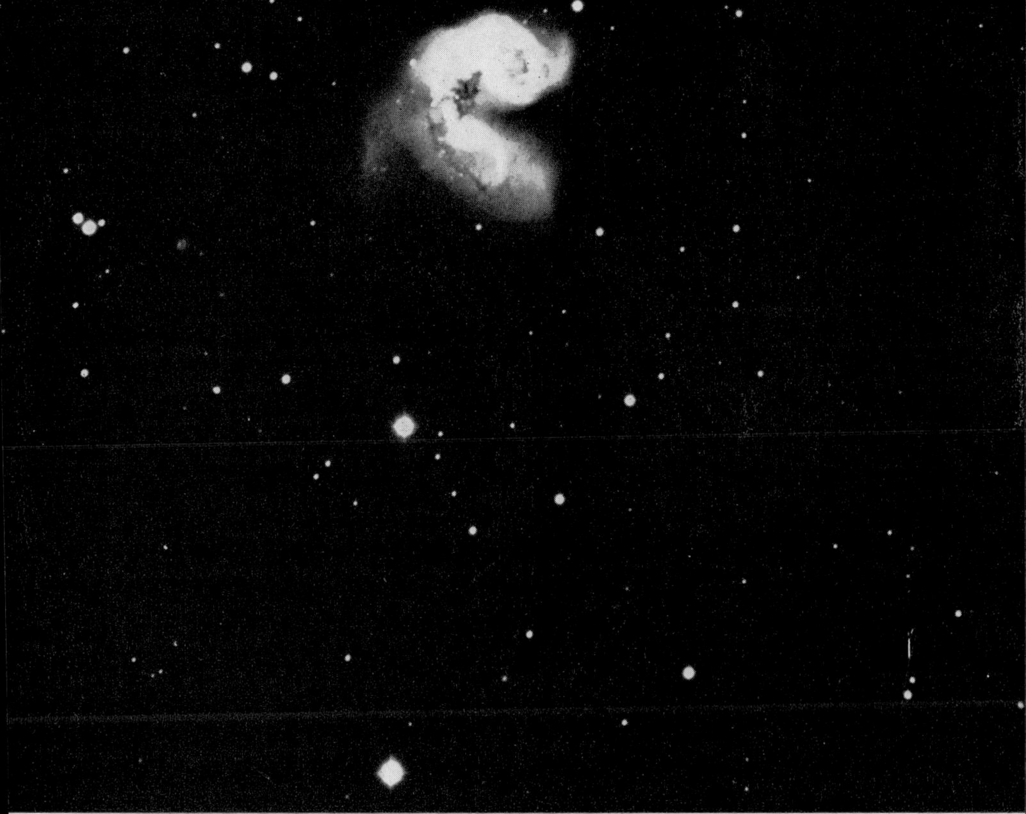

PLATE 13. Two distant galaxies—N.G.C. 4038 and N.G.C. 4039.

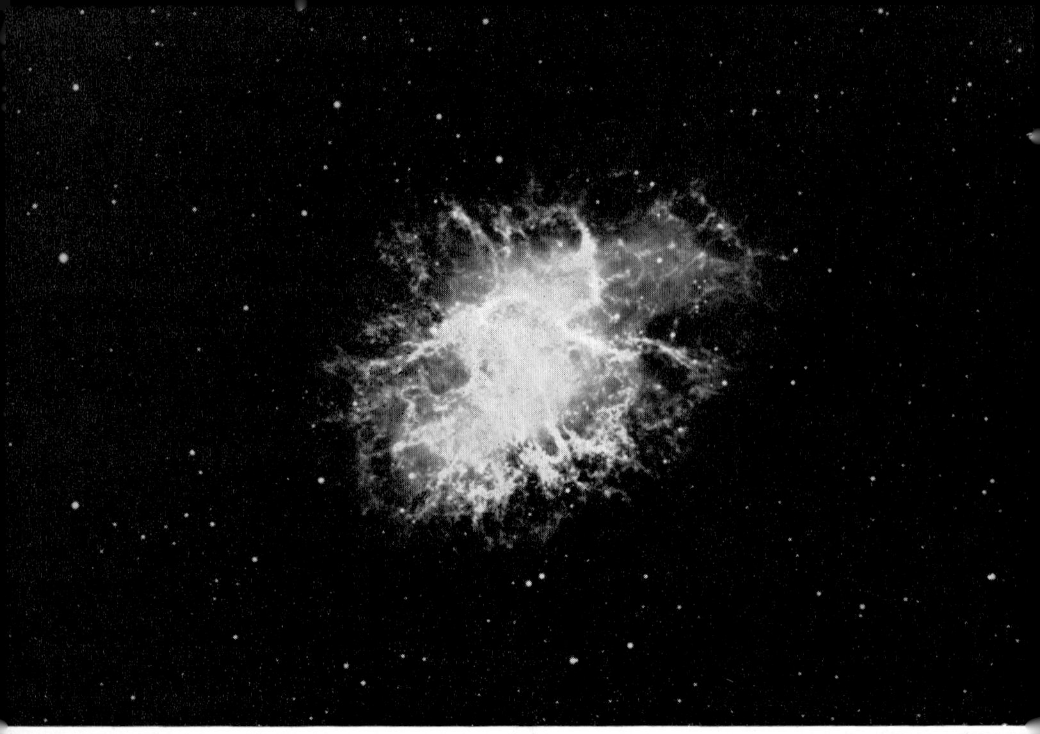

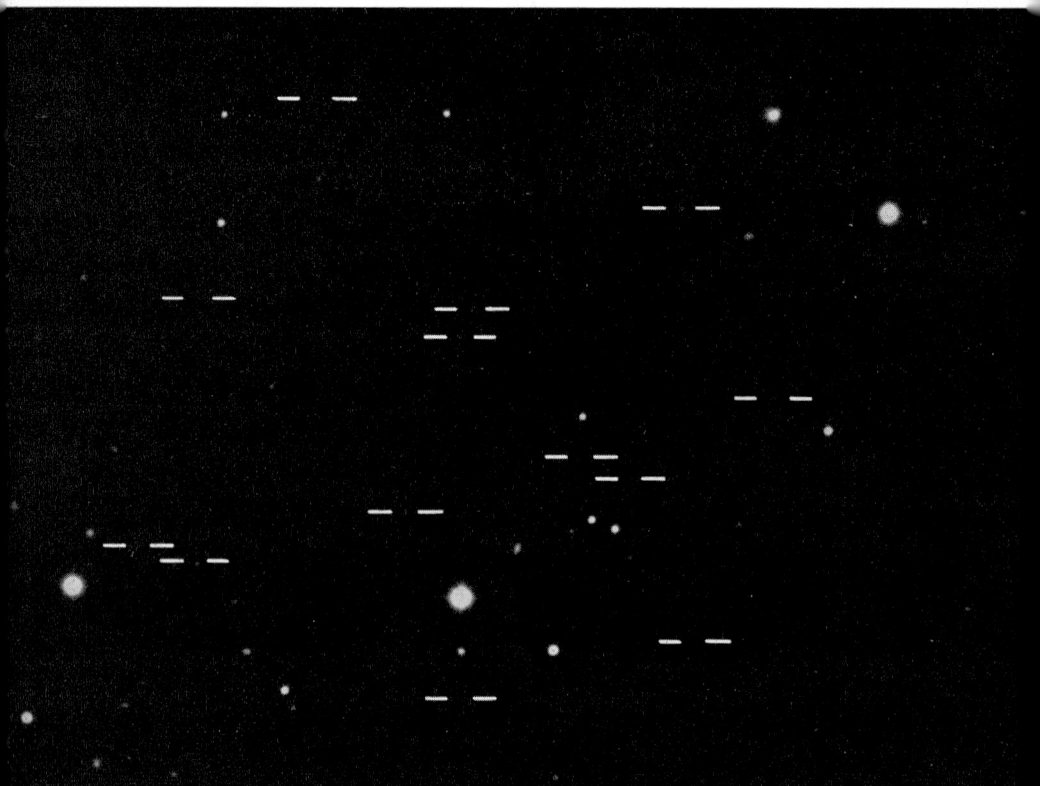

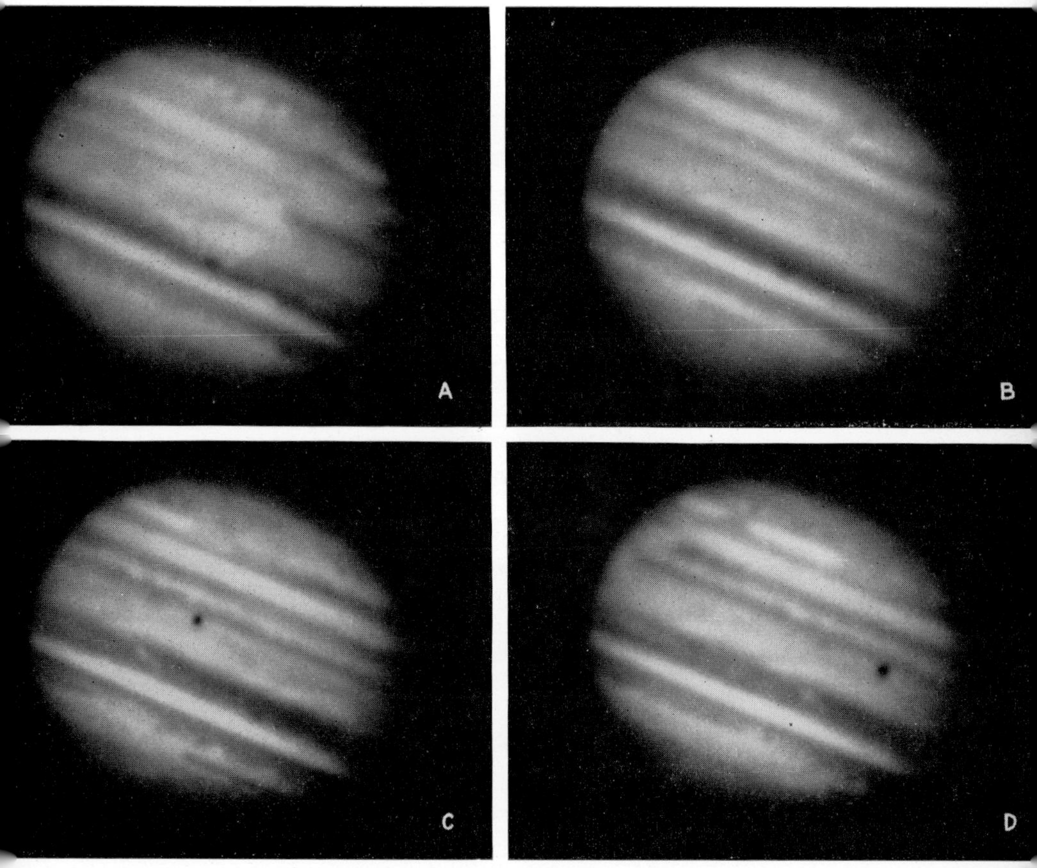

Plate 14. (*Left, above*) The Crab Nebula in Taurus. Messier 1—N.G.C. 1952. (Questions 509 to 517, 785, 868, 870.)

(*Left, below*) The field of distant galaxies in Coma Berenices. The light that registered on this photograph to make the faint spots shown by the white lines travelled from 500 millions to one thousand million years through space. (Question 827.)

Plate 15. (*Above*) Four views of Jupiter. A and B show cloudy streaks. C shows a satellite and its shadow, and D was taken 50 minutes later and shows the shadow only.

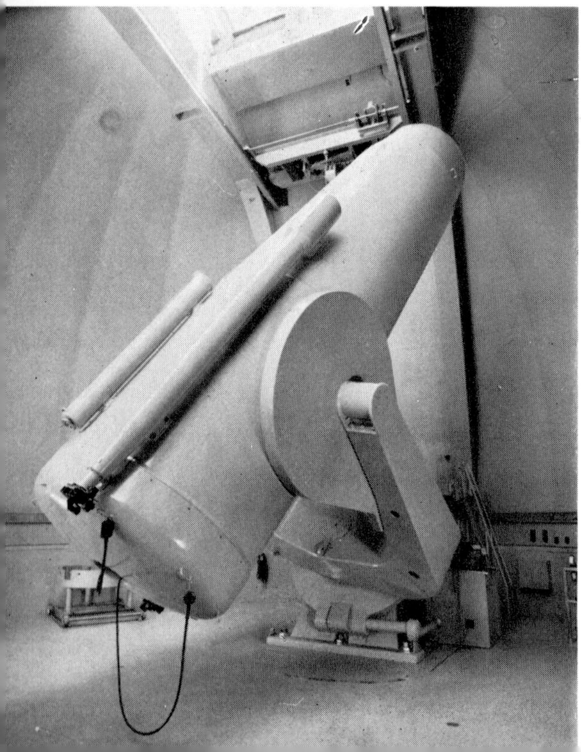

PLATE 16. (*Above*) The 200-inch Hale Reflector at the Mount Palomar Observatory, U.S.A., seen from the east. (Question 961.)

(*Left*) The 48-inch Schmidt telescope at the same observatory.

some violet stars and a few green ones. Every shade and tone of the colours is there. There is no colour that is not represented, but the colours are not striking except for red stars.

474. Do the colours of the stars have any significance? Yes. Colours are indications of star temperatures, and they also tell which of the elements make up the stars. As a rule, the blue and the white stars are the hottest, while the red ones are the cooler stars.

475. How hot are stars? The surface temperatures of stars are measured over a wide range. The hottest stars have a surface temperature on the order of 80,000° Fahrenheit. The coolest visible stars are about 3,000° Fahrenheit at their surfaces. The interior temperatures of all stars must be measured in millions of degrees.

476. How do we know the temperature of a star? An instrument called a thermocouple indicates the amount of radiation we receive from a star. If the star's distance is known, its real temperature can be calculated. The spectroscope will tell us of the behaviour of the atoms that make up a star, and laboratory experiments give us a criterion for the behaviour of atoms of various elements at various temperatures. It is possible to estimate very closely the temperature of any star whose spectrum can be studied.

477. How is the light of a star analyzed? Light is analyzed by means of a spectroscope, one of the most valuable instruments in astronomy. If light is passed through a prism or across the surface

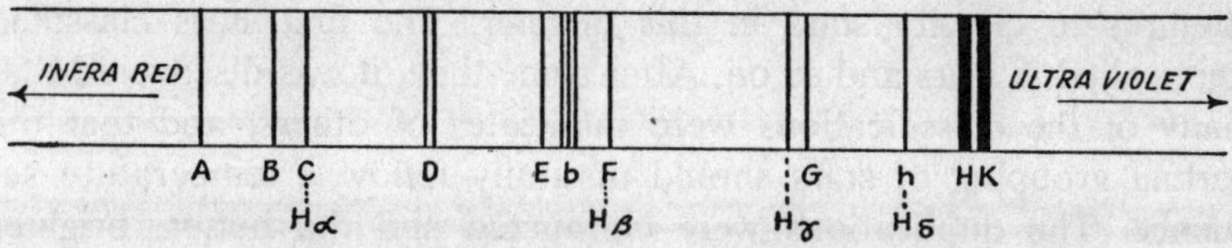

A few of the absorption lines in the solar spectrum. Fraunhofer's designations were alphabetical. Four prominent hydrogen lines are shown.

of a finely ruled grating, it will be refracted or bent. The longer waves of light, which appear red to the eye, will be refracted to a lesser degree than the shorter waves, which are violet. A beam of white light, when it is passed through a prism, emerges as a band of colours with red at one end of the band and violet at the other, and all the rest of the colours between them. Such a band of colours produced by a spectroscope is a spectrum. In stars there is almost always a very hot central region surrounded by atmospheres of cooler gases. The action of the atoms in the star's atmosphere absorbs energy from the hot centre of the star and thus partially removes certain very narrow bands of certain colours from the spectrum. The location of these bands, which appear darker than the rest of the spectrum, identifies absolutely the element whose atoms produced them. Certain minute variations in the appearance and position of these bands, or lines, tell much about the material of a star, its temperature and motion, even though the star is, as all of them are, at an incomprehensible distance from us.

478. Can stars be classified according to their spectra? Yes. Stars have been placed in seven major divisions according to the intensity and arrangement of the lines in their spectra. Each of these major divisions is subdivided into ten minor divisions. The major divisions are given letters of the alphabet and the minor divisions are represented by numerals from 0 to 9 which follow the letter of the major division. The letters indicating the major divisions, in order of their brilliance and other contributing factors, are O, B, A, F, G, K, and M. An easy way to remember them is to learn the sentence "O, be a fine girl, kiss me!" There are, in addition, four other divisions: W at the violet end of the spectrum and R, N, and S, which are at the red end. These minor divisions are side branches of the major divisions. The apparently arbitrary selection of letters grew out of pioneer attempts to classify stars in this manner. The first stars classified were called A stars and so on. After some time, it was discovered that many of the classifications were duplicates of others, and that the normal grouping of stars should naturally follow a temperature sequence. The duplications were eliminated and the hotter, brighter stars were placed first in line. Hence the seven major and three minor classifications as they are now.

479. What is the sequence of spectral divisions? The order of spectral classification runs, generally, from hot to cool; from white to red. The lines indicating many of the elements appear and disappear as we go down the series. In the cooler stars, some molecular compounds begin to appear.

480. What are the spectral characteristics of the O-type stars? These stars show the lines of ionized helium, oxygen, nitrogen and hydrogen and must, therefore, be intensely hot. Very few O-type stars are visible to the naked eye because they are generally at vast distances from us.

481. What are the spectral characteristics of the B-type stars? The B-type stars show lines of ionized oxygen and carbon; neutral helium lines begin to appear and the hydrogen lines are stronger. Beta (β) Orionis (Rigel) and Alpha (α) Virginis (Spica) are examples of B-type stars. They are blue-white and very hot.

482. What are the spectral characteristics of the A-type stars? The hydrogen and ionized magnesium lines are very strong in the A-type stars. Lines of helium and ionized oxygen cannot be seen. Weak metal lines of iron, titanium, etc., begin to appear. Alpha (α) Lyrae (Vega) and Alpha (α) Canis Majoris (Sirius) are typical A-type stars.

483. What are the spectral characteristics of the F-type stars? In F-type stars, the hydrogen lines are weaker and the metal lines are stronger. Calcium lines are quite strong. Alpha (α) Carinae (Canopus) and Alpha (α) Canis Minoris (Procyon) are F-type stars.

484. What are the spectral characteristics of the G-type stars? In G-type stars, the metal lines are very prominent. Calcium lines are stronger and the hydrogen lines are fainter. The Sun and Alpha (α) Aurigae (Capella) are G-type stars.

485. What are the spectral characteristics of the K-type stars? The metallic lines are stronger than the hydrogen lines in K-type stars. The H and K calcium lines reach their greatest intensity. Some molecular lines begin to appear in this type of star, e.g., titanium oxide, cyanogen. These stars are definitely reddish. Alpha (α) Tauri (Aldebaran) and Alpha (α) Boötis (Arcturus) are typical K-type stars.

486. What are the spectral characteristics of the M-type stars? In this type of star the molecular bands are strong, with titanium oxide strongest, and the violet end of the spectrum is weak. These stars are definitely orange or orange-red. Alpha (*a*) Orionis (Betelgeuse) and Alpha (*a*) Scorpii (Antares) are M-type stars.

487. What are the characteristics of the W-type stars? These are extremely hot stars which show bright lines in their spectra rather than the darker lines characteristic of energy absorption. Most of the atoms in these stars have been ionized. They are closely associated with the O-type stars and there are none near enough to be visually prominent.

488. What are the spectral characteristics of the N- and the R-type stars? They are the deep orange-red stars. In them, the molecular bands are strong, with carbon compounds predominating. These stars are cool and not visually bright, so that examples of them would necessarily be obscure and unknown generally, except to astronomers. They are offshoots of the M-type stars.

489. How can the motion of a star be told from its spectrum? The speed of a star in the line of sight will change the wave lengths of energy which are coming from the star. If the star is moving away from us, the waves of energy will be lengthened and all the darker lines will be shifted slightly toward the red end of the spectrum. If the star is moving toward us, the waves will be compressed and the lines will be shifted toward the violet end of the spectrum. The amount of this shift away from the normal is proportional to the velocity of the star, or to the combined velocity of the star and the observer. The motion and direction of the Earth are known, so that the resulting shift will give the speed of a star within about ¼ mile per second. This shift of the spectral lines is known as the Doppler effect.

490. Has the Doppler effect any analogy in sound? Yes. The pitch of a train's whistle, for example, will be raised if the train is coming toward the listener. This is because the sound waves are being compressed and made shorter, and so raised in pitch. If the train is moving away, the sound waves from its whistle will be drawn out and lengthened, and the pitch of the whistle will be lowered. The sudden drop in pitch of a train's whistle as the train passes the listener is a common sound.

491. In what area of the sky are there the most stars? The region of the Milky Way in the upper part of the constellation of Cygnus, The Swan, is considered to be the most densely populated area of the sky.

492. How big are stars? The range of star sizes is extreme. The smallest known stars are about the size of planets—less than 10,000 miles in diameter. The largest stars are hundreds of times the diameter of the Sun, which is an average star. For example, the star Omicron_2 Eridani B, which is a white dwarf star, is about 16,500 miles in diameter—a little more than twice the diameter of the Earth. The variable star VV Cephei is 1,220 times the diameter of the Sun—about one thousand million miles in diameter. There are other bodies that may not be technically classifiable as stars which are nevertheless spheres of gas, which may be even larger. One vast body, part of a system whose designation is Epsilon (ε) Aurigae, may be about 3,000 times the diameter of the Sun.

493. What is the largest star visible to the naked eye? We cannot yet be certain. One very large star is Alpha Herculis—*Ras Algethi*—which is a red giant. Possibly the largest star known is the fainter component of the binary system Epsilon Aurigae, near Capella in the sky. The primary in this system is a very luminous yellow supergiant, at least 60,000 Sun-power. The fainter component cannot be seen at all, because it radiates only in the infra-red, but we can measure its size by noting what happens when the infra-red star passes in front of the yellow star and blots out part of its light. The diameter is estimated as over 2,000 million miles. Recently there have been suggestions that the object is something different—a very old star, called a collapsar or black hole—but as yet the evidence is not conclusive.

494. What is the "main sequence" of stars? Ejnar Hertzsprung and Henry Norris Russell arranged all of the known types of stars upon a diagram which had for its vertical co-ordinate the absolute magnitude of the stars and for its horizontal co-ordinate the spectral classification of the stars. The spectral classification also gave temperature and colour. It was found that the vast majority of the stars fell within the boundaries of a broad line which ran from the upper left-hand corner of the diagram—which is known as the Hertzsprung-Russell diagram—to the lower right-hand corner. This broad line is

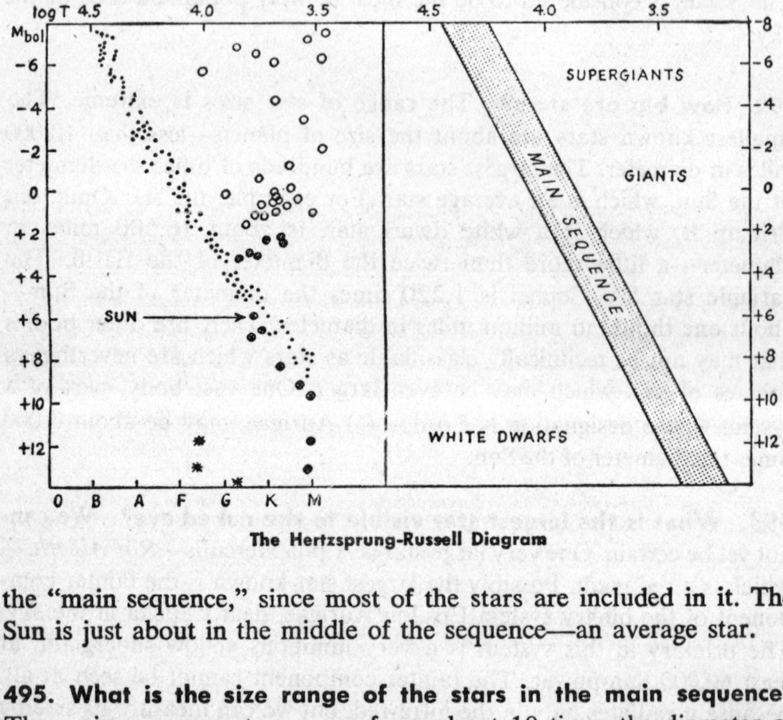

The Hertzsprung-Russell Diagram

the "main sequence," since most of the stars are included in it. The Sun is just about in the middle of the sequence—an average star.

495. What is the size range of the stars in the main sequence? The main-sequence stars range from about 10 times the diameter of the Sun to about 1⁄10 of its diameter.

496. What is a giant star? A giant star is one whose diameter falls somewhere within the general size range between 10 times the diameter of the Sun and 100 times its diameter.

497. Where do the giant stars appear in the Hertzsprung-Russell diagram? The giant stars fall above the main-sequence stars and to the right of the centre of the diagram.

498. Are the giants massive stars? The masses of the stars fall into a very narrow range. The largest star masses known are only about 100 times the mass of the Sun, and such stars are very rare. The giants are large, cool and fairly tenuous stars whose masses start at about the mass of the Sun and go from that point up to about 10 and sometimes 20 times the Sun's mass.

499. Are the giant stars hot stars? Not as a rule, at least at their surfaces. There are very few extremely hot stars that are as large as ten times the diameter of the Sun.

500. What is a supergiant star? Stars which are larger than giants are supergiant stars. A supergiant star is one whose diameter is more than 100 times the diameter of the Sun.

501. Where are the supergiant stars in the Hertzsprung-Russell diagram? The supergiant stars are in the far upper right-hand corner of the diagram.

502. Are the supergiants massive stars? They are tremendously distended and extremely tenuous. The masses of the supergiants will rarely exceed 30 times the mass of the Sun, although the volumes of such stars may be thousands of times the volume of the Sun.

503. Are the supergiant stars hot stars? No. As a rule, the supergiant stars are all of a low surface temperature—about 3,000° C.

504. What is the size range of the supergiant stars? The largest stars are about 3,000 times the diameter of the Sun. From this extreme, the supergiants range in size down to about 100 times the Sun's diameter. In miles, this is from about three thousand million miles in diameter down to about 100 million miles.

505. What is a white dwarf? A white dwarf is a star in what astronomers believe to be the last stages of stellar evolution. The materials which compose a white dwarf have degenerated through a combination of exceedingly high temperatures, lack of internal pressure, scarcity of hydrogen and unfailing gravitational pressure. White dwarfs are intrinsically faint, although they are still at high surface temperatures. This is because they are very small, sometimes planetary in size. These stars are unbelievably dense, for, in spite of their small sizes, they retain stellar masses.

The first star of this type to be discovered was the faint companion of Sirius. Since Sirius is Alpha (α) Canis Majoris A, the smaller, fainter star is Alpha (α) Canis Majoris B. Because Sirius is the Dog Star, Alpha (α) B is also sometimes known as The Pup. The Pup gives off only $\frac{1}{400}$ the light of the Sun, while Sirius produces 26 times the light of the Sun. The Pup has about the same mass as the Sun, although its volume is $\frac{1}{27,000}$ of the Sun's volume. The density of the smaller star, then, must be 27,000 times the density of the Sun, or about 40,000 times the density of water. A pint of water, which weighs about a pound, takes up 20 cubic inches of space. An equal volume of the material of which this star is made would weigh 40,000 pounds, or 20 tons. That is one ton per cubic inch.

The gases of the white dwarf stars are believed to be completely ionized. (See question 1007.) All of the electrons of the atoms of which they are composed have been torn away from their nuclei, so that only the nuclei are left. Since the atom is largely empty space, with its electrons forming the boundary of the volume of the atom, the absence of the electrons permits the nuclei to be in contact, thus doing away with most of the space which the normal atom occupies. The nuclei possess most of the mass of the atom, so that in a white dwarf star, we find most of the original mass of the star compressed into a very small fraction of its original volume.

In spite of their small size and consequent low luminosity, large numbers of these degenerate stars have been found. Typical is Van Maanen's star. Its absolute magnitude is 14.4. It is about the size of the Earth and its specific gravity is 300,000. A cubic inch of its material would weigh 20 tons. The champion of them all, at this writing, is a star listed in the Astrographic Catalogue as +70°8247. It is a thirteenth-magnitude star and is about half the size of the Earth. Its specific gravity is 36 million—620 tons per cubic inch.

In spite of these incredible densities, these stars are made of gas. In order that their gases may be so completely ionized, such stars must have endured tremendous temperatures. Perhaps many of the white dwarfs have gone through the shattering experience of being supernovae. It is believed that the next step for them is the final stage in stellar evolution—complete cessation of all atomic process, and extinction. When this has happened, the star will remain in the heavens for the rest of time, totally invisible, but with most of its gravitational powers remaining, since its mass will have diminished very little. It is possible that there are now as many stars in the sky whose careers have come to an end, as there are bright stars.

506. Where are the white dwarf stars in the Hertzsprung-Russell diagram? The white dwarf stars are very faint but of high temperatures. These qualities place them in the lower left-hand corner of the diagram.

507. What is a nova? The word *nova* is part of the Latin phrase *stella nova,* a "new star." At rare intervals through the years, a bright star has flared up in the skies where no visible star had ever been before. To the people living upon the Earth long ago, these were obviously new stars. With the growth of astronomy, it is now known that about a score of these "new stars" appear every year which are, even at their brightest, too faint to be seen without the help of telescopes. It is only about once every 500 years, on the average, that one flares up so brightly that it attracts universal attention.

These are not new stars. They are stars which have been in their places since stars began, but which have met with stellar catastrophes. They have exploded. We do not yet completely understand the process by which a star becomes a nova. Suddenly such a star will brighten by as much as ten to twelve magnitudes. It will remain thus brighter for a short time and then slowly, irregularly, it will fade until it regains its original dimness. It is believed that some internal explosion, deep within the star, has taken place. The cause of such explosions is not known, but they do release tremendous amounts of matter and raise the temperature of the star's surface sometimes by thousands of degrees. Most of the time, such outbursts, even though they are spectacular, do not damage the star to any great extent, and it returns to its original state.

508. What makes a star become a nova? The cause of the explosion of a star is not completely understood. It is believed to lie in the system of stellar evolution. When a star has reached that point in its career when its hydrogen is about used up, its internal temperature increases suddenly by a tremendous amount. This triggers an explosion deep within the star. In the case of ordinary novae, the star goes through a relatively short period of extreme brightness and high temperature and returns to its original state. In a few cases, stars have become novae more than once, and apparently have been the better for their celestial binges. In the case of supernovae, the explosion is cataclysmic. The entire structure of the star is involved and much of the mass of the star is expelled. In such instances, the transformation of the star is complete, and it probably becomes the strange, degenerate object known as a white dwarf.

509. What happens to the material ejected from a nova? It is thrown into space, sometimes dissipating quickly so that little or no trace of it can later be seen near the ejecting star. Sometimes, as in the case of the Nova Persei, 1901, it remains as a cloud around the star. Most of the matter is set free in space, however, and contributes to the vast quantity of dust and gas that seems to fill interstellar space.

510. What is the origin of planetary nebulae? It was at first believed that planetary nebulae were shells of gases expelled from a star which had become a nova, as most planetary nebulae have a bright central star whose radiation makes these gases visible. The expansion of nova shells, however, is now known to be much more rapid than the expansion of planetary nebulae. Planetary nebulae may result from a slow dispersion of stellar material into space from the central star, which may be in the last stages of stellar evolution.

511. What is a planetary nebula? A planetary nebula is a cloud of gas and dust in space with a more regular form than the ordinary gaseous nebulae. Planetary nebulae were so named by Sir William Herschel because, when seen through a telescope, they show a perceptible disk, somewhat like that seen when the more distant planets are observed. The name does not mean that they are in any way connected with planets, but refers entirely to their appearance. Photographs show that there is really no similarity of appearance between a planet and a planetary nebula, so that the name is now used to identify a class of gaseous nebula of a certain origin. Typical planetary nebulae are the Ring Nebula, in Lyra, M 57; the Owl Nebula in Ursa Major, M 97.

512. Why is the Crab Nebula so called and where is it? The Crab Nebula is a supernova remnant in the constellation of Taurus. It is about one degree from the star Zeta (ζ) Tauri, slightly north of and preceding that star. Its co-ordinates are 5 hours, 31.5 minutes' right ascension, +21°59′ declination. To a person of vivid imagination, a photograph of the Crab Nebula might look vaguely like a crab without claws. (See question 517.)

513. What are some well-known planetary nebulae? Here is a list of the planetary nebulae in Messier's catalogue with their names, the constellation in which they are found and Messier's number.

Name	*Constellation*	*Messier Number*
Dumbbell Nebula	Vulpecula	M 27
Ring Nebula	Lyra	M 57
Owl Nebula	Ursa Major	M 97
——	Sagittarius	M 75

514. How many novae are there? Astronomers find about 20 novae annually, usually on routine patrol plates taken nightly at the big observatories. It is also a part of the duties of some amateur astronomical groups to make a systematic search of the skies at regular intervals for novae. The vast majority of novae, even at their brightest, are still too faint to be seen by the unaided eye.

515. By how much does the light of a nova exceed the normal light of the star? In ordinary novae, an increase of about eight magnitudes is not uncommon. This means an increase of about 16,000 times in the luminosity of the star. In supernovae, the increase must be measured in a factor of millions.

516. What is a supernova and how many of them have there

been? A supernova is a nova which reaches such a degree of brilliance that it attracts general attention. There have been three stars in the historical past that have attained the rank of supernovae. They became so bright that they could be seen even in the daytime. These were the supernova of 1006; the supernova of 1054, whose shell is now known as the Crab Nebula; Tycho's Star, in 1572; and Kepler's Nova in 1604.

517. How do we know of the Supernova of 1054? The existence of a nebula, known as the Crab Nebula, in the constellation of Taurus, led astronomers to learn of the Supernova of 1054. The Crab Nebula is a cloud of dust, gas and debris which is irregularly oval in outline. During the last 30 years, photographs of this cloud show that it is expanding in all directions from a central point. Astronomers have been able to calculate the rate of its expansion, which is about 8,000,000 miles a day. With this knowledge, the probable date at which its growth began has been placed about 900 years ago. Astronomers sought for some record of a nova in the various astronomical annals of that epoch. Both Chinese and Japanese annals told of a "guest star"—one that suddenly appeared in the location in which the Crab Nebula now is, in 1054. There is no doubt that the Crab Nebula is the débris of the 1054 supernova. As well as sending us visible light, the Crab is a source of radio waves and X-rays. In the gas-cloud is one of the remarkable objects known as pulsars, which are rapidly-varying radio sources. A pulsar is believed to be a small, incredibly dense star made up chiefly of neutrons; it represents the final stage in the evolution of a massive star. The Crab pulsar, or "the Crab's powerhouse", is the only pulsar identified with an optical object, though over 60 pulsars are now known by their radio emissions.

The Crab Nebula is 6,000 light-years away, so that the supernova outburst took place 6,000 years before the Chinese and Japanese observed it. The pulsar is the remnant of the actual supernova. The Crab was the first object in Messier's famous catalogue of nebulae, and so is known as M.1.

518. How do we know of the Supernova of 1572? We know of the Supernova of 1572 from an account of it which was written by the great Danish astronomer, Tycho Brahe (*q.v.*). His detailed story of that star is in his work *De Nova Stella*. From this account, this supernova has always been known as Tycho's Star.

519. How do we know of the Supernova of 1604? Johannes Kepler (*q.v.*) has left the principal account of the Supernova of 1604, although Galileo and several other contemporary astronomers kept records of it. It is most often known as Kepler's Star.

520. What are some famous novae? Here is a list of novae which became visible with the naked eye.

Date	*Star*	*Maximum Magnitude*
1054	Crab Nebula—Taurus	-4
1572	Tycho's Star—Cassiopeia	-4
1600	P Cygni	3.5
1604	Kepler's Star—Ophiuchus	-2.5
1670	Nova Vulpeculae	3
1866	T Coronae Borealis	2
1876	Q Cygni	3
1901	GK Persei	0
1918	Nova Aquilae	-0.7
1920	Nova Cygni	1.8
1925	RR Pictoris	1.1
1934	Nova Herculis	1.3
1936	Nova Lacertae	2
1936	Nova Sagittarii	4.5
1942	Nova Puppis	0.4
1946	T Coronae Borealis	2
1950	CP Lacertae	6
1960	DQ Herculis	5
1963	Nova Herculis	3.2
1966	HR Delphini	3.6
1967	Nova Vulpeculae	4.8

521. Has any star ever become a nova more than once? Yes. There are several stars which have become novae more than once. One variable star, T Coronae Borealis, was seen to explode in 1866 and again in 1946. Objects of this kind are called recurrent novae.

522. Are there many double stars? In the list of the 20 brightest stars, seven are double and one is multiple. In our stellar neighbourhood, this ratio of about 40% seems to hold. There may be other regions, however, in which the ratio of double stars is different.

523. Can we see all double stars as pairs? No. Most double stars are so far away that we see them only as single points of light. Some are near enough so that a telescope will separate the two components. Such double stars are called telescopic binaries. Others are so remote that no telescope will resolve them, and their duplicity is known only through spectroscopic analysis. These are called spectroscopic binaries.

524. Are all double stars linked together gravitationally? No. Two stars may appear to be double simply because they are in the same direction as we look at them, although one may be much farther from us than the other. Such a pair of stars is called an optical double, because if it could be seen from some other viewpoint, it would not be a double. If the motions of a double star show that the two components are gravitationally connected, they are known as a physical double. If the two stars are describing orbits about each other and about a common centre of gravity, they are known as a visual binary, and each star is a binary star.

525. What are binaries? Binaries are systems of stars in which two stars are describing orbits about each other and about a common centre of gravity. Not all double stars are binary stars. Stars may be double only because one happens to be in our line of sight with the other, although they are actually light years apart, having no physical connection with each other. Such a pair is called an optical double. If, however, two stars are in one system, connected gravitationally, they are a binary. If the two members of a binary can be separated telescopically, the system is known as a visual or a telescopic binary. If the stars are too far away to be resolved by any telescope, but make their duplicity known only by a doubling and shifting of their spectral lines, the system is a spectroscopic binary. Some binaries—about 1,500 are catalogued—have stars that revolve about each other so directly in our line of sight that they eclipse each other, causing a variation in the light of what we may see as a single star. Such a system is an eclipsing binary.

526. What is an eclipsing binary? When the plane of the orbit of a binary pair of stars is edge on, or almost edge on, to our line of sight, the two stars will mutually eclipse each other, totally or partially, as they revolve about each other. This eclipsing will cause a variation in the light of the system, which we usually see as a single star. The rea-

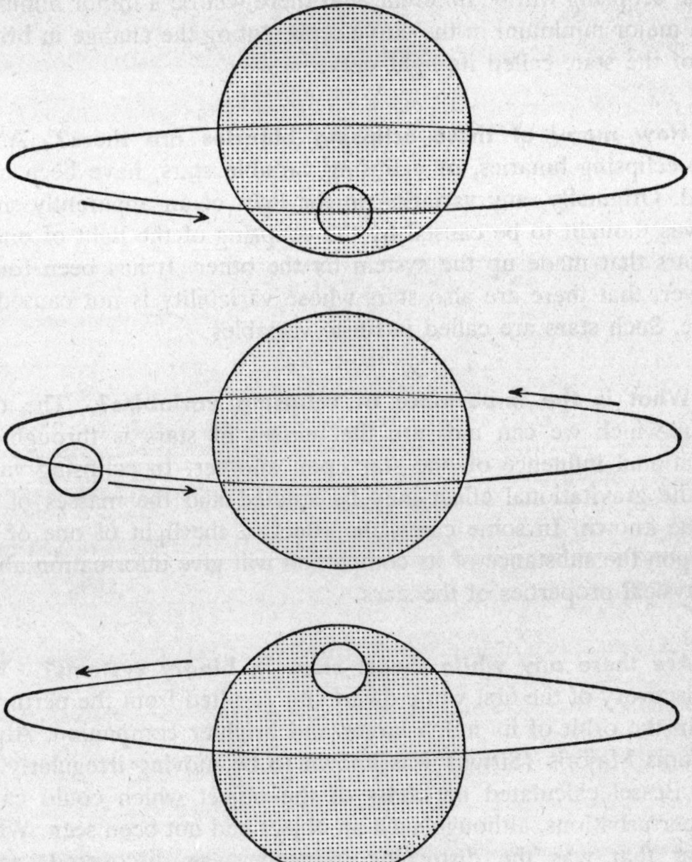

An eclipsing binary system. The orbits of the two stars are almost edge-on to our line of sight. At top, one star completely hides the other, giving minimum brightness. Centre, both stars completely uncovered, giving maximum brightness. Bottom, smaller star covers part of larger star giving a secondary minimum.

son for the variation is that when the stars are not eclipsing each other, to our point of view, we can see the light of both of them. When one eclipses the other, we see only the light from the star nearer to us. If the eclipse is not total, we see all the light from the nearer star plus some of the light from the farther one. If the two stars are not of the same brightness, as is often the case, the dimmer periods of their mutual eclipsing will be different, and there will be a minor minimum and a major minimum in the curve representing the change in brightness of the star, called its light curve.

527. How many of these eclipsing binaries are there? About 1,500 eclipsing binaries, or eclipsing variable stars, have been catalogued. Originally, any variation in the light of an apparently single star was thought to be caused by the eclipsing of the light of one of the stars that made up the system by the other. It has been found, however, that there are also stars whose variability is not caused by eclipse. Such stars are called intrinsic variables.

528. What is the importance of eclipsing variables? The only way in which we can measure the masses of stars is through the gravitational influence of one star upon another. In eclipsing variables, the gravitational effect may be studied and the masses of the stars be known. In some cases, the effect of the light of one of the pair upon the substance of its companion will give information about the physical properties of the stars.

529. Are there any white dwarf stars in binary systems? Yes. The discovery of the first white dwarf star resulted from the perturbations in the orbit of its much larger and brighter companion. Alpha (α) Canis Majoris (Sirius) was known to be moving irregularly. In 1844, Bessel calculated the mass of the object which could cause these perturbations, although such an object had not been seen. When the star that was the disturbing influence was discovered, some years later, it was found to be a white dwarf—very small but of enormous specific gravity. There are several instances of binary stars in which one of the components is a white dwarf. (See question 505.)

530. If there are more than two stars in one system, what is the system called? Strictly speaking, a star system made up of three stars should be called a ternary system; of four stars, a quaternary system; and so on. Actually, if more than two stars are linked together gravitationally, the system is called a multiple system.

531. Is there any other way of detecting a distant double star except by means of a spectroscope? Yes. If the plane of the orbit of the stars in the system is edge on, or nearly so, to our line of sight, there will be a variation in the light of the apparent single star as we see it. This change will be caused by the stars' eclipsing each other as they pass around their orbits. This change, combined with a shifting of the lines in the spectrum of the system, tells us that we are looking at an eclipsing binary.

532. How many stars in a cluster? There is no arbitrary low limit to the number of stars that must be associated in order to be called a cluster. Some galactic, or open, clusters contain relatively few stars, while some of the great globular clusters have many hundreds of thousands of stars.

533. What is a globular cluster? A globular cluster is an aggregation of stars which appears to be spherical. There are about 100 globular clusters which have been identified in our Galaxy. Astronomical opinion holds that there are about 100 more which are so situated that they cannot be seen from the Earth. Globular clusters are tremendously concentrated. Where galactic clusters may contain anywhere from 10 to perhaps 1,000 stars, globular clusters will be packed with two or three hundred thousand stars.

The globular clusters in our Galaxy are at tremendous distances. Even the nearest is more than 20,000 light years away. Most of the clusters visible to us seem to be concentrated in the vicinity of Sagittarius and Scorpius—that part of the Milky Way which contains the hub of the great wheel of stars, dust and gas of which our Sun is a very minor member. The location of so many of these great spheres of stars, all at about the same distance from where the centre of the Galaxy was believed to be, gave Harlow Shapley the clues from which he constructed the theoretical size and shape of our Galaxy about 40 years ago. His picture has been accepted almost universally by astronomers everywhere.

The individual stars that lie in the globular clusters must be tremendously large and bright to be seen at all, even through the agency of photographs of long exposure made with the great telescopes; such are the distances of these clusters. The absolute magnitudes of the stars must be on the order of —10. The brightest stars in the globular clusters are red stars. This means that the individual stars are even larger since the total luminosity is governed by the amount of surface giving out the light. Red light is weaker light, and stars of enormous surface area must fill the globular clusters in order to make them as bright as they must be to cast their light as far as they do. Many of the galactic clusters, by contrast, are filled with hot, blue stars, and can make a better showing with considerably less star area.

The motions of the individual clusters seem to indicate that they pursue rather eccentric orbits about the centre of the galaxy. The stars within a globular cluster must also be moving in orbits about the centre of the cluster. Since the clusters are spherical, or nearly so, the orbits of the stars must be, in many cases, of extremely high inclination. There is no evidence of much dust or obscuring matter in the globular clusters. In this respect, they are also most unlike the galactic clusters, which are, more often than not, highly involved in nebulosity.

When the distribution of the globular clusters in our own Galaxy was analyzed, and the shape and dimensions of the Galaxy determined from that distribution, a search was made in other galaxies for a similar distribution of globulars. Only those galaxies relatively near to ours provided any chance of finding globular clusters in them. They have been seen in the Great Spiral in Andromeda and in some of the other near galaxies. In these distant neighbours, the globular clusters follow the same distribution pattern, confirming the theory of Shapley and others about the size and shape of the Milky Way Galaxy.

534. What is meant by a galactic cluster? A galactic cluster of stars is so called because it partakes of the general rotational motion of the galaxy. They are also known as open clusters or coarse clusters because the stars in them are loosely associated and haphazardly placed within the general outline of the cluster. There is no specific form or shape to a galactic cluster. Indeed, no two of them are alike. Their locations are, in most cases, within the spiral arms of our own and of other galaxies. The stars in a galactic cluster seem to have a common motion in space and, since they are generally of the same spectral class in each cluster, they may also have had a common origin. There are about 300 galactic clusters known within the part of our own Galaxy visible to us, and there are probably many more so situated that they cannot be seen.

Galactic clusters are discernible in the Great Spiral in Andromeda and in other galaxies which are near enough so that some detail of their structures can be observed. Typical examples of galactic clusters in the Local Galaxy are the Pleiades, Praesepe, and the double cluster in Perseus.

535. Some stars have double roman letters as designations. What does this mean? The double roman letters and single roman letters beginning with R are used to designate variable stars. Bayer's use of Roman capitals did not extend beyond Q in any of the constellations, so variable stars were given the rest of the Roman capitals from R to Z. Then, if there are more variables in a given constellation, the letters are doubled and used again, as RR, RS to RZ, SS to SZ and so on, until ZZ is reached. After that, the designation for variables begins again at AA and goes to AZ and so with all the letters, except J, until QZ is reached. This system provides for 334 variable stars in each constellation. If there are more than 334 variable stars in any one constellation, the letter V followed by a number from 335 on is used, as V335, V336 and so on. Each of these designations is followed by the genitive form of the constellation name, as RU Ophiuchi; SS Cygni. If a star had already been given a Greek letter designation by Bayer, and is a variable, its original designation was not changed. Betelgeuse, Alpha (α) Orionis, for example, is a variable, but it has not been given any other than Bayer's original designation.

536. What are variable stars? Variable stars are stars whose brightness is not constant, but whose light changes from time to time.

537. What is the cause of variation in the brightness of a star? Some stars seem to change in brightness because they are members of a system of two or more stars whose revolution about each other is in the plane of our line of sight so that, as we see them, they mutually eclipse each other. These stars are variable only because of position and motion and are known as extrinsic variables. With other stars, the cause of the variability lies within the star and is due to some unbalance of the star's atomic structure which is not yet fully understood. Such stars are intrinsic variables.

538. Are all pairs of stars eclipsing variables? No. The double star must be so far from us that we can see it only as a single point of light, and the two stars that make up the double must be revolving about each other in orbits whose plane is more or less in our line of sight so that the two stars mutually eclipse each other completely or partially.

539. Are all eclipsing variables the same type of star? Almost every type of star has been found in eclipsing variables. There are extremely massive stars and stars of very small mass; there are supergiants, giants and main-sequence stars.

540. Do all eclipsing variables vary in the same manner? No. There is almost infinite variety in the periods and in the amount of variation. If the two stars chance to be equal in size and in luminosity, the variation will be even and regular. This is seldom the case, however. The components are more often widely different in size, in luminosity, and the variations in the planes of the orbits of eclipsing variables also make eclipses of different degrees. Sometimes the mutual eclipses are total, sometimes partial. The relative sizes, luminosities, motions, and angles of the pairs of stars can give a tremendous range of variability.

541. What are the typical eclipsing variable stars? The best known eclipsing binary star is probably Beta (β) Persei (Algol). Its variability was discovered in 1669 and since then many details have been learned about it. The main binary is a system of two stars moving in orbits about a common centre of gravity and about each other in such a manner that they partially eclipse each other as we see them. One of the stars is about three times the diameter of the Sun and is three magnitudes brighter than its companion, which is, nevertheless, about $\frac{1}{5}$ larger than the brighter star. The centres of the two stars are 13 million miles apart, but since the radii of the stars take up about three million miles, their surfaces are about 10 million miles apart. Every 2 days, 20 hours and 49 minutes the mutual orbit of the two stars is completed. For 2 days and 11 hours, the single point of light that we see is constant in magnitude 2.3, except for a very slight fall and rise about halfway in this period. Then, in about five hours, Algol will drop almost one magnitude, to 3.5. In the next five hours, it will rise again to 2.3. Thus it loses and regains almost 2½ times its normal brightness during each cycle.

One of the strangest of the eclipsing binaries is Epsilon (ϵ) Aurigiae. Here, the period is 27 years. During all this time, the spectrum of the brighter star remains visible, but it is shining through the very tenuous body of its enormous component, which may be the largest star known. This larger star, which does not contribute any light to the system and must, therefore, be so cool as to be dark to our vision, has a diameter which has been estimated to be about 3,000 times the diameter of the Sun, or almost three thousand million milcs.

The star Beta (β) Lyrae shows spectroscopic effects that indicate the passage of stellar gases between the two components of the star system. The two stars are probably so close to each other that the attraction between them has distorted them until they are pear-shaped, with their smaller ends pointing toward each other. Gases from one star are being drawn across the gap between them.

542. In a pair of stars, how is the motion of the individual stars measured? The distance of the two component stars of a double must be determined and their orientation must also be found. Then, a series of observations, often covering years of time, must be made to determine their motions.

543. How can we measure the distance between two of a pair of stars? The distance between the stars which constitute a double can be determined by means of a filar micrometer. The micrometer is a circular instrument that is fitted into the field of a telescope. Across the micrometer are two parallel wires (often called cross hairs) whose distance from each other can be changed by means of a finely threaded screw. The direction of the pair of wires can also be rotated about the field of the telescope. One of the two wires is placed upon the image of one of the stars and the wires are then opened or closed until the other wire is over the other star. The distance between the two wires is then read from a scale in seconds of arc.

544. How do we determine the orientation of a double-star system? The micrometer is set so that one of two cross hairs in it points to the north. The other cross hair is then rotated until it lies over the line joining the two components of the double star. The angle made by the two cross hairs is then measured from the north point counter-clockwise and read in seconds of arc. This angle is known as the position angle.

545. What is the plane of the orbit of a double star? We always see the motion of a double star as though it were in the plane of the sky. Actually, however, it must be regarded as being in three dimensions and can be at any angle to our line of sight. It is very seldom that the plane of the orbits of a pair of stars is so oriented that it is really at right angles to our line of sight.

546. What is the shape of the orbits of a pair of stars? The shape of the orbit of any celestial body is an ellipse.

547. What is meant by the line of apsides? The line of apsides is the major axis of an ellipse, or any extension of the major axis beyond the border of an ellipse. The point at which the major axis intercepts the border of an ellipse is an apse. The line upon which these points of interception lie is the line of apsides.

548. What must be known about the orbits of the components of a double star in order to find out the facts about the stars? There are eight essential pieces of information which are needed to determine the facts about stars in a double system. These facts are called the elements of the relative orbit. They are:

1. The semi-major axis of the relative orbit. This is one half of the largest diameter of the ellipse in which the star travels. It is expressed in angular measure in seconds of arc.
2. The time of periastron. This is the time at which the stars in the system will be nearest together. It is expressed by the symbol T.

3. The eccentricity of the relative orbit. This is the amount by which the elliptical shape of the orbit differs from a circle. It is expressed by the symbol e.
4. The inclination of the plane of the relative orbit from the plane of the sky. This is the amount, in degrees, minutes and seconds of arc, by which the plane of the orbit is tilted away from or toward a plane which is at right angles to our line of sight. It is expressed by the symbol i.

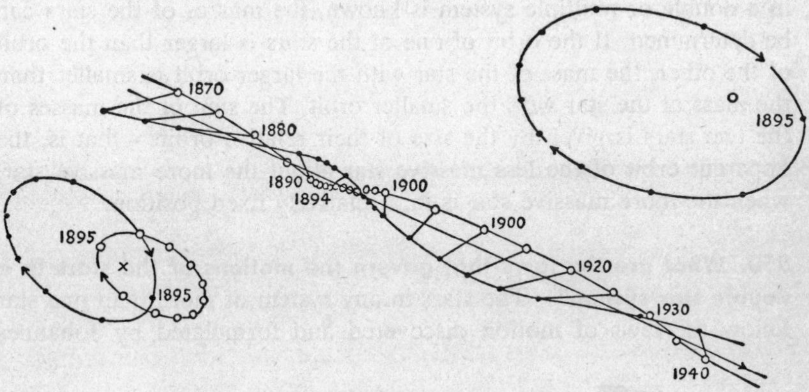

The orbits of a binary system. Lower left; both stars showing fixed orbits of each. Center; motions of stars as seen from Earth. Upper right; orbit of fainter component shown about brighter component considered as fixed.

5. The position angle of the node. This is the point at which the plane of the orbit of the star which is considered to be in motion cuts the plane of the sky—that theoretical plane which is always at 90° to our line of sight. This position angle is in degrees, minutes and seconds of arc and is always measured around the plane of the orbit of the second star counterclockwise. The figure for the angular distance must lie between 0° and 180°. The symbol for this element is a capital omega (Ω).
6. The angle between the node and the periastron point measured in the direction in which the star is moving. This figure determines the orientation of the relative orbit. It is expressed in degrees, minutes and seconds of arc and its symbol is a small omega (ω).

7. The period of complete revolution, expressed in years. Its symbol is the letter P.
8. The mean motion of the star considered to be in motion. This is obtained by dividing 360° by P, the period of revolution, and the answer will be so many degrees per year. This element is written 360°/P.

549. What can be learned about the stars from the laws governing their motion in a multiple system? If the size of the stars' orbits in a double or multiple system is known, the masses of the stars can be determined. If the orbit of one of the stars is larger than the orbit of the other, the mass of the star with the larger orbit is smaller than the mass of the star with the smaller orbit. The sum of the masses of the two stars is given by the size of their relative orbits—that is, the apparent orbit of the less massive star about the more massive star, when the more massive star is in a relatively fixed position.

550. What are the laws that govern the motions of the stars in a double star system? The stars in any system of more than one star follow the laws of motion discovered and formulated by Johannes

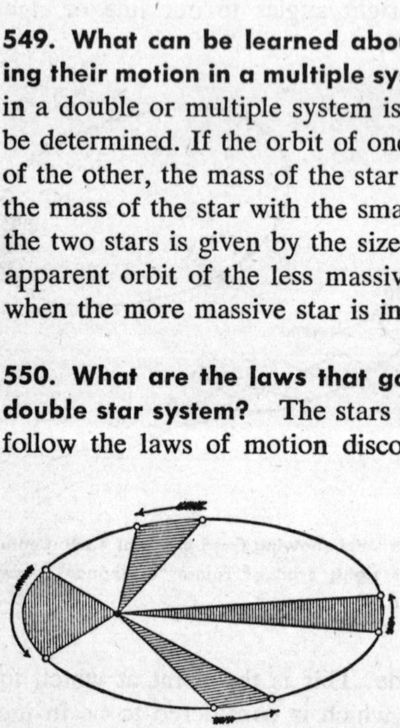

Kepler's Law of Equal Areas. Each area is swept out in the same period of time by the radius vector—the line between the focus of the elliptical orbit and the planet. Hence, the planet must move most rapidly when it is nearest the Sun, which is at one focus of its orbit.

Kepler for the motions of planets in revolution about the Sun. Each star travels in an ellipse about the other star, with the second star at one focus of the ellipse. Both stars move in ellipses about the common centre of gravity of the system. There are thus three ellipses, and each has the same eccentricity. The radius vector of each star sweeps out equal areas in equal periods of time; in other words, the line from the focus of the elliptical orbit to the star covers areas of equal size in equal times. The square of the time taken by each star to complete one circuit of its orbit is proportional to the cube of the mean distance of the stars from each other, divided by the sum of the masses of the two stars.

551. What are intrinsic variable stars? Intrinsic variable stars are stars whose change in brightness is not caused by an eclipsing companion star, but is due to some action within the single star itself. Such stars pulsate, growing brighter and then dimmer with astonishing regularity in most instances, and without any regard for regularity in others.

552. Do the intrinsic variable stars have any common characteristic? No. They are of all kinds, though most of them have fairly late-type spectra.

553. What is the most famous of the regularly intrinsic variable stars? The star Delta (δ) Cephei is the prototype of the regularly periodic intrinsic variable stars. Such stars are called "Cepheid variables" from Delta (δ) Cephei.

554. What is a Cepheid? A Cepheid variable star is a star whose brightness varies in a regular period of time, because of some intrinsic quality in the physical make-up of the star. The reason for the variation of Cepheid variables is not yet fully understood. It has been discovered that the actual brightness of a Cepheid variable star is directly related to the period of its variation. The longer the time required for the star to go from maximum brightness through its faintest phase and back to maximum again, the brighter the star is. This fact was discovered by Miss Henrietta Leavitt at Harvard Observatory about 1911 and the formula relating brightness and the period of variability was written by her and Dr. Harlow Shapley and is the period-luminosity law. This peculiarity of the Cepheid variable stars is of tremendous value to astronomy. Since the period of variation will disclose the true luminosity, or absolute magnitude, of the star, and since we know that the intensity of light diminishes with the square of the distance, an application of the law of inverse squares will give the distance of the star in question. Cepheid variable stars have been found in almost every part of the sky and in all of the distant clusters and in all galaxies near enough for any individual stars to be resolved. Thus, our knowledge of the distances of such specific objects and our conceptions of stellar and galactic distances have been tremendously aided by Cepheid variable stars. The prototype of such stars is one in the constellation of Cepheus, in the northern skies. This particular star bears the astronomical designation of Delta (δ) Cephei; hence the general term of Cepheid variable for all such stars.

555. What is peculiar about the Cepheid variables? The length of the period of a Cepheid variable is directly related to its absolute magnitude. The longer the period of the star—the length of time the star takes to go through its entire cycle of variability, from maximum brightness to maximum brightness again—the brighter is the star.

556. What is the value in knowing the relation between the period and brilliance of Cepheid variable stars? The period of change in a Cepheid variable star discloses its true brightness or absolute magnitude. The difference between its absolute magnitude and its visual magnitude—its brightness as we see it—will give us its distance from us. Cepheid variables are found in many distant organizations of stars, such as clusters and galaxies. The strange property of these stars has enabled astronomers to measure distances in space with much greater accuracy than was possible before the secret of the Cepheid variables was discovered.

557. Do Cepheid variables pulsate as they vary in brightness? Cepheid variables change in size in the same period of time in which they change in brightness, but their size changes do not keep perfect step with their changes in luminosity. The maximum brightness of a Cepheid variable occurs at a point of time in which the star is growing in size, but before it has reached its maximum dimensions. The minimum brightness occurs when the star is in the process of shrinking, but before it has reached its minimum size. Actually, the star is at about the same size when it is brightest as when it is faintest, but in one case its size is increasing and in the other it is decreasing.

558. How are Cepheids used to measure distance? The strange property of a Cepheid variable star is that the star's period—the length of time the star takes to pass from one phase of its cycle of variability through all its changes and arrive again at the same phase —is directly proportional to the real brightness or absolute magnitude of the star. The longer the period of such a star, the brighter the star is. When the length of the period of such a star has been determined, its absolute magnitude is known. We also know that the intensity of the light of any object is inversely proportional to the square of our distance from that object. Hence, if we know the difference between the absolute magnitude of a star and its visual magnitude—its light as we see it—we know the distance to that star. The periods of Cepheid variable stars, which are found in many parts of space, disclose their real brilliance. The difference between their real brilliance and their apparent brilliance measures the distance from the star to the observer.

559. Is it necessary to know anything further about a Cepheid variable than its absolute magnitude in order to use it for a measure of distance? In order to use the Cepheid variables as reasonably accurate indicators of distance, astronomers must also know what is called "the zero point"—that is, exactly what the absolute magnitude of a Cepheid variable of a certain period is. The discovery of the period-luminosity relation was made about 1911 by Henrietta Leavitt during a study of the Cepheid variables in the Magellanic Clouds. At that time, the distance to the Magellanic Clouds was considered to be about 75,000 light years. The zero point of the classical Cepheid was based upon that distance. Recent developments, however, indicate that 75,000 light years is not sufficiently distant for the Magellanic Clouds. In 1952, Walter Baade, at Mount Palomar, revised the zero point for the classical Cepheids and thereby doubled, at least, the distances to objects lying outside of our own local galaxy.

560. Is there more than one kind of Cepheid variable star? Short-period variables are generally divided into two large classes, according to their periods of variation. Those with very short periods—not longer than one day—are often called "cluster variables," because such stars were first discovered in globular star clusters, though they are now more generally termed "RR Lyrae stars". Cepheids with periods of from one day up to 45 days are called "galactic Cepheids" or "classical Cepheids" or "typical Cepheids."

561. What are the distinguishing characteristics of the classical Cepheids? The classical Cepheids are found, generally, in the vicinity of the Milky Way. Their most frequent period is about five days. They are generally very consistent both in the length of their periods and in the manner in which they complete their periods. They increase in brightness in a much shorter time than they take to diminish, and the shape of their light curves varies slightly with a slight difference in their periods. The range of their variation is about one magnitude. They are yellow supergiant stars. Alpha (α) Ursae Minoris (Polaris, the Pole Star) is a classical Cepheid.

562. Which is the best known of the irregular intrinsic variable stars? The most famous of this type of star is probably Alpha (α) Orionis (Betelgeuse). The brightness of Betelgeuse varies from magnitude 0.5 to 1.1. This variation follows a fairly regular basic pattern of about six years, but there is a secondary variation of anywhere from 150 to 300 days imposed upon the longer period. At the same time, the diameter of this tremendous star undergoes a change which involves about 150 million miles. At times, the diameter of Betelgeuse is about 530 times that diameter of the Sun. At other times, its diameter shrinks to about 360 times the Sun's diameter.

563. Are there any lost stars which were once visible but which have now disappeared? It is possible that there are. Tycho's Star, which in 1572 was one of the brightest of the supernovae, cannot now be identified with any star, though it has been identified with a radio source. There are stars whose luminosity has altered during historical time, but the span of human history is not long enough for man to have knowingly witnessed the complete natural extinction of any star. The process is undoubtedly going on and there may come a time when man will be able to record the death of a now bright star. It is possible that there are now almost as many dark, invisible and unknown stars in space as there are bright stars.

564. For what is Barnard's Star notable? Barnard's Star, in the constellation of Ophiuchus, is famous for its large proper motion or tremendous velocity in space. It is moving at about 300 miles per second, and is the star with the greatest known proper motion. However, Barnard's Star is only displaced, from our point of view, by 10″.25 of arc a year. At that rate, it would take the star 352 years to change its position by 1°—twice the apparent diameter of the full moon. It seems to be attended by at least one planet.

565. What is Betelgeuse and where is it? Betelgeuse is the Arabic name for the star Alpha (α) Orionis, visually the brightest star in the constellation of Orion. It is the northeastern star of the four bright stars that outline the main figure of Orion. The name *Betelgeuse* means "The Armpit of the White-Belted Sheep" and is probably a combination of the Arabic translation of the name given to the star by Ptolemy in order to identify it by location in the conventional constellation figure of his time and of the more ancient Arabic name for the entire constellation. Betelgeuse is one of the best known stars in all the heavens and holds a commanding position in a commanding constellation, for Orion dominates the winter skies.

Betelgeuse is a member of a class of stars which are called supergiants because their diameters are more than 100 times the diameter of the Sun. The diameter of Betelgeuse is not constant, however. The star shrinks, at times, until its diameter is a mere 360 times the diameter of the Sun, and then swells until it is about 530 times the Sun's diameter—170 million miles larger. Betelgeuse changes thus irregularly, with no certain cycle, and its brightness changes with its size. When its size is increasing, it is brightest with a magnitude of about 0.1, but when its diameter is decreasing, its brightness diminishes to 0.8—a change of half a magnitude or eight times over in brightness. The reason for this change in both dimension and brilliance is not fully known, but seems to be fairly common among the stars of which Betelgeuse is a typical example. These great red stars are fairly cool—about 3,000° C. at their surface—and it is possible that molecular changes are taking place inside them which cause their variation in brightness.

Betelgeuse is also typical of the red supergiant stars in being extremely tenuous. Its mass is only about 15 times the mass of the Sun. It is a red-hot vacuum. If a piece of Betelgeuse measuring 1 yard on a side—a cubic yard of the material of which the star is made—could be brought to Earth and placed upon scales, it would weigh 1/30 of an ounce.

Betelgeuse was one of the first stars whose diameter was actually measured. This was done in 1920 by Dr. Adams who used an interferometer mounted on the 100-inch Hooker Reflector at Mt. Wilson. (See question 977.)

566. Where is the star Canopus? Canopus is the brightest star in the constellation of Carina, "The Keel" (of Argo Navis, "The Ship *Argo*"). It lies just over 52° south of the celestial equator, about on a line with the bright star Sirius, in Canis Major, "The Greater Dog," but some 35° farther to the south. The astronomical designation of Canopus is Alpha (α) Carinae. The star's position is 6 hours 23.1 minutes of right ascension and 52°40′ south declination. Canopus is magnitude —0.73, second only to Sirius in visual brightness. It is a supergiant star, 210 times the Sun's diameter, and it is intrinsically one of the brightest of stars. Its luminosity is at least 5,200 times that of the Sun, but because it is about 100 light years away from us, its brilliance is considerably reduced by distance. *Canopus* was also the name of a great and fabulous city in ancient Egypt.

567. Which of the stars have been Pole Stars? The star Alpha (α) Draconis (Thuban) was the Pole Star at the time when man began to record his observations of heavenly objects, about 3,000 B.C. The star Beta (β) Ursae Minoris has the Arabic name *Kochab,* which means "The Pole Star." The line of the circle of precession brought the extension of the Earth's north pole past Kochab, but not very near to it, from about 1,500 B.C. until about 300 B.C. Today Alpha (α) Ursae Minoris (Polaris) is the Pole Star, and has been for about 2,000 years. The extension of the North Pole into space will be much nearer to Polaris 100 years from now than it is today.

568. In what part of the sky are the Yeds? The Yeds are two stars in the constellation of Ophiuchus, the Serpent Bearer. Ophiuchus is in the southern part of our skies during late July and August, and represents, in Greek mythology, the figure of the first physician. Across the large figure of Ophiuchus cuts a second constellation, Serpens, the Serpent, which is divided in two parts, Serpens Caput, the Head of the Serpent, which is to the west of Ophiuchus, and Serpens Cauda, the Serpent's Tail, to the east. Ophiuchus is pictured, in the old drawings of the constellation, as grasping the serpent. The Yeds represent his hand. *Yed* is part of the Arabic name for "hand." The upper of the two stars is called Yed Prior, a combination of Arabic and Latin, which means "the first star in the hand," and the other is Yed Posterior, "the rear star in the hand." The astronomical designations of the two stars are Delta (δ) Ophiuchi and Epsilon (ϵ) Ophiuchi, respectively.

569. Why don't the stars appear larger when viewed through a telescope than when seen with the naked eye? The stars are all so far away that no apparent star disk can ever be seen. All that we get from even the nearest star is light—in the form of a beam which we see as a point with only one dimension—length. This point of light is not capable of being magnified. A telescope can increase its intensity, since the function of a telescope is to increase the power of the eye alone to collect light. Thus, while we can never see more of a given star through a telescope than we can without one, we can see more stars, for the telescope collects light from stars so faint that our eyes alone cannot collect their light.

570. What are navigational stars? The navigational stars are 57 bright stars. The positions of these stars on the celestial spheres in relation to the Sun are given in the *Nautical Almanac* for every hour of every day in the year. This information is furnished for the use of navigators.

571. How are these stars listed? The navigational stars are listed in order of their position in right ascension and also in alphabetical order of the popular names of the stars. The magnitudes of the stars are given in round figures, and have not yet been brought up to date with the latest findings. The astronomical designations of the stars are not given.

572. Why are the particular 57 stars selected for navigational stars? The stars chosen are bright—all of them are brighter than third magnitude—and therefore are easily found. The positions of the stars are such that some of them will be in the sky from any point on the surface of the Earth at some time on any night in the year.

573. What are the navigational stars? Here is the list of 57 stars. To this list has been added the astronomical designation of each star. The magnitudes of the stars have also been brought up to date.

Names	*Vis. Mag.*	*Right Ascension* h	m	*Declination* °	′	*Designation*
Alpheratz	2.06	00	06.3	28	52	Alpha (α) Andromedae
Ankaa	2.39	00	24.3	—42	31	Alpha (α) Phoenicis
Schedar	2.16	00	38.2	56	19	Alpha (α) Cassiopeiae
Diphda	2.02	00	41.6	—18	12	Beta (β) Ceti
Achernar	0.51	01	36.2	—57	26	Alpha (α) Eridani
Hamal	2.00	02	04.9	23	16	Alpha (α) Arietis
Acamar	2.92	02	36.7	—40	28	Theta (θ) Eridani
Menkar	2.54	03	00.2	03	56	Alpha (α) Ceti
Mirfak	1.80	03	21.5	49	43	Alpha (α) Persei
Aldebaran	0.86v	04	33.6	16	26	Alpha (α) Tauri
Rigel	0.14v	05	12.6	—08	15	Beta (β) Orionis
Capella	0.05	05	13.7	45	58	Alpha (α) Aurigae
Bellatrix	1.64	05	23.0	06	19	Gamma (γ) Orionis
El Nath	1.65	05	23.8	28	35	Beta (β) Tauri
Alnilam	1.70	05	34.2	—01	14	Epsilon (ε) Orionis
Betelgeuse	0.71v	05	53.0	07	24	Alpha (α) Orionis
Canopus	—0.73	06	23.1	—52	40	Alpha (α) Carinae
Sirius	—1.43	06	43.4	—16	40	Alpha (α) Canis Majoris
Adhara	1.48	06	57.1	—28	55	Epsilon (ε) Canis Majoris
Procyon	0.37	07	37.2	05	20	Alpha (α) Canis Minoris
Pollux	1.16	07	42.9	28	07	Beta (β) Geminorum
Avior	1.97	08	21.7	—59	23	Epsilon (ε) Carinae
Suhail	2.24	09	06.5	—43	16	Lambda (λ) Velorum
Miaplacidus	1.67	09	12.8	—69	33	Beta (β) Carinae
Alphard	1.98	09	25.6	—08	29	Alpha (α) Hydrae
Regulus	1.36	10	06.2	12	10	Alpha (α) Leonis
Dubhe	1.81	11	01.3	61	58	Alpha (α) Ursae Majoris
Denebola	2.14	11	47.0	14	48	Beta (β) Leonis
Gienah	2.59	12	13.7	—17	19	Gamma (γ) Corvi
Acrux	1.39	12	24.4	—62	53	Alpha (α) Crucis
Gacrux	1.69	12	28.9	—56	53	Gamma (γ) Crucis
Alioth	1.79	12	52.3	56	11	Epsilon (ε) Ursae Majoris
Spica	0.91v	13	23.1	—10	57	Alpha (α) Virginis
Alkaid	1.87	13	46.0	49	31	Eta (η) Ursae Majoris
Hadar	0.63	14	01.0	—60	11	Beta (β) Centauri
Menkent	2.04	14	04.3	—36	10	Theta (θ) Centauri
Arcturus	—0.06	14	13.8	19	23	Alpha (α) Boötis
Rigil Kentaurus	—0.27	14	36.9	—60	40	Alpha (α) Centauri
Zubenelgenubi	2.76	14	48.5	—15	50	Alpha (α) Librae
Kochab	2.04	14	50.8	74	19	Beta (β) Ursae Minoris

Alphecca	2.23v	15	33.0	26	51	Alpha (α) Coronae Borealis
Antares	0.92v	16	26.9	−26	21	Alpha (α) Scorpii
Atria	1.93	16	44.4	−68	57	Alpha (α) Tri. Aus.
Sabik	2.46	17	08.1	−15	41	Eta (η) Ophiuchi
Shaula	1.60	17	30.9	−37	05	Lambda (λ) Scorpii
Rasalhague	2.09	17	33.1	12	35	Alpha (α) Ophiuchi
Eltanin	2.21	17	55.7	51	30	Gamma (γ) Draconis
Kaus Australis	1.81	18	21.5	−34	24	Epsilon (ε) Sagittarii
Vega	0.04	18	35.6	38	45	Alpha (α) Lyrae
Nunki	2.12	18	52.8	−26	21	Sigma (σ) Sagittarii
Altair	0.77	19	48.8	08	46	Alpha (α) Aquilae
Peacock	1.95	20	22.5	−56	52	Alpha (α) Pavonis
Deneb	1.26	20	40.1	45	08	Alpha (α) Cygni
Enif	2.31	21	42.2	09	41	Eta (η) Pegasi
Al Na'ir	1.76	22	05.7	−47	09	Alpha (α) Gruis
Fomalhaut	1.19	22	55.4	−29	50	Alpha (α) Piscis Austrini
Markab	2.50	23	02.8	14	59	Alpha (α) Pegasi

574. Where in the sky is the Beehive? The Beehive is the open cluster of stars in the constellation of Cancer. (See question 611.) It is also known as Praesepe, M 44, and N.G.C. 2632, depending upon whether you consult the ancient Romans, Messier's Catalogue of Star Clusters and Nebulae, or the New General Catalogue.

575. What is the Coalsack and where is it? The Coalsack is the name given to a region of the southern sky lying adjacent to the Southern Cross in which a cloud of cosmic dust hides the light of many stars. The Coalsack appears as a well-defined black spot and was first reported by the crews of Magellan's expedition on their return from the first voyage around the world. For many years the Coalsack and similar regions in many parts of the sky were considered to be places where the distribution of the stars thinned out. Sir William Herschel called such places "holes in the sky." Improved methods and instruments, however, show that all of these dark areas are caused by clouds of dust, dark and inert, matter in its most primitive form. The individual grains that make up these vast clouds are estimated to be about $\frac{1}{100,000}$ of an inch in diameter. It has been estimated that about 10% of the mass of our Galaxy is made up of dust and gas, and dark matter in many other galaxies shows plainly, usually as a dark line through the centre of a galaxy seen edgewise, and as streaks and spots in galaxies presented at other angles.

576. Where is Charles's Wain? Charles's Wain is a name for the group of seven bright stars in the constellation of Ursa Major, The Great Bear, which is known in the United States as The Big Dipper. The name *Charles's Wain* originated in England more than 300 years ago, and the Charles referred to was Charles I of England. The name arose from the resemblance of this group of stars to a cart without wheels, but with a shaft to which horses could be hitched; it is now popularly known as the Plough. Similarly, the Japanese call The Great Bear "The Emperor's Carriage."

IX. THE CONSTELLATIONS

577. What is a constellation? A constellation is a group of stars within a specific region. The modern constellations are units of area in the celestial sphere. There are 88 recognized constellations in the heavens now, whose boundaries were surveyed and established by a commission appointed in 1925 by the International Astronomical Union. The chairman of this commission was M. E. Delporte of the Royal Observatory of Belgium. The Report of the Delporte Commission was published in 1930, and the constellation boundaries it recommended have been standard ever since.

Most of the constellations which lie north of a line running about 45 degrees south of the celestial equator are ancient, since this is the part of the sky which was visible to the parts of the Earth where the ancient civilizations flourished. Originally, the constellations did not necessarily border each other and parts of the sky belonging to no official constellation were left between the more striking regions. These vacant places have been filled within the last 300 years by modern constellations which, from the very inconspicuous nature of the regions originally left unnamed, are faint and usually small. The names of 23 constellations in the extreme Southern heavens are considered modern and were named by various leaders in astronomy during the last 200 years. One of the original constellations, Argo Navis, lying mainly in the southern sky, was found to be so big and to contain so many stars that it was impossibly unwieldy. It was broken up into three smaller constellations, each having a name denoting a part of a ship, as Carina (The Keel), Puppis (The Deck) and Vela (The Sails).

Constellations should be considered in the same light as geographical entities. They designate specific areas on the celestial sphere, and are the addresses of all stars and other objects permanently within their borders and of planets, comets, meteors, the Sun, Moon and other temporary visitors to their areas. Constellations are infinite in depth, for objects known to be at vast distances beyond the stars are considered to be within the confines of a constellation. For example, the Great Spiral in Andromeda, which lies at least 2,200,000 light years away from the Sun, is located as being in the constellation of Andromeda, as are the stars in the same region which are mere hundreds of light years away.

For details of the various constellations, see the questions dealing with each.

578. How many constellations are there, and what are they? There are 88 constellations now recognized in modern astronomy. Their names, together with the conventional pronunciations of the names and their meanings, are given below.

Andromeda	An drom′e da	Andromeda
Antlia	Ant′li a	(Air) Pump
Apus	A′ pŭs	Bird of Paradise
Aquarius	A kwar′ i ŭs	Water Bearer
Aquila	Ak′ wi la	Eagle
Ara	A′ ra	Altar
Aries	A′ ri ez	Ram
Auriga	O rī′ ja	Charioteer
Boötes	Bo o′ tez	Herdsman
Caelum	Sē′ lum	Chisel (Sculptor's)
Camelopardalis	Ka mel′ o pard′a lis	Giraffe
Cancer	Kan′ ser	Crab
Canes Venatici	Ka′ nez Ve nat′ ī si	Hunting Dogs
Canis Major	Ka′ nis Ma′ jer	Greater Dog
Canis Minor	Ka′ nis Mī′ ner	Lesser Dog
Capricornus	Kap′ rī kor′ nŭs	Horned Goat
Carina	Kar īn′ a	Keel
Cassiopeia	Kas′ ī o pē′ ýa	Cassiopeia
Centaurus	Sen to′ rŭs	Centaur
Cetus	Sē′ tŭs	Whale
Chamaeleon	Ka mē′ le ŭn	Chameleon
Circinus	Sŭr′ sī nŭs	Compass (dividers)
Columba	Ko lŭm′ ba	Dove (Noah's)
Coma Berenices	Ko′ ma Be re nī′ sēz	Berenice's Hair
Corona Australis	Ko ro′ na os tra′ lis	Southern Crown
Corona Borealis	Ko ro′ na bo rē al′is	Northern Crown
Corvus	Kor′ vŭs	Crow or Raven
Crater	Kra′ ter	Cup
Crux	Kruks	Cross (Southern)
Cygnus	Sig′ nŭs	Swan
Delphinus	Del fēe′ nŭs	Dolphin
Dorado	Do ray′ do	Swordfish

Draco	Dray′ ko	Dragon
Equuleus	E kwo oo′ le ŭs	Colt
Eridanus	E rid′ a nūs	Eridanus (a river)
Fornax	For′ naks	Furnace
Gemini	Jem′ i nī	Twins
Grus	Groos	Crane
Hercules	Her′ ku lēz	Hercules
Horologium	Ho ro lo′ ji ŭm	Clock
Hydra	Hī′ dra	Water Snake (female)
Hydrus	Hī′ drŭs	Water Snake (male)
Indus	In′ dŭs	Indian
Lacerta	La ser′ ta	Lizard
Leo	Le′ o	Lion
Leo Minor	Lē′ o mī′ nor	Lesser Lion
Lepus	Le′ pūs	Hare
Libra	Lee′ bra	Balance
Lupus	Loo′ pūs	Wolf
Lynx	Lingks	Lynx
Lyra	Lī′ ra	Lyre
Mensa	Mayn′ sa	Table
Microscopium	Mī kro sco′ pi ŭm	Microscope
Monoceros	Mon o′ ser os	Unicorn
Musca	Moo′ ska	Fly
Norma	Nor′ ma	Ruler
Octans	Ok′ tans	Octant
Ophiuchus	Of ee u′ kŭs	Serpent Bearer
Orion	O ri′ on	Orion
Pavo	Pah′ vo	Peacock
Pegasus	Peg′ a sus	Pegasus (winged horse)
Perseus	Per′ soos	Perseus
Phoenix	Fee′ niks	Phoenix
Pictor	Pic′ tor	Painter
Pisces	Pie′ seez	Fishes
Piscis Austrinus	Pie′ sis os treen′ us	Southern Fish
Puppis	Pŭp′ is	Deck
Pyxis	Piks′sis	Mariner's Compass
Reticulum	Re tik′ u lŭm	Net
Sagitta	Sa jit′ a	Arrow
Sagittarius	Sa jit aīr′ i ŭs	Bowman
Scorpius	Skor′pi ŭs	Scorpion

Scutum	Sku′ tŭm	Shield
Serpens (Caput)	Ser′ pens (Kap′ ut)	Serpent (Head)
(Cauda)	(Kow′ da)	(Tail)
Sextans	Seks′ tans	Sextant
Taurus	To′ rŭs	Bull
Telescopium	Te le skop′ i ŭm	Telescope
Triangulum	Trī an′ gul ŭm	Triangle
Triangulum Australe	Trī ang′ u lŭm os tra′ le	Southern Triangle
Tucana	Too ka′ na	Toucan
Ursa Major	Ur′ sa Ma′ jer	Greater Bear
Ursa Minor	Ur′ sa Mī′ ner	Lesser Bear
Vela	Vay′ la	Sails
Virgo	Vŭr′ go	Virgin
Volans	Vo′ lans	Flying Fish
Vulpecula	Vŭl pek′ ŭ la	Little Fox

579. On what nights are the various constellations in the best position to be seen? Here is a list of the nights on which the different constellations culminate, that is, on which they reach the highest part of their course across the sky. Some of them cannot be seen from latitudes too far north or south, but unless you live on the equator, this will always be the case.

January 3	Reticulum	March 21	Pyxis
January 14	Taurus	March 30	Vela
January 15	Caelum	April 8	Sextans
January 27	Orion	April 9	Leo Minor
January 28	Lepus, Mensa	April 10	Antlia
January 30	Pictor	April 15	Chamaeleon, Leo
January 31	Dorado	April 25	Ursa Major
February 1	Columba	April 26	Crater
February 4	Auriga	April 29	Hydra
February 6	Camelopardalis	May 12	Corvus, Crux
February 16	Canis Major	May 14	Centaurus, Musca
February 19	Gemini, Monoceros	May 17	Coma Berenices
February 22	Puppis	May 22	Canes Venatici
February 28	Canis Minor	May 26	Virgo
March 4	Volans	June 14	Circinus
March 5	Lynx	June 16	Boötes
March 16	Cancer	June 23	Libra, Lupus
March 17	Carina	June 27	Ursa Minor

July 3	Corona Borealis, Norma
July 5	Apus
July 7	Triangulum Australe
July 8	Draco
July 18	Scorpius
July 21	Serpens
July 25	Ara
July 26	Ophiuchus
July 28	Hercules
August 14	Corona Australis
August 15	Scutum
August 18	Lyra
August 21	Sagittarius
August 24	Telescopium
August 29	Pavo
August 30	Aquila, Sagitta
September 8	Vulpecula
September 13	Cygnus
September 14	Delphinus
September 18	Microscopium
September 22	Capricornus, Equuleus
October 9	Aquarius, Piscis Austrinus
October 12	Grus, Lacerta
October 16	Pegasus
November 1	Tucana
November 10	Sculptor
November 11	Pisces
November 13	Cepheus
November 18	Phoenix
November 23	Andromeda, Cassiopeia
November 29	Cetus
December 7	Triangulum
December 10	Hydrus
December 14	Aries
December 17	Fornax
December 22	Perseus
December 25	Eridanus, Horologium

Circumpolar Octans. Because this constellation lies all around the South Celestial Pole, it will be in the southern sky every night.

580. Was there ever a constellation called Cynosura? Cynosura was the ancient name for the constellation which we now call Ursa Minor, The Little Bear. The primitive Greeks imagined that the stars in this region of the sky represented a dog, and the word *Cynosura* comes from two Greek words which mean "The Dog's Tail." Because this part of the sky contains the north celestial pole, it is the central point about which everything else seems to revolve. By implication, the modern word *cynosure* means "the focus of attention, the point toward which all eyes are directed."

581. How did the constellations get their names? The present names of the constellations are given, in most countries, in Latin. Latin is used as the nearest to a universal language, for astronomy embraces all civilized nations and languages. The names have come to us either from classic Greek civilization, which, in turn, took them from earlier sources, or have been given to various associated stars by relatively modern astronomers.

Every ancient civilization had its own names for various groups of stars. The Babylonian constellations were, for the most part, much larger than the present constellations, taking in more territory for each star group than did the Greek or modern astronomers. The Babylonians, Egyptians, Arabs and Greeks have all contributed to the present names of the constellations. In many cases, a group of contiguous constellations are associated in Greek mythology; for example, the Perseus-Andromeda group, and it is difficult to tell whether the myth or the constellation came first. In a surprising number of cases, constellations bear the same name in the languages of several ancient cultures that could not possibly have had any communication with each other. The Greeks, the Finns and the American Indians all called the stars of Ursa Major "The Bear." The first orderly catalogue of the constellations was that of Ptolemy, who listed 48 constellations. He admitted that he had altered some of these from more ancient shapes in order that certain stars might be more conveniently located within them. Ptolemy's constellations did not necessarily touch each other, nor did they cover the entire sky. There were gaps between the more striking arrangements of stars in which there was nothing visually notable. Such spaces Ptolemy left blank. The extreme Southern regions, which were not visible from the eastern end of the Mediterranean, were also omitted. Ptolemy did, however, list 108 individual stars which were not assigned to any of his constellations but which were left orphans in these various unclaimed areas.

So great was Ptolemy's influence that even the Arabs, who had a religious restriction against depicting the image of any human being anywhere, even in the heavens, gradually overcame that prejudice and accepted, at least in part, the Greek tradition.

Astronomers of the sixteenth and seventeenth centuries used some of Ptolemy's "unformed" stars as a basis for new constellations to fill in the vacant areas. Hevelius, for example, placed Canes Venatici, Lynx and Leo Minor in the unoccupied regions near Ursa Major. He also added Lacerta, Scutum, Sextans, and Vulpecula in various strategic areas. Bayer contributed 13 new constellations in 1603, most of them well to the south, along the horizon of his world. Bertschius added Camelopardalis, Monoceros and Columba; and

Lacaille invented no less than 14 constellations, all in the deep south, during an expedition to the Cape of Good Hope in 1750–54.

The 88 modern constellations contain 15 men and women, one head of hair, 9 birds, one insect, 22 land animals, 10 water creatures, one serpent, sometimes divided into two parts, and 30 inanimate objects. For details of each, see the question dealing with each constellation.

582. Who was Andromeda, for whom the constellation was named? The story of Andromeda includes not only the constellation named for her but several neighbouring constellations. The legend begins with Perseus who was, at the moment, returning from his conquest of the Medusa. Perseus was one of the great heroes of Greek mythology, son of Jupiter and of Danaë, whom Jupiter had managed to visit in the guise of a shower of gold. Perseus was sent to rid the earth of the Medusa, an unprepossessing creature whose hair was a nest of venomous serpents and whose glance was so frightful that anyone who beheld her was turned to stone. Perseus borrowed the winged sandals of Mercury, met the Medusa and, by means of catching the reflection of her head in the polished surface of his shield, managed to kill her and to decapitate her without looking directly at her.

The blood of the Medusa spilled upon the ground and as it soaked into the earth, it was transformed into Pegasus, a great horse with wings sprouting from its shoulders. Perseus captured Pegasus and rode homeward from his adventure. As he was soaring above the Mediterranean, over the coast of Africa, he saw, far below him, a beautiful girl chained to a rock on the shore. In the sea, churning its way toward her, was a tremendous sea monster. The maiden was Andromeda and the monster was Cetus.

Andromeda had got into her predicament through no fault of her own. She was the daughter of Cassiopeia and Cepheus, queen and king of Ethiopia. Cassiopeia had been so ill advised as to boast of her beauty, comparing it favourably with that of the sea nymphs. The sea nymphs, by way of putting Cassiopeia in her place, sent Cetus, the monster, to ravage the coast of Ethiopia. Cepheus appealed to one of the oracles of his day and was told that he must sacrifice Andromeda to the monster. Accordingly, he chained Andromeda to the rock and reconciled himself to the worst. In the finest tradition, Perseus, riding upon his newly acquired flying horse, carrying the head of the Medusa, arrived in the nick of time. He tricked Cetus into taking a good look at the Medusa's head, putting him out of the running, rescued Andromeda and married her. The whole cast of characters is in this region of the skies. Pegasus and Cetus are there, west and south of Andromeda. The rest of the *dramatis personae*—father, mother, daughter and son-in-law (Cepheus, Cassiopeia, Andromeda, and Perseus)—are there, wheeling nightly above us.

583. Where is Andromeda? Andromeda is a constellation in the northern sky lying generally between 20° and 50° north declination and from the twenty-third hour of right ascension to between the second and third hours. It culminates—reaches the meridian—at 9 P.M. on November 5. From the latitude of 40° north, the central part of Andromeda lies across the zenith. The most prominent group of stars which will help to locate and identify Andromeda is an easily-found enormous square, the Great Square in Pegasus. There are four bright stars that make up the Great Square. The star in the northeast corner of the Square is the brightest star in Andromeda. Andromeda may be found on Map 1.

From the bright star in the northeast corner of the Great Square, two more bright stars indicate a line curving to the northeast. Above this line of stars is the main area of the constellation. There is a second line of fainter stars converging with the line of bright stars. The stars in this second line have an interesting agreement of position with the stars in the lower and brighter line—each bright star seems to have a counterpart in the fainter line. A part of Andromeda lies directly above the Great Square, extending westward across the top of the Square.

584. What are the notable stars and other objects in Andromeda? The star in the upper corner of the Great Square in Pegasus is Alpha (α) Andromedae (Alpheratz). This star is of visual magnitude 2.06 and is about 90 light years distant. The next star eastward in the bright line of stars is Beta (β) Andromedae (Mirach). This is a giant star with a visual magnitude of 2.02, about as bright as Polaris, the North Star, and is about 76 light years away.

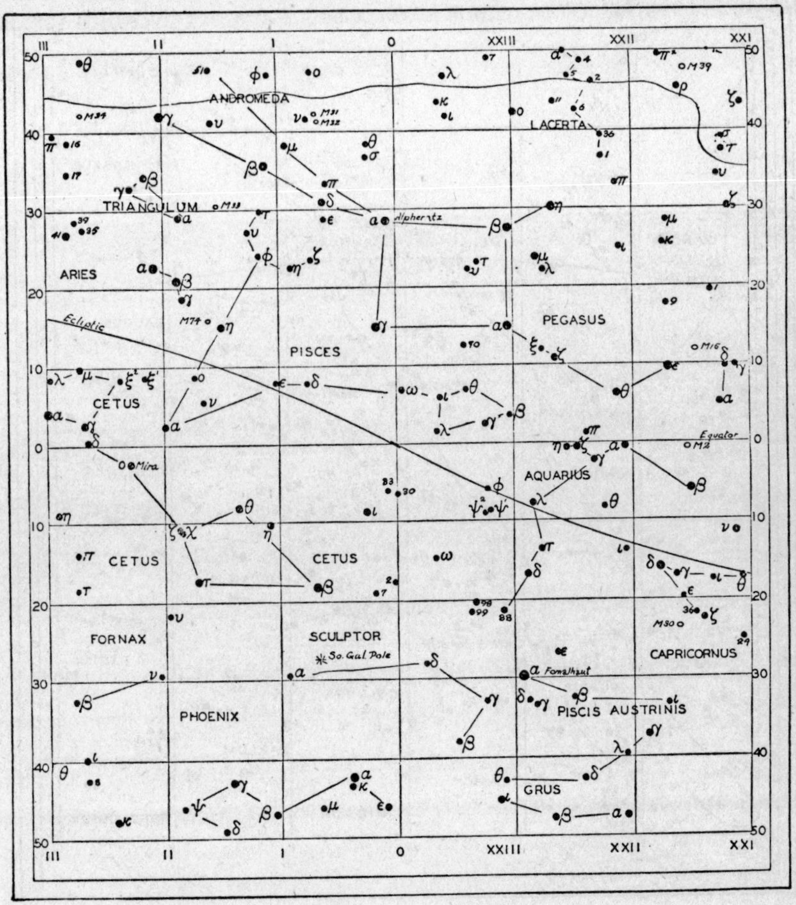

MAP I

MAGNITUDES 1 2 3 4 5
CLUSTERS–NEBULAE
GALAXIES

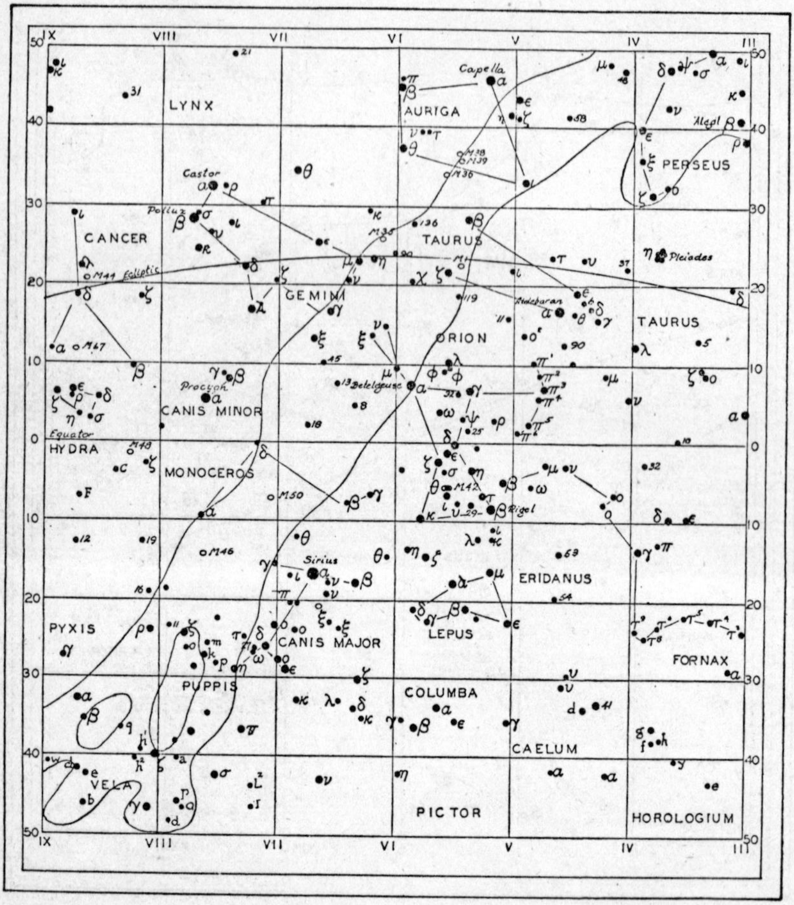

MAP 2

MAGNITUDES 0 1 2 3 4 5

NEBULAE-CLUSTERS

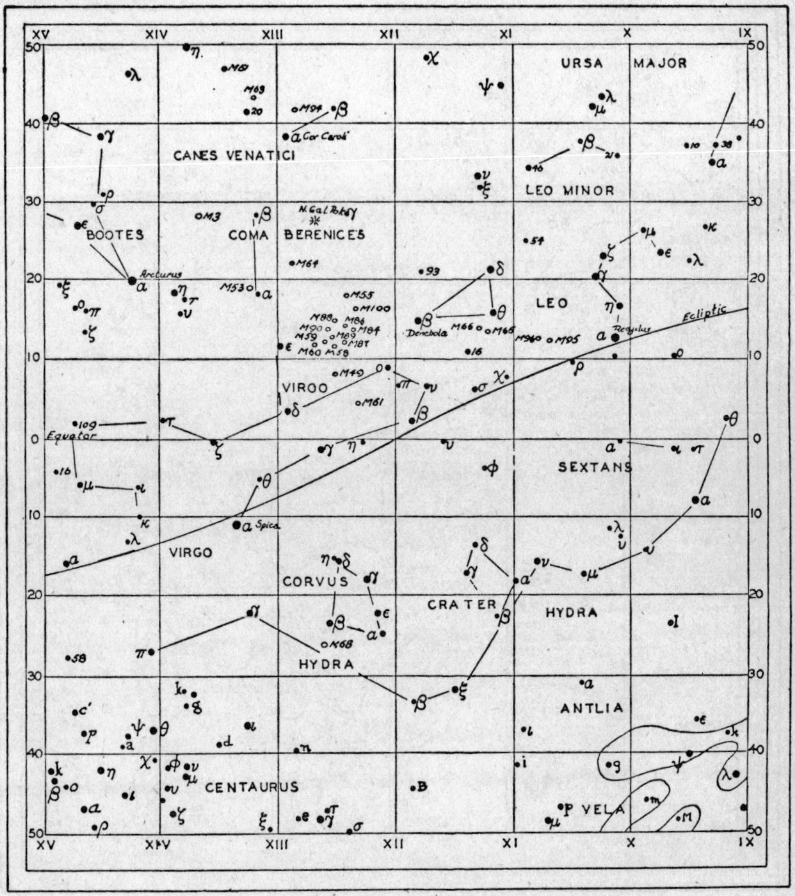

MAP 3

MAGNITUDES 1 2 3 4 5
NEBULAE-CLUSTERS
GALAXIES

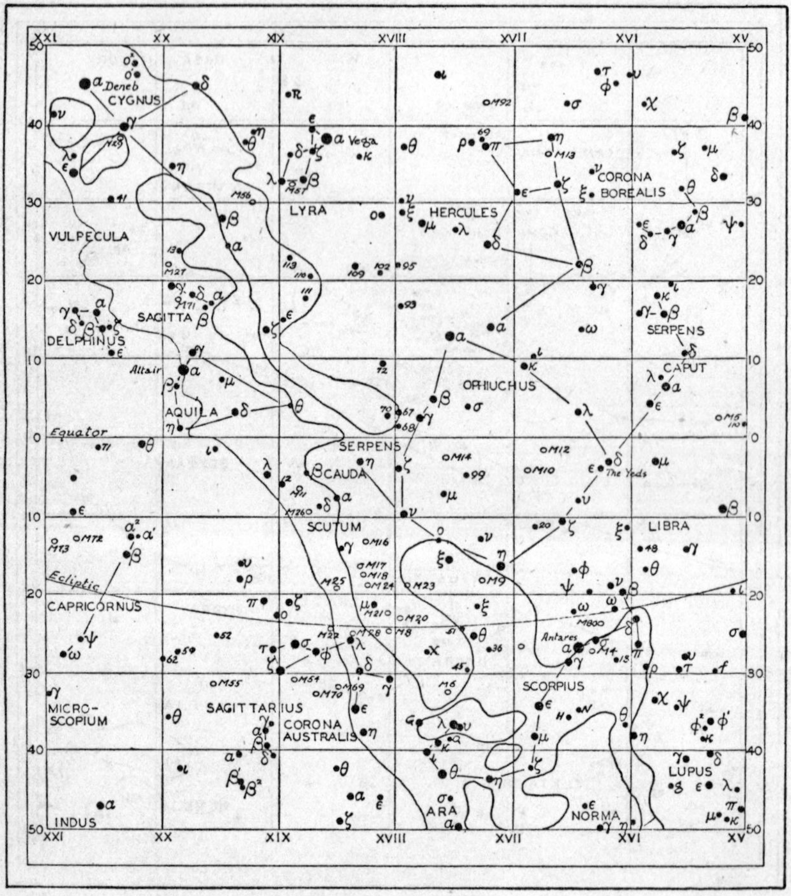

MAP 4

MAGNITUDES 1 2 3 4 5
NEBULAE-CLUSTERS

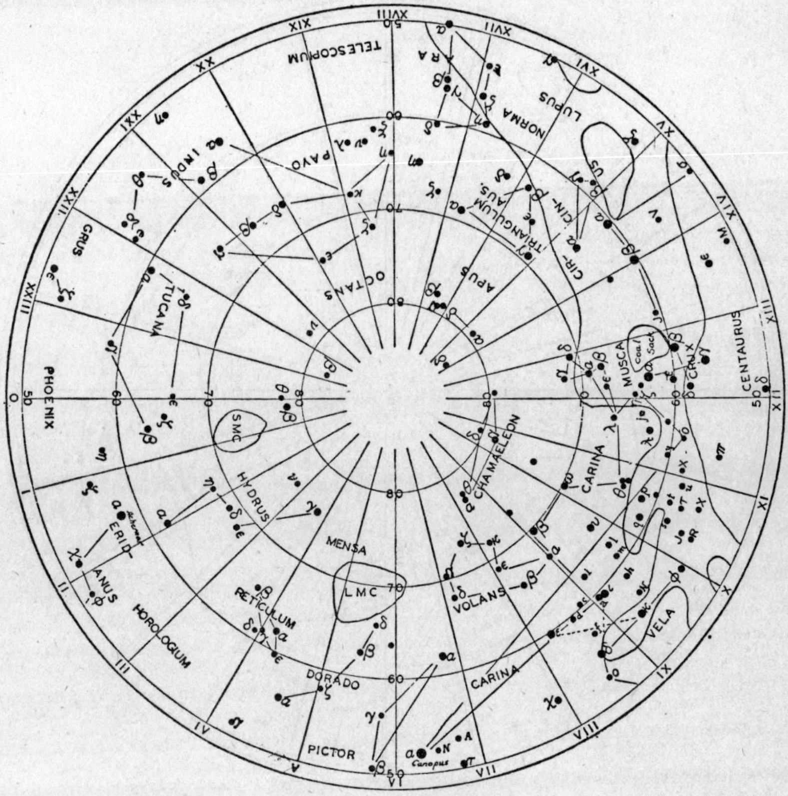

MAP 5

MAGNITUDES 0 1 2 3 4 5

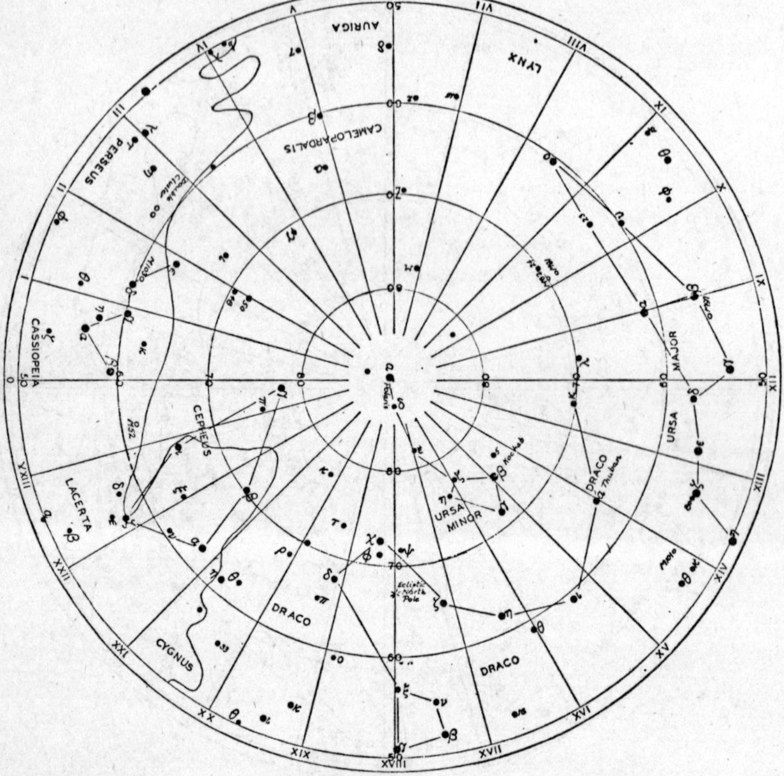

MAP 6

MAGNITUDES 2 3 4 5
NEBULAE-CLUSTERS
GALAXIES

The star at the eastern end of the line of bright stars is Gamma (γ) Andromedae (Almach), a famous multiple star. As we see it, it is one star whose visual magnitude is 2.14. Actually, the light we see from Almach comes from three stars. The brightest component of this triad is yellow and is a giant. One of its companions is greenish with a magnitude of 5.4 and the third is blue and of magnitude 6.2. This system of stars is 260 light years distant. The names of these stars illuminate slightly the confusion between Arabic star names and the constellation place names assigned to the stars by Ptolemy. *Alpheratz* is Arabic for "the horse's navel," and is a translation of Ptolemy's locational name for this star in the figure of the winged horse, now represented by Pegasus. Alpheratz was originally considered as belonging to both Andromeda and Pegasus. Mirach is a continuation of the equine figure and means "the loins." Almach has nothing whatever to do with the Greek mythological story of Andromeda, but probably stems from an independent Arabic conception of these stars. *Almach* means "the Earth-Kid," an Arabic name for a badger.

Slightly above and to the west of Mirach is The Great Spiral in Andromeda. This is numbered M 31 in Messier's Catalogue and N.G.C. 224 in the New General Catalogue. M 31 is an external galaxy, another vast star system quite similar in general form to our own Milky Way galaxy. Apart from the two Magellanic Clouds, which can be seen only from the Southern Hemisphere, this is the only external galaxy visible to the naked eye. It takes a very good eye and a fine, clear moonless night to see it. M 31 then appears as a slightly luminous, oval haze. The visual magnitude of the Great Sprial is 5.0, just above the lower limit of naked-eye visibility. The light that strikes the eye when the Great Spiral is seen has been travelling through space for rather over two million years. The Great Spiral Nebula in Andromeda is a vast organization of stars and of every variety of celestial object that we find in our own galaxy—star clusters, gaseous nebulae, variable stars and, undoubtedly, dark nebulae. We see the Andromeda Galaxy at an angle, not exactly edgewise or in plane, and that point of view gives it an oval appearance. Actually, it is circular in form, generally flat with considerable thickening at its hub. It is about half again as large as our own galaxy and contains about 150 thousand million stars—judging from its mass—and a considerable amount of dust and gas. It is about 150,000 light years in diameter and its absolute magnitude, taken as a whole, is —17.5, which makes it 1,600,000,000 times as bright as the Sun.

The Great Spiral has a smaller companion galaxy, M 32, which appears in the same field as the Great Spiral in a small telescope. This is an elliptical galaxy without any trace of arms or dust, and it may have lost its dust by passing completely through the larger galaxy. It was once thought that collisions of galaxies were quite common. The galaxies involved in such encounters could slip easily through each other because the great distances between the stars in each permit their mutual passage without any danger of star collision. The interstellar matter in each, however, would collide and is probably expelled with considerable force and tumult by such a passage. However, the idea of colliding galaxies has now been given up.

585. Why was Antlia so named? Antlia is a modern constellation, added some 200 years ago to fill in some of the area left vacant between the ancient constellations. In the vicinity of Antlia there was once the largest of all the constellations, Argo Navis, the ship Argo, which carried Jason and his fellow adventurers on their quest for the Golden Fleece. Argo Navis was too cumbersome for modern astronomical handling and was broken up into the smaller constellations of Carina, The Keel; Vela, The Sails; and Puppis, The Deck. In the region near these constellations, modern astronomers added several more prosaic constellations, presumably to make sure that the Argonauts should not suffer for lack of equipment. Antlia Pneumatica, the air pump, was one of these less happily named late additions.

586. Where is Antlia? Antlia extends from about halfway between the ninth and tenth hours of right ascension to just beyond the eleventh hour, and runs from about 23° south declination at its western edge in a series of steps so that its eastern border extends from 35° to 40° south, although its western border reaches from 23° south to 40° south. Hydra touches Antlia on the north and east; Vela is to its south; and Pyxis on its west.

Theoretically, the entire constellation of Antlia should be visible from latitude 40° north, just skirting the southern horizon on April 5 at 9 P.M. Actually trees, houses and the like probably stand in the way almost everywhere except on the ocean or on a barren plain. There is not much there to see, anyway. The brightest star in Antlia is fourth magnitude and it and one companion slightly fainter are the finest luminaries in the constellation. Antlia is singularly devoid of objects of interest. Antlia may be found on Map 3.

587. Where did Apus get its name? Apus means "The Bird of Paradise," and was the name given to this hitherto unassigned region in the skies by Bayer in his catalogue of 1603. Apart from the obvious celestial implication in the name of the bird, it is not known why Bayer christened it so.

588. Where is Apus? The constellation of Apus, the Bird of Paradise, extends from the eighteenth hour of right ascension to the fourteenth hour and from 70° to about 85° south declination. It is, in overall shape, a truncated triangle with a small bay in the lower eastern corner. There are just four stars of fourth magnitude in Apus and none brighter. Three of these stars form a tiny triangle near the centre of the constellation and the fourth is some distance to the west. One of the very faint stars in the constellation is an irregular variable, but there is nothing further of interest within the boundaries of Apus. Apus may be found on Map 5.

589. Where did Aquarius get its name? Aquarius, the Water Carrier, is one of the oldest of the constellations. This pattern of stars has been represented in the mythologies of several ancient civilizations as a man pouring water from a jar. In Egypt, the rainy season began when the Sun rose at the same time as the stars of Aquarius. This coincidence of seasons, which occurred over much of the territory in which early civilization was cradled, may have had something to do with the association of this constellation with a water bearer. The Greeks took the ancient legend and crystallized it by identifying Aquarius with Ganymede, who was one of the cup-bearers to Jove.

590. Where is Aquarius? Aquarius extends from the zero line of right ascension on the east to about halfway between the twentieth and twenty-first hour lines on the west, and runs from about 25° south to about 5° north in declination. Cetus is to the east of Aquarius; Piscis Austrinis takes up most of its southern border;

and Capricornus appropriates much of its western edge. Delphinus, Equuleus and Pegasus cover most of its northern side with Pisces finishing off the last easternmost portion. Aquarius culminates at 9 P.M. on October 5. Aquarius is a zodiacal constellation and its stars form the background for the Sun during the period from February 21 to March 21. Aquarius may be found on Map 1.

591. What are the notable objects in Aquarius? In spite of the large area covered by Aquarius, there are no very bright stars in the constellation. In the western region is Beta (β) Aquarii (Sadal Suud). This last name is Arabic and means "The Luckiest of the Lucky." The star is 2.86 visual magnitude. It is very much like the Sun in its chemical make-up, but it is a supergiant and is more than 1,000 light years distant. Northeast of Beta (β) is Alpha (α) Aquarii, Sadal Melik, "The Lucky Star of the King." This star is another supergiant at a distance of about 1,000 light years. It is magnitude 2.96. Just east of Alpha (α) are three stars which form a small triangle which is the identifying asterism of Aquarius. These three stars represent the jug out of which the Water Carrier is pouring the water.

Just above Beta (β) Aquarii is a globular cluster of stars which has the designation M 2, or N.G.C. 7098. The visual magnitude of the cluster is 5 and it is almost 45,000 light years away. It was discovered by Maraldi in 1764. Far down in the southeast corner is N.G.C. 7009, a planetary nebula which Sir William Herschel found in 1782. It is elliptical in cross section and is about three thousand million miles in its longest diameter. Its visual magnitude is 8 and its superficial resemblance to Saturn and its rings in overall shape led Herschel to christen it the Saturn Nebula. Just a little below and to the west of the Saturn Nebula is M 72—N.G.C. 6981—another globular cluster, very small and faint, lying tremendously far out in space. Near this one is M 73—N.G.C. 6994. This may mark the location of one of those comets which Messier was forever seeking, for there is nothing in this region of the skies now but four faint stars.

592. Why is Aquila so named? In the roughly triangular outline of Aquila, the ancients imagined an eagle with outstretched wings. There has been a remarkable consistency of opinion here. Every culture of the past has identified Aquila with some sort of bird. The

Greeks again consolidated and fixed the ancient legend by identifying the constellation with the eagle that carried Ganymede up to Olympus.

593. Where is Aquila? Aquila occupies the space bounded roughly by the nineteenth hour of right ascension on the west and the constellation extends eastward to about halfway between the twentieth and twenty-first hour lines. It runs from about 10° south of the celestial equator to about 16° north. Sagittarius and Capricornus bound it on the south; Aquarius and Delphinus on the east; Sagitta on the north, and Hercules, Ophiuchus, Serpens Cauda and Scutum on the west. It is on the meridian at 9 P.M. on August 25. In general outline, Aquila is an acute triangle without a clearly defined base. It can be found on Map 4.

594. What are the noteworthy objects in Aquila? The brightest star in Aquila, Alpha (α) Aquilae (Altair) is a bit to the east of the apex of the triangle formed by the stars of the constellation. The Arabic name for the entire constellation was "The Flying Eagle," and *Altair* is a part of the Arabic name. Altair is one of the nearer stars to the Sun, since it is only 16.5 light years distant. It is also one of the visually bright stars, as its magnitude is 0.77. It is a normal Main Sequence star.

Altair is flanked by two fainter stars, one to the northwest and one to the southeast. The upper star—to the northwest—is Gamma (γ) Aquilae (Tarazed). This is 340 light years away. It is a supergiant and its magnitude is 2.67. The star to the southeast is Alshain, and is an orange-coloured star whose magnitude is 3.9. The names of both these stars come from the Persian name for the constellation, which was *Shahini Tarazad,* "The Soaring Falcon." *Alshain* is an Arabic corruption of *Shahini* and *Tarazed* is a variant of *Tarazad.*

Directly below Altair, lying almost on the celestial equator, is Eta (η) Aquilae, a famous variable of the pulsating type. At its brightest, Eta is of visual magnitude 3.7. In a period of time of seven and a fraction days, the star dims to magnitude 4.4 and then brightens. It stays bright for 40 hours and takes 66 hours to diminish to its minimum, where it remains for 30 hours before rising again to maximum to complete its cycle. Like all Cepheid variable stars, Eta is a supergiant. It is yellow, and it is also a spectroscopic binary.

Above and to the west of Altair is Zeta (ζ) Aquilae, a double star. The system is composed of a fairly bright star of magnitude 2.99 with a companion of the twelfth magnitude.

Aquila has been the locale of several novae. One of the earliest was recorded as appearing not far from Altair in 389 A.D. Two more novae were seen in Aquila in 1936 and another in 1899. In 1918, on the western edge of the constellation, the brightest nova in 300 years blazed up on June 18. It was brighter than first magnitude.

595. How did Ara get its name? Ara is the Altar. Even though the constellation does lie far to the south, it is, nevertheless, one of the very ancient constellations. Three thousand years ago, when man first began to record his observations, the precession of the equinoxes had arranged matters so that the region of the heavens in which Ara lies was easily visible from the southern shores of the Mediterranean. Thuban, in the constellation of Draco, about 15° from Polaris, was then the North Star. The name of Ara probably came from a fancied resemblance in the pattern of its stars to the conventional shape of a sacrificial altar.

596. Where is Ara? Ara lies just below the southern horizon as seen from 40° north latitude and culminates at 9 P.M. on July 20. Its western border is midway between the sixteenth and seventeenth hour lines of right ascension and it extends eastward to about the eighteenth hour line. Its northern boundary is 45° south declination and it goes south to about 68°. It lies below Scorpius, which is its northern neighbour, while Telescopium and Pavo bound it on the east. Norma and Triangulum Australe border it on the west. There are only four stars in Ara brighter than fourth magnitude, and none of them is remarkable in any way. Ara may be found on Map 5.

597. Why is Aries so named? Aries, the Ram, has been called the Ram or the Sheep in many ancient cultures. As in so many cases, its present name was bestowed upon it by the Greeks, who pictured it as the ram with the Golden Fleece which was the goal of the quest of the Argonauts. Aries is not a conspicuous constellation, but it is probably the most famous of the 12 constellations of the zodiac. Some thousands of years ago, when the zodiac was first organized, the vernal equinox lay within the boundaries of Aries, and the Sun stood before the stars of this constellation at the moment when it crossed the equator on the first day of spring. Since then, the precession of the equinoxes has moved the vernal equinox westward about 30°, so that it now lies in Pisces. The point where the Sun crosses the equator on its way north, however, is still called "The First Point of Aries" in spite of its new location.

598. Where is Aries? Aries extends from a little west of the second hour line of right ascension to about halfway between the third and fourth hour lines, and from 10° to 30° north declination. It culminates early in December. Triangulum and Perseus border it on the north; Taurus on the east; Cetus and Pisces on the south; and Pisces on the west. The more prominent portion of Aries is a much elongated triangle of three fairly bright stars with its base running northeast-southwest in the western part of the constellation area. Other than this figure, there is little or nothing to see in Aries. It may be found on Map 1.

599. What are the noteworthy objects in Aries? The brightest star in Aries is Alpha (α) Arietis (Hamal). This is the uppermost and most eastern star of the flattened triangle that is the most prominent portion of Aries. Hamal is magnitude 2 and is about 76 light years distant. Its name is the Arabic word for "Sheep." The apex of the triangle of stars is Beta (β) Arietis (Sheratan) magnitude 2.7 and 52 light years distant. The third star of the triangle is a famous double star. Its two components are very nearly equal in magnitude, which makes it very easy to separate in a small telescope. The magnitudes of the two stars are 4.2 and 4.4. This star is Gamma (γ) Arietis (Mesartim). Mesartim is part of the Arabic name for the entire constellation which meant "The Exceedingly Fat Ram." Sheratan is the Arabic word for "The Sign" and probably referred to the point at which the vernal equinox once stood, thousands of years ago.

600. Why is Auriga so named? The origin of the name *Auriga* is completely lost. It is obviously the name of a person, but who that person was no one seems to know. The accepted traditional figure of the constellation shows a bearded man carrying a goat on his shoulder and, by convention, the man is further occupied in driving a chariot. The Assyrians pictured this constellation as a chariot and the Greeks bestowed upon it the personality of Hephaestion, a lame demigod, who rode because it was difficult for him to walk. Hephaestion is more commonly known as Vulcan. The goat seems to have been added to the picture to fit in with the name for the brightest star in Auriga, Capella, which is Latin for "The Little She-goat." Capella was the goat that suckled the infant Jupiter.

601. Where is Auriga? Auriga is one of the northern constellations. It lies between 20° and 50° north declination and extends from a little west of the fifth hour of right ascension to beyond the seventh hour. On the west, it is bounded by Perseus; on the north by Camelopardalis and a part of Lynx, which also covers most of its eastern border. Gemini and Taurus bound it on the south and Gemini also covers a bit of its eastern side. The bright stars of Auriga outline a very symmetrical kite-shaped figure which does not, however, take in a large area of the constellation lying to the east in which there are no bright stars. Auriga culminates at 9 P.M. on January 20. It may be found on Map 2.

602. What are the notable objects in Auriga? The brightest star in Auriga is one of the best known of all stars. It is Capella, Latin for "The Little She-goat." Capella is on the western end of the crossbar in the kite-shaped figure made by the brighter stars of Auriga. Capella ranks number 6 on the list of the brightest stars with a magnitude of 0.09. It is also one of the nearer stars, at a distance of 45 light years, and is about 120 times as bright as the Sun. It is also the same sort of a star that the Sun is, although Capella ranks as a giant with a diameter about 12 times the diameter of the Sun. Capella is a yellow star, but many ancient records list it as definitely red. This might mean that it has increased in temperature since it was first observed, as star colours play a large part in the rough estimation of star temperatures.

Just below Capella, there is a group of three much fainter stars which are known collectively as "The Kids" from their proximity to "The Little She-goat." These stars are Epsilon (ϵ), Zeta (ζ) and Eta (η) Aurigae. Epsilon Aurigae is an amazing star. It is an eclipsing binary with a period of 27 years. During this time, its light diminishes by one half a magnitude and returns again to its original brightness. What actually happens is that a small star of magnitude 3 is revolving about an enormous star whose temperature is too low to produce visible radiation. Its radiation is infrared. This tremendous body is about 3,000 times the diameter of the Sun, but does not contribute any visible light to the system. We know of its presence only because it interferes with the light of the brighter, smaller star revolving about it. Recent suggestions—that the object may be a "collapsar" or "black hole", made up of a small, super dense star surrounded by a region from which no light can escape—remain to be confirmed. Zeta Aurigae is a similar system, but the sizes of the two stars are not so extreme. The smaller of the two is about seven times the diameter of the Sun, while the other is a supergiant about 200 times the Sun's diameter. The two stars are so directly in our line of sight as they revolve about each other that they eclipse one another. The greater size of the cooler of the two stars compensates for its lesser brilliance per unit of area so that the two stars are of about the same brightness. The larger, cooler star, however, has an extensive atmosphere. As the smaller, hotter star passes behind it, this atmosphere absorbs some of the energy from the hot star, producing dark lines in that star's spectrum. These lines gradually fade and the character of the spectrum changes as the larger star more and more covers the smaller one. Next, we can see only the spectrum of the larger star and, finally, the process reverses itself as the smaller star emerges on the other side of the larger star. This whole cycle takes nearly three years.

The tip of the kite of Auriga is Beta (β) Aurigae (Menkalinan). Menkalinan is a part of the star's rather complicated Arabic name, which means "The Shoulder of the Rein Holder," and was probably the locational name of the star in Ptolemy's catalogue. Menkalinan is a spectroscopic binary. What we see is a star of magnitude 1.86, about 88 light years distant.

Theta (θ) Aurigae marks the other corner of the kite and is a double star. The magnitudes of its two components are 2.67 and 7.14, and we see it as a star of magnitude 2.65. It is a little more than 100 light years from us.

There are three very interesting clusters of stars within the boundaries of Auriga. M 36—N.G.C. 1960—is an open cluster of stars lying well below the centre of the general figure of the constellation. It is rather difficult to separate the stars of the cluster from the general stellar population of the region. M 37—N.G.C. 2099—a little below and to the east of M 36, is a similar group of stars but it is more concentrated and easier to define. M 38—N.G.C. 1912—is farther south and west and has a definite crosslike form. All of these clusters are at tremendous distances, lying some 2,000 light years out beyond the brighter stars of the constellation.

603. Why is Boötes so named? The word *Boötes* is Greek and means a "Herdsman." The root of the word can be found in many English words which refer to cattle. Boötes is one of the oldest-established constellations and is mentioned in the Odyssey. From its position in the celestial sphere, following Ursa Major around the pole, it was known as the Bear Keeper. The name of its brightest star, Arcturus, is Greek for "The Bear Keeper," and may at one time have been the name for the entire constellation. Boötes is generally represented by the figure of a man holding a spear or goad in one hand. In his other hand is a leash to which are fastened two dogs, which figure in a neighbouring constellation, Canes Venatici, "The Hunting Dogs." There are several versions of the identity of the individual in the constellation. Among these is Atlas, the giant who is holding the world on his shoulders. Boötes gained this personality from its proximity to the Pole, about which the heavens seem to revolve nightly.

604. Where is Boötes? Boötes is rather irregular in outline. It lies generally between 10° and 55° north declination and extends from midway between the thirteenth and fourteenth hours of right ascension almost to the sixteenth hour in its most northern region. In Boötes, as in many constellations, the shape outlined by the brighter stars is that of a kite. The symmetry of the figure is marred by a bend in the west side of the kite, as though it has struck a telephone wire. Boötes may be found on Map 3.

605. What are the notable objects in Boötes? The brightest star in Boötes is Alpha (α) Boötis (Arcturus), "The Bear Keeper." This is another famous star. It was once ranked sixth among the stars in order of brightness, but a recent reclassification of the brighter stars has elevated it to fourth place. Its magnitude is –0.06. It is one of our nearer neighbours, since it lies about 45 light years away. Arcturus is a normal giant with a diameter of about 30 times the diameter of the Sun and it is 100 times as bright as the Sun.

About 250 years ago, Edmond Halley, the astronomer after whom Halley's comet was named, discovered that three of the brighter stars, among them Arcturus, had changed their positions from the locations given for them in older catalogues. This was the first inkling that man had that the stars might not be really "fixed." Now it is taken for granted that all the stars are in motion, most of them at incredible speeds. We also suspect now that the motions of some of the fast-moving stars are more apparent than real and that they may be almost standing still while our solar system rushes past them. The motions of objects in space, including the Earth, are amazingly complicated. From our point of view, Arcturus is one of the fastest-moving stars, spinning along at about 85 miles a second, but much of this may be due to changes in position on the part of the Sun. Even at this angular displacement, which amounts to about 2″ of arc a year, Arcturus' position changes by one Moon's breadth only after an interval of 900 years. Arcturus is typical of most giant stars in that it is extremely tenuous. Its specific gravity is about $1/5{,}000$. Compare that with the Sun's specific gravity of 1.41. Halfway up the figure of the kite of Boötes on the east side is Epsilon (ϵ) Boötis, Pulcherrima. This is a lovely double star, often described as one of the most beautiful in the sky. One of the two stars that make the pair is a bright orange colour and is magnitude 2.47, while the other is a rich blue-green of magnitude 5.04. The star's name is modern and was given to the star because of its appearance. *Pulcherrima* is the Latin superlative of "beautiful."

The easternmost of the three stars that form the top of the kite of Boötes is Delta (δ) Boötis. It is a double star and its components are magnitudes 3.47 and 7.84, respectively. Above and to the west of it is Beta (β) Boötis (Nekkar), an Arabic word meaning "The Drover." Nekkar is magnitude 3.48. Still farther west and a bit to the south is Gamma (γ) Boötis (Seginus) completing the top of the kite. Seginus, the origin of whose name is unknown, is magnitude 3.05. Above Delta, going east again, is Mu (μ) Boötis (Alkalurops). This name is a strange combination of Arabic and Greek. The Greek portion of the name means a "Shepherd's Crook." To this, the Arabs added their definite article *al*. Alkalurops is a complicated double star that we see as a single star of magnitude 4.3. There are actually four stars contributing to this light: a double of magnitudes 4.0 and 6.5 plus another visual binary whose components are 7.2 and 7.8.

The four stars Mu (μ), Delta (δ), Gamma (γ) and Beta (β) form a slightly cockeyed rectangle which the Arabs used to call "The She-wolves."

606. Where is Caelum and why is it so called? The constellation of Caelum is far to the south, below the western corner of Lepus. It lies between Eridanus on the west and Columba on the east. Pictor and Dorado form its southern border. Caelum extends from about 28° south declination irregularly south almost to 50° and lies between the fourth and fifth hours of right ascension. It may be found on Map 2.

The constellation is one of those that was named by Lacaille, in modern times, and where or why he got the name is not known. *Caelum* is Latin for a "Chisel," particularly a sculptor's chisel. The constellation of Sculptor lies some five hours to the west of Caelum, and there may be a connection between the Sculptor and one of his implements.

Caelum contains only two stars brighter than fifth magnitude, and there is nothing of especial interest in the constellation.

607. Why is Camelopardalis so named? To start with, there does not seem to be complete agreement as to the exact name of this constellation. The name used here is from the *Report of the Delporte Commission,* created in 1930 to redefine and correct the boundaries of the constellations. In many star atlases, the name is given either as *Camelopardus* or *Camelopardalus.* In any event, the word is Latin for "Giraffe." The region now covered by Camelopardalis was once unclaimed space between the older constellations of Cassiopeia, Perseus, Auriga, Ursa Major and Ursa Minor. Bertschius, a Dutch astronomer of the seventeenth century, decided to give this region an identity. He also attempted to give it a religious tone. His intention was to call his new, synthetic constellation "The Camel," to memorialize one of the camels that carried Rebecca to Isaac. In Latinizing the word, however, a couple of extra syllables were added, and the name came out *Camelopardalis,* which is a far cry from Bertschius' original idea.

608. Where is Camelopardalis? Camelopardalis is far to the north. Its irregularly triangular outline extends from 52° to 85°

north declination and from the third hour of right ascension to the halfway point between the fourteenth and fifteenth hours. This makes it sound much larger than it really is, because it is full of strange little bays and projections which cover many of the crowded polar hour lines. Camelopardalis is bounded by Perseus and Cassiopeia on the west and south and also by a short stretch of Cepheus on the west, as well. Lynx, Ursa Major, Draco and Ursa Minor complete its most irregular boundaries.

In all its area, there are seven stars brighter than fifth magnitude. The efforts of early charters of the skies to stretch this scanty material into the resemblance of a giraffe are wonderful to behold. There is nothing remarkable in Camelopardalis. Camelopardalis may be found on Map 6.

609. Where did Cancer get its name? Cancer is Latin for "The Crab." The crab conception comes from the Greeks, who called this group of stars *Grapsias,* which also means "The Crab." The idea of the crab probably comes from the fact that the constellation once held the summer solstice, when the Sun, having climbed to its most exalted position above the northern part of the Earth, began to go backwards, like a crab, on its way south again.

The Egyptians called Cancer a beetle and the people of India called it a tortoise, but the Chaldeans pictured it as a crab, which is probably where the Greeks got the idea.

610. Where is Cancer? Cancer is in the northern sky and lies generally between 7° north and 33° north declination. It extends from about the eighth hour of right ascension to 9 hours and 20 minutes. It is bounded on the north by Lynx; on the west by Gemini and part of Canis Minor, which also forms part of its southern boundary; on the south by Hydra and on the east by Leo. Cancer culminates at 9 P.M. on March 21. It may be found on Map 2.

611. What are the notable objects in Cancer? Cancer is one of the faintest of the constellations from a visual standpoint and it is of importance chiefly because it is in the zodiac. Thousands of years ago it marked the background for the Sun when that star reached the summer solstice—its greatest elevation over the northern hemisphere. The Sun then stood 23½° north of the celestial equator and was overhead at the same distance north of the terrestrial equator. The line around the Earth at that distance north of the equator was called the Tropic of Cancer. The precession of the equinoxes (see question 59) has moved this point westward until it now lies very close to the western edge of Gemini.

None of the stars in Cancer is much brighter than fourth magnitude. There are two interesting clusters within its borders. One of these, north of the centre of the constellation, is *Praesepe,* "The Manger." Praesepe is an open cluster of stars covering about three square degrees. It contains nearly 400 stars ranging from magnitude 6 to magnitude 12. Most of these stars really belong to the cluster, having a common motion, while the rest just happen to be in the neighbourhood. There are many double, triple, and multiple stars in the cluster and they are, in the main, yellowish white. Another name for Praesepe, obviously from its apparent shape, is The Beehive. Praesepe can be seen without a telescope on a clear night and it does appear to have the conventional domed shape of a beehive, although none of its individual stars are visible to the naked eye. Praesepe was first mentioned by Hipparchus about 150 B.C. The other cluster in Cancer is a telescopic object lying somewhat to the south and slightly east of Praesepe. It is M 67—N.G.C. 2682—and is a faint, rather closely set open cluster.

612. Why is Canes Venatici so called? Canes Venatici, "The Hunting Dogs," is another modern constellation created to fill in space which was not occupied by any of Ptolemy's original constellations. Johannes Hewel (1611–1687), a Dutch astronomer who is better known by the Latinized version of his name, Hevelius, published a star catalogue in 1690 in which he named seven areas which had not officially belonged to any constellation before then. Canes Venatici is supposed to represent two hunting dogs who are pursuing the bear as it circles the pole. Hevelius called the more northern of the two dogs "Asterion" and the more southern "Chara." Hevelius was probably inspired to give this area the name it bears by the much older Arabian convention which named the regions to the south—in Coma Berenices and in Virgo—"The Retreat of the Howling Dogs."

613. Where is Canes Venatici? Canes Venatici lies directly below the handle of the Big Dipper in Ursa Major, and culminates at 9 P.M. on May 20. It extends from about 12° to about 33° north declination and from the twelfth hour to halfway between the thirteenth and fourteenth hours of right ascension. Ursa Major bounds it on the north; Boötes on the east and part of the south; Coma Berenices on the south and Ursa Major again on the west. Canes Venatici may be found on Plate 3.

614. What are the noteworthy objects in Canes Venatici? The brightest star in Canes Venatici, Alpha (α) Canum Venaticorum, is called *Cor Caroli,* "The Heart of Charles." There is some confusion among the experts as to which Charles was intended, but it does appear that the star was given this name by Sir Charles Scarborough in memory of Charles I. Charles I was, in many ways, more worthy of such a memorial than was Charles II, for whom some say the star was named when he established the Greenwich Observatory. Halley, in 1725, set Cor Caroli apart as a one-star constellation, but the star has always been considered a part of Canes Venatici. Cor Caroli is a famous double star. Its components are easy to separate. One of them is magnitude 2.9 and the other 5.6. The brighter of the two is again a spectroscopic binary.

Just at the edge of Canes Venatici, near the border of Boötes, is M 3—N.G.C. 5272—a globular cluster. M 3 is a tremendous sphere of giant stars about 65 light years in diameter. It must contain well over 100,000 stars and is about 60,000 light years distant. Many of the stars in the cluster are Cepheid variables. M 3 is typical of more than 100 globular clusters that may be found in many parts of the sky, and is one of the finest in the northern hemisphere. In a small telescope, it looks like a tiny piece of luminous cotton wool.

In the upper part of Canes Venatici, very near to Alkaid, the last star in the handle of the Great Bear, is M 51—N.G.C. 5194—the Whirlpool Nebula. This is one of the most famous of the spiral nebulae. It is a galaxy lying at some six million light years' distance. It is one of the very few galaxies that are so situated that we are able to see it in plane. In long-exposure photographs it appears as an almost circular figure with every appearance of being in rotational motion, showing a distinct spiral arm construction. At the end of one of its two arms there is a satellite formation, at some distance from the main body.

"Two round, condensed nebulae, of a very similar appearance in good glasses; quite easily picked up with a good finder on a dark night by alignment with Eta (η) Ursae Majoris, the star at the end of the Bear's Tail. In large telescopes, the more southerly of the two is seen to be surrounded by a faint halo, which takes on a spiral aspect in apertures approaching 12 inches. This was the first spiral nebula to be detected as such, an accomplishment of the third Earl of Rosse. A glimpse of some of the stellar condensations can be detected by averted vision in instruments of about 10 inches aperture or over." (Steavenson)

Just about in the centre of Canes Venatici is M 63—N.G.C. 5055—another external galaxy. This one is considerably fainter and is over seven million light years distant. It is described as a bright object—its magnitude is 9.6—lying at a very small angle to the line of sight, greatly foreshortened, with many arms.

Farther north and near the eastern edge of the constellation is N.G.C. 4258, which has a magnitude of 9.2. This is still another distant galaxy, almost nine million light years out in space.

615. Why was Canis Major so named? The constellation of Canis Major, and particularly its brightest star, Sirius, has been identified with a dog by many ancient cultures. Here again is the strange association so often encountered in constellation names, among races that could not possibly have had any communication with each other. In Greek mythology, Canis Major was the hunting dog of Orion. In ancient Scandinavia, the constellation was the Dog of Sigurd. In India, Canis Major was the Deerslayer. The stars may also have represented the dog Anubis to Egyptians. Originally, there was just one constellation called the Dog, and this was it.

616. Where is Canis Major? Canis Major is a winter constellation and culminates at 9 P.M. on February 18. It is bounded on the north by the constellation of Monoceros; on the west and south by Puppis; and on the south and a part of the east by Columba. The remainder of the eastern side is touched by Lepus. Canis Major is small and regular in shape, running from 11° south declination to 33° south, and from just west of the fourth hour line to west of the seventh hour line of right ascension. The Milky Way runs across the northwestern corner of Canis Major, but misses the greater part of the constellation. Because Canis Major lies in the fringe region of this river of stars, it contains many bright stars in itself. Canis Major may be found on Map 2.

617. What are the noteworthy objects in Canis Major? Sirius, Alpha (α) Canis Majoris, is not only the brightest star in Canis Major, but it is the brightest star in the sky in point of visual magnitude. Sirius looks so bright because it is one of the nearest stars in the list of the 20 brightest stars—at a distance of only 8.7 light years from the Sun. Its absolute magnitude—the brightness it would have if it were placed at a theoretical distance of 32.6 light years—is 1.45, but its visual magnitude is —1.43. The absolute magnitude of the Sun is 4.9—quite bright enough to be visible on a clear night to an exceptionally good eye—and Sirius is just about 26 times as luminous as the Sun.

The name *Sirius* has several possible derivations. There is a Greek word, sounding very much like *Sirius,* which means "The Scorcher." The ancient Egyptian name for the Nile, however, was Siris, and since Sirius rose at the same time as the Sun each year when the annual flooding of the Nile began, the Egyptians associated that life-giving event with the appearance of the bright star in the morning twilight, just before sunrise. It is quite possible that its name may have come from that association. Sirius is so situated that it can be seen from almost anywhere on Earth. Because of that, and its outstanding brilliance, it was bound to attract attention and legend from the very beginning.

Sirius is a double star. Its duplicity was discovered mathematically by Friedrich Wilhelm Bessel (1784–1846), a very great Prussian astronomer and mathematician. In 1834, Bessel's observations of the motion of Sirius in its orbit convinced him that its regular progress through space was being perturbed by some other body near it. He applied Kepler's laws to the motion and to a projected orbit of Sirius and calculated the probable mass and distance of such a companion. No one was able to find any such star, however, for nearly 30 years. Then, in 1862, Alvan Clark, one of the greatest builders of telescopes, was testing a new objective lens which he had just finished. The lens was 18 inches in diameter and through it, close to Sirius, Clark saw a very faint star which had not been placed on any previous catalogue. The distance of this star from Sirius and the perturbations it was causing in the orbit of Sirius made it possible to determine the mass of this newly found star.

It was about 50 years after its discovery, however, before its size was determined. When its known mass was applied to its size, astronomers gasped. The result was unbelievable. This star was only about 26,000 miles in diameter—slightly more than three times the diameter of the Earth—but its mass was $\frac{9}{10}$ of the mass of the Sun. Its density, then, had to be 53,000 times the density of water, and its mass was nearly 300,000 times the mass of the Earth. This strange star was the first of the white dwarfs to be discovered. Its magnitude is 8.66, which is not very dim, but its proximity to the blazing brilliance of Sirius makes it very difficult to isolate except with very large telescopes and at the most favourable times in its orbit, when it swings wide of its brilliant companion. For all its extreme density, this star is composed of gas at a high temperature. The gases in such a star must be completely ionized—all the electrons of all the atoms have been torn away from the nuclei of these atoms—and the nuclei must be pressed close together. This condition would cause the material in the star to be condensed to an extent beyond anything with which we are familiar. Most of the volume of an atom is empty space. When the electrons which form the boundaries of the atom are removed, the nuclei of the atom can be moved together, and since the nuclei contain practically all of the mass of the atom, the tiniest fraction of the volume of such a star will contain almost all of the star's original mass. The designation of this amazing star—a teaspoon full of which would weigh about a ton—follows the astronomical rule for denoting the fainter components of bright stars. It is Alpha (α) Canis Majoris B. Sirius itself is Alpha (α) Canis Majoris A. This pioneer white dwarf is familiarly known as The Pup.

To the west and south of Sirius is Beta (β) Canis Majoris (Murzim), a bright star which, in almost any other constellation, would attract much more attention than Murzim does in the exalted company it is keeping. Murzim is of magnitude 1.9 and is an extremely brilliant luminary, since it is about 750 light years distant. The name is a variant of the Arabic word *Mirzam*, "The Proclaimer." The star was so named because it appears above the horizon very shortly before Sirius itself rises and was considered a herald of the brighter star.

Below Sirius and slightly east of it is Epsilon (ϵ) Canis Majoris (Adhara). This is a double whose brighter component is magnitude 1.48 and whose fainter component is 7.5. Adhara is from the Arabic name once given to the entire group of fairly bright stars in this region, but which has persisted only in the case of this one star. It means "The Virgins."

Slightly above and to the west of Adhara is Delta (δ) Canis Majoris, magnitude 1.85, but actually a tremendously bright star, for it sends its light to us over 2,100 light years. Its real brilliance is absolute magnitude —7.1. Many of the stars in this part of the sky are really tremendous luminaries. It is pure astronomical chance that Sirius, which appears to be the brightest of the stars in Canis Major, is actually outstripped, and far outstripped, by many of its companions. The Arabic name for Delta is Wezen, "The Weight," which may mean that this star is supposed to serve as a balance to Sirius. Below and east of Wezen is Eta (η) Canis Majoris, magnitude 2.46, much fainter, visually, than the other stars, but again a supergiant of the most luminous class, for Eta is 2,700 light years distant and has the same absolute magnitude as Delta.

Just south of Sirius is M 41—N.G.C. 2287—an open cluster of about 50 stars. This is probably the faintest object of which mention has come to us from ancient times. Aristotle speaks of this cluster, and it certainly requires an atmosphere totally free from any impurities, with no Moon to interfere, and a superb pair of eyes to see it. The cluster has one red star near its centre, but is not otherwise a distinguished group. It is not as distant as are many of the individual stars in Canis Major as it is only 1,300 light years from us.

618. Why is Canis Minor so called? Both Canis Minor and Canis Major were traditionally the hunting dogs belonging to Orion, The Hunter. Canis Minor, "The Lesser Dog," is so named to distinguish it from the larger and more densely populated constellation of Canis Major. As in the case of Canis Major, the smaller group has been identified with a dog by many diverse ancient civilizations.

619. Where is Canis Minor? Canis Minor lies roughly between the seventh and eighth hours of right ascension although its eastern border does extend beyond the eighth hour line in one place. Canis Minor reaches from the celestial equator north to about 13° north declination. On the west, it is bordered by Monoceros and by a tiny extension of Gemini, and Gemini covers most of its northern edge. The eastern end of the northern side of Canis Minor touches on Cancer, and Cancer and Hydra cover the eastern border. Monoceros, again, extends along its entire southern boundary. Canis Minor culminates at 9 P.M. on February 18. The constellation may be found on Map 2.

620. What are the notable objects in Canis Minor? The brightest star in Canis Minor is Alpha (α) Canis Minoris (Procyon). The name means "The Leading Dog" and obviously refers to the fact that Procyon rises some time before Sirius and thus foretells its coming. Like Sirius, Procyon is near the Earth and appears as bright as it does for that reason. Its visual magnitude is 0.37 and it is only 11.3 light years distant. It is less than six times as luminous as the Sun and is not unlike the Sun in many of its characteristics. Procyon is twice the size of the Sun but is considerably less dense, for its mass is only 1.1 times the mass of the Sun.

Procyon, by pure coincidence, has a faint companion as has its brighter neighbour Sirius. The faint companion of Procyon is Alpha (α) Canis Minoris B, a white dwarf extremely dense, tiny and massive. Alpha (α) Canis Minoris B is magnitude 10.7.

Northwest of Procyon is Beta (β) Canis Minoris (Gomeisa), of visual magnitude 2.9, but considerably more luminous than is Procyon. Gomeisa is 210 light years distant and has an absolute magnitude of —1.1. The name *Gomeisa* means the "Weeping One" and comes from an involved legend which tells the story of the marriage of Suhail (the star Canopus) to Rigel (a star in Orion). Domestic trouble occurred and Suhail killed Rigel and fled to the south. Sirius followed, but Procyon remained behind and mourned. Gomeisa is helping Procyon with the mourning.

621. Why is Capricornus so called? Capricornus is another of the very ancient constellations which has come down to us substantially without change from the beginnings of history. The Chaldeans spoke of the constellation as a goat because the goat is a climber and can scramble up the hills. Capricornus was the constellation before which the Sun stood when it began its climb up the sky from the low point it had reached in the winter solstice. Like Cancer, Capricornus is devoid of spectacular objects, but is considered an important zodiacal constellation because its boundaries once held the point in the sky which marked the turn of the seasons. The Greeks, for a time, identified Capricornus with the great nature god, Pan, who was pictured as half man, half goat, and hence did not lose entirely the goatish implication. Capricornus is the westernmost constellation of a group which has a watery significance—Aquarius, The Water Carrier; Piscis Austrinus, The Southern Fish; Pisces, The Fishes; Cetus, The Whale, and so on. For that reason, Capricornus has assumed a fishy character and is widely known as the Sea-goat. It is sometimes represented as a beast with the head of a goat and the tail of a fish. The word *Capricornus* really means "The Goat's Horn" or "The Horned Goat" and has no marine significance.

622. Where is Capricornus? Capricornus culminates on September 20 at 9 P.M. It lies along the southern horizon when seen from latitude 40° north. It extends from about 8° south declination to about 28° south and lies between the twentieth and twenty-second hours of right ascension. The stars of Capricornus form the outline of a deep-keeled yacht hull. They are not particularly bright, but there are no very bright stars in the vicinity to make them suffer by contrast. On the west, Capricornus is bounded by Sagittarius, which also takes up a bit of its southern edge, and by Aquila, which turns the corner to the north. Aquarius takes care of the remainder of the northern boundary and of all of the eastern side, while Piscis Austrinus and Microscopium account for the greater part of the southern border. Capricornus may be found on Maps 1 and 4.

623. What are the notable objects in Capricornus? The stars in Capricornus are not overwhelming in their brilliance. Alpha (α) Capricorni (Gaedi) is high in the upper western corner of the constellation. Its name is Arabic and means "The Kid." Gaedi is a naked-eye double. The brighter star of the two is magnitude 3.2 and is called Prima Gaedi. In itself, this star is a double with a ninth-magnitude companion. The other star of the pair, Secunda Gaedi, is magnitude 3.8, and is a telescopic double with a faint companion of magnitude 11.

Beta (β) Capricorni (Dabih) is just below it. Dabih is a close double whose components are third and sixth magnitudes. *Dabih* is part of the Arabic name for the star and means "The Lucky Star of the Slayer"; it has nothing to do with the goatish character of Capricornus. The western end of Capricornus is Delta (δ) Capricorni, Deneb Algedi. *Deneb* may be the one name borne by the greatest number of stars. It is the Arabic word for "Tail." *Deneb Algedi* is, of course, "The Tail of the Goat," and it is good to know which way the goat is heading. Deneb Algedi is a variable star whose magnitude varies from 2.88 to 2.95.

Some distance below Deneb Algedi and a trifle to the west of it is M 30—N.G.C. 7099. This is a bright globular cluster which has been seen, on fine nights, without a telescope, although from northern latitudes, it would be very near to the horizon and more difficult to pick up because of the denser atmosphere through which it must be seen. In a telescope, M 30 appears to be foreshortened, which is common in an extragalactic spiral nebula but unusual in a globular cluster.

624. Why is Carina so called? *Carina,* "The Keel," was once a part of the enormous southern constellation Argo Navis, "The Ship *Argo,*" which originally occupied this and much of the neighbouring region of the sky. Modern astronomers found that Argo Navis was far too cumbersome to handle as a single constellation. The old constellation was broken up into three smaller constellations: Carina; Puppis, "The Deck"; and Vela, "The Sails." Carina is the most southern and hence the lowest part of Argo Navis.

625. Where is Carina? Carina is too far to the south ever to be seen from the latitudes of the United States. It culminates at 9 P.M. on March 5. Directly north of it are Puppis and Vela, the Deck and the Sails, both of which were once a part of Argo Navis. To the west is Pictor and to the south, Volans and Chamaeleon. Centaurus and Musca fill up the eastern border. Carina is irregular, but extends roughly from 50° south declination to 75° south, and from the sixth hour of right ascension to just past the eleventh hour. Carina may be found on Map 5.

626. What are the notable objects in Carina? Alpha (α) Carinae (Canopus) is the second brightest star in visual magnitude, ranking after Sirius. Canopus is magnitude —0.73, but unlike Sirius, it is really brilliant, for its absolute magnitude is —3.1. It is 98 light years distant and is about 5,200 times as luminous as the Sun.

There is a strange and interesting collective impostor lying largely within the boundaries of Carina. The stars Epsilon (ϵ) and Iota (ι) Carinae, together with Delta (δ) and Kappa (κ) Velorum, form the False Cross, an asterism with a striking resemblance to the Southern Cross. The False Cross is pointing at an angle of about 45° from the direction in which the long axis of the Southern Cross points. Since the long axis of the Southern Cross points almost exactly to the south celestial pole, it would be embarrassing, to say the least, for a navigator to choose the False Cross as a guide. The star Eta (η) Carinae, once known as Eta (η) Argûs, is a famous and puzzling variable. It is not shown on most conventional modern charts because its magnitude now is about 7.6, far below naked-eye visibility. In old charts, it is shown as a fourth-magnitude star, but in 1827, it began to grow brighter until, in 1843, it was second to Sirius. It held that exalted position for about 10 years and then gradually declined to its present condition.

Halfway up the eastern edge of Carina is N.G.C. 2808, described as a "large, rich globular cluster, 5′ in diameter, of 13th or 15th magnitude stars like the finest dust. The centre is a blaze of closely packed stars."

627. Why was Cassiopeia so named? The Greek mythological story of Cassiopeia is related in question 582. The Arabs first held this constellation to be a great hand with fingers outspread. Later, a kneeling camel was identified with the stars of Cassiopeia, but finally they too came to see here a woman in a chair. The name *Cassiopeia* may have come from the Phoenicians. In some of their ancient tablets, the constellation is called "Kasseba" and is depicted as a goddess of plenty with a sheaf of corn in her arms.

628. Where is Cassiopeia? Cassiopeia is one of the more familiar northern constellations, conspicuous principally for the striking arrangement of its stars rather than for any remarkable single object. The brighter stars in Cassiopeia make a very definite though slightly crooked W when Cassiopeia is below the pole in the northern winter, and an equally distinct M in the summer when the constellation is beyond the pole in what is known as lower culmination. Cassiopeia is a circumpolar constellation from latitude 40° north. It is almost opposite the Great Bear in relation to the north celestial pole. Cassiopeia extends from just below 50° north declination to about 78° north. Its outline is irregular so that its southern boundary extends from just west of the twenty-third-hour line eastward to halfway between the first and second hour lines of right ascension. The northern half of Cassiopeia works its way eastward in a series of steps in such a manner that the upper portion lies between about 0 hours, 30 minutes and 3 hours, 30 minutes of right ascension. Cepheus borders Cassiopeia on the north and along most of its western side. Lacerta and Andromeda complete the western edge and Andromeda extends all across the southern boundary. Perseus and Camelopardalis form the eastern boundary. Cassiopeia culminates at 9 P.M. on November 20. It may be found on Map 6.

629. What are the notable objects in Cassiopeia? There are five bright stars in Cassiopeia that make up the figure of the W. The first of these, beginning from the west, is Beta (β) Cassiopeiae, Caph. The word *Caph* is Arabic for "The Hand" and goes back to the time when the Arabs professed to see a great hand in the stars of this region. Caph is of magnitude 2.26 and is one of the nearer stars, 45 light years distant. The first lower joint of the W is Alpha (α) Cassiopeiae, Schedar. With this name, we come into the present conception of Cassiopeia, for *Schedar* means "Breast" and refers to the anatomy of the lady herself. Schedar is magnitude 2.16 and is slightly variable. Gamma (γ) Cassiopeiae is the center peak of the W and is a variable star whose brightness ranges from 1.6 to 2.9. It has a faint companion of magnitude 8.18. The second lower joint of the W is Delta (δ) Cassiopeia, Ruchbah, which is suspected of being an eclipsing variable, as its light changes slightly over a period of 759 days. It is usually of magnitude 2.67. The star at the eastern end of the W is Epsilon (ϵ) Cassiopeiae, which is considerably more luminous than most of its neighbours. We see it as a faint-magnitude 3.3, but it is more than 500 light years distant and its absolute magnitude is —2.7.

A little above and to the east of Beta is the spot where Tycho's

Star appeared in 1572. (See question 578.) The Milky Way passes through the lower part of Cassiopeia and contributes many small clusters which are scattered among the brighter stars. M 103—N.G.C. 581—is an open cluster slightly to the west and above Delta. M 52 —N.G.C. 7654—is somewhat west of and above Beta and is another open cluster, rather difficult to distinguish because it lies in so rich a field of stars.

630. Why is Centaurus so called? Centaurus takes its name from the creature of Greek mythology who is depicted as having the torso of a man and the body and legs of a horse. The particular Centaur with whom the constellation is identified is Chiron, the greatest of the Centaurs. Chiron was skilled in many arts and was appointed teacher to many of the lesser gods and heroes of Greek mythology. He was killed by Hercules in an accident, and was accorded a memorial by being placed among the stars as a constellation.

631. Where is Centaurus? Centaurus is an enormous constellation, very irregular in outline. It extends from about the eleventh hour line of right ascension eastward to the fifteenth hour line, in part, and it reaches from 30° south declination to 65° south, again in part. Hydra makes the northern boundary of Centaurus. Lupus and Circinus touch it on the east; Musca, Crux and Carina on the south; and Carina, Vela and Antlia on the west. It culminates at 9 P.M. on May 14. Centaurus may be found in part on Map 3 and in part on Map 5.

632. What are some of the notable objects in Centaurus? Centaurus is famous principally for one star, Alpha (α) Centauri, which is the nearest star in all space to the Sun. Alpha (α) Centauri is 4.3 light years away—a matter of 26 billion miles. The Sun, at a distance from us of 93 million miles, is so near that its light takes 8 minutes to reach the Earth. Alpha (α) Centauri is 272,000 times as far away. All the other stars are still farther away. Actually, Alpha Centauri is just about as bright as the Sun and resembles the Sun in many other characteristics. If you wish to see what the Sun would look like from a distance of 4.3 light years, look at Alpha Centauri. Its visual magnitude is —0.27, which makes it the third brightest star by that standard. Alpha Centauri is a double star whose two components are easily separated by a small telescope. One of these is of magnitude 0.01 and the other of magnitude 1.4. Not far from this pair is a very faint, tiny object whose motion seems to indicate that it is swinging around the main pair in a tremendous orbit and belongs to the same system. At enormous intervals of time, this star passes between the Sun and the brighter pair of stars and becomes, technically, the nearest star to the Sun. From that peculiarity it has been given the name of *Proxima,* which is Latin for "nearest." It is of magnitude 11.3. Alpha Centauri has been given a name, probably an Arabic translation of Ptolemy's place name in his charts. This name is *Rigil Kentaurus* and means "The Centaur's Knee."

The discovery that Alpha Centauri was our nearest neighbour was made by Henderson in 1839. He suspected that the star might be a candidate for the post because of its proper motion—almost four seconds of arc a year—which was an indication of possible nearness. To the west of Alpha Centauri is Beta (β) Centauri (Hadar). Beta Centauri differs from its visually brighter neighbour in that it is legitimately bright. It is about 490 light years distant, has an absolute magnitude of —5.2 and is 12,000 times as bright as the Sun—or as Alpha Centauri.

There is a famous globular cluster in Centaurus which is easily visible to the naked eye. Theoretically, it should be seen from the latitude of New York, as it is not quite 50° south of the celestial equator. This cluster is Omega (ω) Centauri, a closely packed sphere of more than 100,000 stars which is about 30,000 light years distant. Omega Centauri also bears the designation N.G.C. 5139.

633. Why is Cepheus so named? The story of Cepheus in Greek mythology can be found in question 582. Before that, indeed more than 5,000 years ago, the constellation was a unit much as it is today. In spite of its lack of brilliant stars, Cepheus is one of the oldest constellations. Cepheus has a possible association with several famous historical monarchs. One was Cush, king of Ethiopia and husband of Cassiopeia. It is as this personality that he figures in the Greek story. There is also a theory that Cepheus may be Cheops, the famous pharaoh of Egypt who built the Great Pyramid. To the Arabs, the constellation was a flock of sheep, complete with shepherd and dog.

634. Where is Cepheus? Cepheus is another circumpolar constellation that never sets at latitudes north of 40°. Cepheus culminates on October 5 at 9 P.M. The northern point of the constellation almost touches Polaris. The western boundary of Cepheus lies mainly between the twentieth and twenty-first hour lines of right ascension. Its western edge starts, at the southwestern corner, at about the twenty-second hour line and goes north and west in a series of steps, finally reaching as far east as the eighth hour line near the Pole. The southern boundary of the constellation lies irregularly along the line of 55° north. From there, Cepheus stretches to about 88° north. On the west, it is bounded by Cygnus and Draco; on the south, by Cygnus, Lacerta, Cassiopeia and Camelopardalis, which also border its southern edge because of its irregular shape. The small northern boundary reaches Ursa Minor. The five brighter stars of Cepheus outline a figure that resembles a square with a triangle set on top of it, the apex of the triangle pointing a bit to the east of north. The western line of the W of Cassiopeia points directly at the centre of Cepheus. You can find Cepheus on Map 6.

635. What are the noteworthy objects in Cepheus? Alpha (α) Cephei (Alderamin) is the southwestern star of the main figure of the constellation and is of magnitude 2.4. *Alderamin* is Arabic for "The Right Forearm" and refers to the ancient conception of Cassiopeia and certain neighbouring stars as an open hand. Beta (β) Cephei (Alfirk), is above Alpha and is a variable star whose light changes from 3.14 to 3.19 magnitude in less than five hours. This star is a binary system with one of the swiftest orbital motions known. *Alfirk* is a part of the older Arabic name for the constellation—"The Stars of the Flock."

The point of the triangle-above-the-square in Cepheus is Gamma (γ) Cephei, (Er Rai), "The Shepherd." Er Rai is of magnitude 3.2 and is the nearest bright star to Polaris in the constellation of Cepheus. Across from Alpha Cephei is Zeta (ζ) Cephei, magnitude 3.31. Just east of Zeta is the most famous star in the entire constellation. This is Delta (δ) Cephei, an intrinsic variable whose light changes from magnitude 3.51 to 4.42 in 5.4 days. This regular pulsation of Delta makes it the prototype of the Cepheid variable stars whose variation has been discovered to be related to their absolute magnitude. The longer the period between maximum and maximum of such a star, the brighter the star is. (See question 554.) Delta Cephei is the prototype Cepheid variable star. Its magnitude is considered to be 3.96, which is the mean between its two extremes. Delta is also a double and has a companion of magnitude 6.19.

636. Why is Cetus so named? The connection of Cetus with the Andromeda legend is considered in question 582. This region of the sky, however, seems to have been given over to the terror and mystery of the deep ocean long before Andromeda was heard of. The Babylonians considered it as representing the waste of waters and pictured Cetus as a strange and fearful sort of monster—a sea serpent, not a whale. Years later, it was incorporated into the Andromeda-Perseus story.

637. Where is Cetus? Cetus is an enormous constellation ranking fourth in total area covered. Some part of it lies across the meridian for about two months after the middle of November. It begins east of the third hour of right ascension and extends irregularly westward to beyond the zero hour line. It runs from 25° south declination irregularly north to 10° north declination. It is bounded on the east by Taurus, Eridanus and Fornax; on the south mainly by Sculptor; and on the west and part of its northern edge by Aquarius and Pisces. It is not a conspicuous constellation in spite of its size. It can be found some distance south and east of the Great Square in Pegasus, Andromeda and Perseus. The faint stars in the northern area of Cetus form a fairly symmetrical lozenge and the equally faint stars of the southern region spread out in a strangely elongated and irregular triangle difficult to describe. Cetus can be found on Map 1.

638. What are the notable objects in Cetus? The brightest star in Cetus is found in the extreme southwest and is Beta (β) Ceti (Diphda). *Diphda* is magnitude 2.02. Until a few years ago, this star was called Deneb Kaitos, "The Whale's Tail." *Deneb* is a common name among stars—so common, in fact, that there is another Deneb in Cetus. For this reason the name of Deneb Kaitos was changed to

Diphda. The remaining Deneb is Eta (η) Ceti, lying somewhat above and to the west of Diphda. Deneb is magnitude 3.47. Still farther to the east and lower is Tau (τ) Ceti, a faint star of magnitude 3.5, whose chief claim to fame is that it is one of our near neighbours. It is only 12 light years distant.

Almost in the middle of Cetus, about 3° south of the equator, is one of the best-known and least-understood of all the stars. This is Omicron (o) Ceti (Mira). *Mira* means the "Marvellous." This name was bestowed upon the star by Fabricius on August 13, 1596, upon the occasion of his discovery that it was a variable star. Mira has a tremendous range of variability. At its brightest it may reach magnitude 2, though at other maxima it may become no brighter than magnitude 4 or 5. It remains in this exalted position for only about ten days, however, and then begins to fade, very slowly. For about seven months it continues to diminish in brilliance, finally reaching a low point of magnitude 9.6—far below naked-eye visibility and about $\frac{1}{1,500}$ of its brilliance when at maximum. It remains in this state only a short while before it begins its climb to glory, which takes it about three months. The entire period of the star is 332 days. No one knows exactly what happens to Mira to make it behave in this way. It is an intrinsic variable, and the manner of its variation seems to be not uncommon in stars of its type. It is a supergiant—at least 300 times the diameter of the Sun—and a cool star. Mira's temperature changes, too, but nothing like as much as its light, for when Mira is coolest, its temperature is still about one third of what it was at maximum brilliance. It is possible that dense clouds form in the outer atmosphere of this star, cutting off its light but allowing the longer waves of infrared radiation to pass. Mira, in addition to its other accomplishments, is a double star. Its companion, which is about as bright as Mira is at minimum, was discovered only twenty-odd years ago, and seems to be slightly variable in its own right, with the sort of uneasy variability that denotes degeneration; possibly it is on the verge of becoming a white dwarf. This strange star is tiny—about as tiny, compared to Mira, as "The Pup" is compared to Sirius.

Alpha (α) Ceti (Menkar) is magnitude 2.54, but it is the brightest of the four faint stars that make up the lozenge that is popularly known as the Head of Cetus. The name *Menkar* means "The Nostril."

639. How were Chamaeleon, Circinus and Columba named? Chamaeleon was added to the constellation list by Bayer in his catalogue of 1603. A chameleon is a lizard having the power to change the colour of his skin according to the general tone of his surroundings.

Circinus was one of the contributions of Lacaille, who added 14 constellations in 1763, as a result of an astronomical expedition to the Cape of Good Hope. It is, theoretically, part of the navigational equipment for the crew of Argo Navis, the great constellation that once dominated the southern heavens. Circinus means the "Dividers" or a "Drawing Compass."

Columba is memorable as a praiseworthy attempt to inject a religious note into the constellation names. Bertschius attempted twice to do this (see question 608), and both times he failed. *Columba* means "Dove" and Bertschius' original name was *Columba Noae,* "The Dove of Noah." This name is still the official name of the constellation, but the identifying words are almost never used. Bertschius hoped to memorialize the dove that Noah sent out from the ark.

640. Where are Chamaeleon, Circinus and Columba? Chamaeleon culminates at 9 P.M. on April 14. It is shaped like a segment of the side of a cone and extends from just before the eighth hour to a little inside the fourteenth hour of right ascension. Chamaeleon runs from 75° to 85° south declination. It is bordered on the south by Octans, the constellation that contains the South Celestial Pole, on the west by Mensa, on the north by Volans, Carina and Musca, and on the east by Apus. There are no stars brighter than fourth magnitude in Chamaeleon, but two of these stars, Alpha (α) and Theta (θ) Chamaeleontis, are less than one degree apart, visually.

Circinus culminates on June 14 at 9 P.M., and is another deep south constellation. It extends irregularly from 55° to 70° north declination and from slightly west of the thirteenth hour to about 15 hours and 30 minutes of right ascension. Apus borders it on the south; Musca and Centaurus on the west; Lupus and Norma on the north; and Lupus and Triangulum Australe on the east. Its brightest star, Alpha (α) Circini, is of magnitude 3.18 and is a double with a faint component of magnitude 8.61.

Columba is further north, so that its northern region may be seen from latitude 40° N., but there isn't much to see. Columba lies some distance south of Orion, with Lepus between them. Lepus and

Canis Major form its northern boundary; Caelum is to the west; Pictor and Puppis to the south, and Puppis and Canis Major to the east. Columba is roughly between the fifth-hour line and the line of 7 hours, 45 minutes of right ascension and runs from 28° south to 43° south declination. Alpha (α) Columbae is a double star whose components are 2.64 and 12 magnitude and Beta (β) Columbae is an unassuming luminary of magnitude 3.12. Columba may be found on Map 2, and Chamaeleon and Circinus on Map 5.

641. Why is Coma Berenices so named? Coma Berenices is a relatively recent constellation but not quite so modern as some of the southern groups. Berenices was the wife of one of the Ptolemys of Egypt. She promised her hair, which was unusually beautiful, to Venus, if her husband returned safely from the wars. He did, so Berenice cut off her hair and placed it on the altar in the temple of Venus. The lovely offering disappeared and it took all the efforts of the royal astronomer to convince Berenice that her hair had been translated to the skies as Coma Berenices. Eratosthenes mentions this constellation, but Tycho Brahe was the first to give it an official place among the others.

642. Where is Coma Berenices? Coma Berenices is a small constellation lying between Leo and Boötes, which border it on the east and west respectively. Ursa Major touches its northwest corner, Canes Venatici is to the north of it, and Virgo lies to the south. Coma Berenices extends from a little west of the twelfth hour line of right ascension eastward to halfway between the thirteenth and fourteenth hours, and irregularly from about 13° north declination to about 32° north. It culminates on May 17 at 9 P.M. It may be found on Map 3.

643. What are the notable objects in Coma Berenices? As far as stars are concerned, there is nothing as bright as fourth magnitude in Coma Berenices. It is an area, however, which contains several points of tremendous interest. In it is located the North Pole of the Galaxy, whose equator is the Milky Way. The technical location of the North Pole of the Galaxy is at right ascension 190° or 13 hours, 45 minutes, and at declination 28° north.

Within the boundaries of Coma Berenices, there are eight Messier objects, all of them external galaxies. These are only a few of the vast number of galaxies, too faint for Messier, that have been found there since his time. This region is the northern part of the great Coma-Virgo cluster of galaxies. De Vaucouleurs, the French astronomer, has advanced the theory that our Galaxy is a member of a metagalaxy—a galaxy of galaxies—whose overall form is similar to that of our galaxy; a rather flat disk with a thickening at its hub. De Vaucouleurs says that the hub of this metagalaxy lies in this Coma-Virgo Cluster. The Arabs, apparently, must have had some inkling that there was something strange and mysterious about this region, for they called it "The Retreat of the Howling Dogs." (See question 808.)

644. What does Corona Australis mean? *Corona Australis* means "The Southern Crown" and is the name of a tiny constellation far to the south. Corona Australis was visible from the first civilized portions of the world in the early days of astronomy and was so named because its tiny semicirclet of faint stars so closely resembles its brighter northern counterpart, Corona Borealis, the Northern Crown. The curve of Corona Australis is toward the west and that of Corona Borealis toward the south, but apart from that and the general size of the figures and the brightness of the stars, the two constellations are strikingly alike.

645. Where is Corona Australis? Corona Australis lies just south of the Milk Dipper, the prominent asterism in the western part of Sagittarius, which also borders Corona Australis on the east. Telescopium and Ara touch it on the south and Scorpius covers its western side. Corona Australis runs from west of the eighteenth hour to well past the nineteenth hour of right ascension and extends from 37° to 46° south declination. It is a tiny semicircle of faint stars, four of which are evenly spaced, making a symmetrical arc with its curve to the east. None of the stars is brighter than fourth magnitude. Corona Australis can be found on Map 4.

646. What is the origin of Corona Borealis? The name *Corona Borealis,* "The Northern Crown," goes back to the legend of Theseus and Ariadne. When Theseus rescued Ariadne from the Minotaur, he married her, as was expected of him. Shortly afterward, however, he deserted her, which was rather unexpected. Bacchus, out of pity for her and probably out of love for her as well, gave her a jewel-studded crown, and this is it. The Arabs called Corona Borealis "The Broken Dish" because the circle of stars that forms the constellation is incomplete. The Bushmen of Australia call it "The Boomerang" for obvious reasons.

647. Where is Corona Borealis? Corona Borealis culminates on July 3 at 9 P.M. It lies between Boötes and Hercules, which form its western and eastern borders, respectively. These same two constellations also meet to make its northern boundary, while Serpens Caput bounds it on the south. Corona Borealis runs from 40° to 25° northern declination and from about midway between the fifteenth and sixteenth hours to halfway between the sixteenth and seventeenth hours of right ascension. It is a small but beautiful circlet of five fairly bright stars, unequal in brilliance. The curve of Corona Borealis faces generally south. Corona Borealis may be found on Map 4.

648. What are the notable objects in Corona Borealis? Alpha (α) Coronae Borealis (Gemma or Alphecca) is an eclipsing variable with a range of brightness of about $\frac{1}{10}$ of a magnitude and a period of 17.4 days. Its mean magnitude is 2.23. There are two other notable stars in Corona Borealis, neither of which is ordinarily visible to the naked eye. One of these, T Coronae Borealis, is one of the few stars that has a history of becoming a nova twice. (See question 521.) The other, R Coronae Borealis, is the prototype of irregular variables which remain fairly steady at one level of brightness and then, rather suddenly, drop in a brief time to a minimum. These stars remain dim for a short space and then, without warning, return to their brighter phase, there to stay for an indefinite period. These are intrinsic variables—the cause of their variation being contained in the star itself and not due to some eclipsing companion.

The name of the brightest star in Corona Borealis, *Gemma,* is Latin for "Bud," and is the same root word from which we get our word "gem." The Arabic name, *Alphecca,* means "The Broken One" and refers, of course, to the incomplete circle of stars which is the visible part of Corona Borealis.

649. What is Spica's Spanker? Spica's Spanker is a name given to the constellation of Corvus, the Crow. The brighter stars of Corvus outline a shape which resembles the mainsail of a boom-and-gaff-rigged boat. Corvus lies a little to the west of the bright star Spica, in Virgo, the Virgin, and for that reason it is called Spica's Spanker.

650. How did Corvus get its name? Like many other ancient constellations, Corvus has been given the same form and name in many unassociated early cultures. Corvus was a raven to most of the western civilizations. Sometimes Corvus was shown pecking at the coils of Hydra, the Water Snake, above which Corvus is located. Even the Chinese, then almost totally out of touch with the rest of the world, called Corvus "The Red Bird." The figure outlined by the brighter stars of Corvus might, with a lot of help from the imagination, resemble that of a bird in flight. The origin of the bird conception is lost in antiquity, but the Greeks, with their gift for dramatizing all things, fashioned a rather complicated tale about Apollo and a crow that he sent to spy upon a young lady about whom he had certain suspicions. These suspicions must have been correct, for the services of the crow were rewarded by a position among the stars.

651. Where is Corvus? Corvus lies between 11° and 25° south declination and between 11 hours 50 minutes, and 12 hours 50 minutes right ascension. Virgo covers the eastern and northern borders of Corvus. Crater lies to the west and Hydra is to the south. Corvus culminates at 9 P.M. on May 12. It can be found on Plate 3.

652. What are the notable objects in Corvus? The four brighter stars in the constellation of Corvus, the Raven, outline a rather shallow kite-shaped figure which is tipped toward the northwest. The fifth visible star—fainter than any of the first four—oddly enough bears the designation of Alpha (α) Corvi (Al Chiba). Just why this one should have been given the leading designation is not clear, as there does not seem to be any sequence or order in the position of the stars. Al Chiba is fainter than fourth magnitude and lies below the most southwestern of the four other stars. The name *Al Chiba* is Arabic for "The Tent" and was once the Arabic name for the entire constellation. The other stars are none of them startling. They are all between 2.5 and 3.5 magnitude. Delta (δ) Corvi, Al Gorah, is a double star. Its brightest component is of magnitude 2.7 and its companion is 8.26. *Al Gorah* is Arabic for "The Raven."

653. Why is Crater so called? The constellation of Crater, "The Cup," gets its name from relatively recent times. The Greeks won in the face of considerable competition when they named it the Cup of Apollo. In ages past, Crater has been many things. The Chinese pictured the constellation as a dog, and the Arabs saw a stall for a horse in these same stars. The Babylonians and the Egyptians, however, both made some sort of a mixing bowl for wine out of Crater and that is the source of the Greek name.

654. Where is Crater? Crater lies just west of Corvus and is slightly larger than its neighbour. It extends from 7° south declination to 25° south and from slightly west of the eleventh hour line to west of the twelfth hour line of right ascension. Hydra and Sextans border it on the west; Leo and Virgo on the north; and Virgo and Corvus on the west. There are no interesting stars or notable objects within the boundaries of Crater. Crater can be found on Map 3.

655. What does the name *Crux* mean? *Crux* is Latin for "Cross" and is the name given to the constellation which is entirely made up of the Southern Cross—probably the most famous of all the constellations. Crux is a relatively modern group of stars. It may have been seen vaguely in the very remote past because the precession of the equinoxes drew it above the southern horizon of civilization thousands of years ago. It was named in 1679 by A. Royer. The bright stars of Crux form a small but distinct cross whose long axis points almost to the south celestial pole. Crux is the only constellation to have a place on the flag of a nation. Australia uses Crux in the outer portion of its standard.

656. Where is Crux? Crux is just below the visible horizon from the latitudes of Europe and North America. It is perfectly rectangular, and is the smallest in area of all the constellations. Crux lies between 55° and 65° south declination and extends from a little west of the twelfth hour line to west of the thirteenth hour line of right ascension. Centaurus surrounds it on the east, north and west, and Musca bounds it on the south. Crux culminates at 9 P.M. on May 12, and may be found on Map 5.

657. What are the notable objects in Crux? The four brighter stars of Crux form a fairly symmetrical cross whose long axis points just a fraction of a degree west of the south celestial pole. The southern end of the upright axis of the cross is marked by the constellation's brightest star, Alpha (α) Crucis, Acrux. This name is a sort of telegraphese for the astronomical designation of the star. Acrux is a triple system. There are in it two bright stars of 1.39 and 1.86 magnitudes, and a fainter companion of magnitude 4.90. These two that we see as a single star of magnitude 0.87 are about 370 light years from the Sun and are considerable stars in their own right to appear as bright as they do. Gamma (γ) Crucis, opposite in position to Acrux, is of magnitude 1.67 and is a giant star at a distance of 220 light years. Beta (β) Crucis, the eastern end of the arm of the cross, is of magnitude 1.31 and is also a much more imposing star than its appearance indicates. It is extremely bright and is almost 500 light years distant. Delta (δ) Crucis, the western end of the arm, is a variable with a shallow range of from 2.78 to 2.84 magnitude. It, too, is intrinsically very bright and lies about 570 light years away. Apparently all of the bright stars in Crux are actually tremendous suns, very bright and large and at incredible distances.

About halfway between Delta and Alpha, spoiling the symmetry of the cross, is Epsilon (ϵ) Crucis, a third-magnitude star. Mark Twain held a very low opinion of this star. He disliked its appearance in the position in which it was and declared he was going to write to his congressman and urge that it be moved into the very centre of the cross, where it would look much better. It is this star, however, that helps to identify the real Southern Cross, making the distinction between it and the False Cross (see question 805). Compare the Southern Cross with the Northern Cross, made up of stars in the central portion of Cygnus.

Crux lies in the path of the Milky Way, but lying between Alpha and Beta is a famous object. Here is the Coalsack, a black blotch against the splendour of the heavens here. This is so prominent as to be easily noticed and was the first of many such dark places among the stars that aroused the curiosity of astronomers. In the early days of telescopic observation, this and similar regions were considered to be holes in the sky—places where the distribution of stars left blanks. Now it is known that all of these apparently empty spaces are really great clouds of dust and gas, inert and not capable of being excited into brilliance by the radiation of stars near them. This is cosmic dust. The individual particles are microscopic, but this material is considered to be the building blocks of the universe—the substance out of which everything is made.

658. Why is Cygnus so called? *Cygnus* means "The Swan." In this case, it does not put too much of a strain on the imagination to see a swan in the outline of this constellation. The short tail and the long, outstretched neck, the widely spread wings can easily be identified even under the less than ideal conditions that prevail in most northern urban skies. It is easy to picture the men of ancient times looking through clear, unpolluted air, seeing this figure distinctly. Almost every culture has identified Cygnus with a bird. The Arabs once called it The Eagle and at another time, descended to the prosaic figure of a Hen in describing it. Once the Swan figure had been established, Cygnus became, at one time or another, almost every famous swan in history, particularly the swan form that Jupiter assumed when he made his famous conquest of Leda.

659. Where is Cygnus? Cygnus is one of the prominent northern constellations. It culminates at 9 P.M. on September 13 and is directly overhead at the latitude of New York. The southern boundary of Cygnus is irregular, but lies mainly a little south of 30° north declination, while its northern boundary falls between 50° and 60° north. The nineteenth hour line just misses its western edge and the twenty-second hour line covers most of its eastern border. Draco and Cepheus border Cygnus on the north; Lacerta and Pegasus on the east; Vulpecula on the south; and Lyra and Draco on the west. The most conspicuous portion of Cygnus is the Northern Cross, which is formed by five of its brightest stars. The long axis of the Northern Cross is almost 15° long and its arms extend about 10°. Cygnus may be found on Map 4, with a small segment on Map 1.

660. What are the notable objects in Cygnus? The brightest star in Cygnus is Alpha (α) Cygni, Deneb, which marks the tail of the Swan and the upper end of the upright of the Northern Cross. The word *Deneb* is Arabic for "The Tail." It is a name often used for stars whose position seems to warrant it. Deneb ranks as nineteenth in the list of the brightest stars, visually. It is magnitude 1.26, but since its distance from the Sun is tremendous—estimates range from 540 light years to 1,600 light years—and its absolute magnitude is estimated from —4.8 to —7.1, it is actually among the brightest of all the first magnitude stars. The lower of the figures for its absolute magnitude makes it 6,000 times as luminous as the Sun, while the higher figure gives it nearly 30,000 times the Sun's brightness. Deneb must be a tremendously luminous star, pouring out energy in the form of radiation at an astounding rate.

Beta (β) Cygni, Albireo, is at the lower end of the upright of the cross and represents the head of the swan. The name *Albireo* is supposed to mean "The Chicken's Head," but there does not seem to be a word resembling *Albireo* in Arabic or in any related language. It may be a corruption of a word that is so far from the original that it cannot be traced. As a star, Albireo is one of the most beautiful sights in the heavens. It is an optical double. One of its components is golden yellow and is of magnitude 3.07. The companion is greenish, and about two magnitudes fainter. The pair is as lovely a spectacle in a small telescope as the eye of man can hope to see.

Gamma (γ) Cygni, Sadr, is the star that marks the junction of the two arms of the cross. *Sadr* is the same word used as *Schedar* in Cassiopeia (see question 629), and means "The Breast"—the breast of the Swan. Sadr is of visual magnitude 2.22, and it is another of those distant and extremely bright stars. It is 750 light years distant and has an absolute magnitude of —4.6. Delta (δ) Cygni is a distant double. We see it as a star of magnitude 2.87. There are actually two stars there, one of magnitude 2.91 and the other of magnitude 6.44, and the system is 270 light years away. Epsilon (ϵ) Cygni, the other end of the arm of the cross, is fairly near. It is 75 light years distant and has a visual magnitude of 2.46. Cygnus contains many remarkable stars which, while they are not visually startling are, nevertheless, objects of intense interest and study on the part of astronomers.

Chi (χ) Cygni, a little more than halfway between *Albireo* and *Sadr*, was one of the first discovered long-period variable stars. Its brightness changes through about eight magnitudes over a period of about 400 days. The record of variations in this and in many other variable stars shows a regular main cycle of variation and minor irregularities superimposed upon the main cycle which may, when sufficient evidence has been gathered, turn out to be regular in themselves. Chi (χ) Cygni, for example, has been under more or less continual observation since 1686, but its pattern is not yet completely known. So with other variable stars. One of the best-known astronomical societies was founded, in 1911, for the purpose of studying variable stars. This is the American Association of Variable Star Observers, whose members now number in the hundreds and live in every part of the world. They are both professional and amateur. Their work in the observation and recording of the performances of variable stars and the data they collect have proven invaluable to professionals.

Another famous variable star lies within the boundaries of Cygnus. This is SS Cygni (see questions 430 and 629). It is found somewhat east of Deneb. SS Cygni may be the most-observed faint star in the heavens. Its characteristics are a very rapid rise to maximum brightness, sometimes occurring from one night to the next, followed by a short but irregular period during which it remains at maximum, and finally an abrupt drop to minimum. The decline in brightness of this star is not, as a rule, as rapid as its rise. After the descent to minimum, there follows a long period of quiescence at minimum, during which every true and loyal variable observer watches breathlessly, night after night, for it to break out again. SS Cygni is typical of a class of variables that show tendencies of becoming novae. They have sometimes been called "dwarf novae," and appear to undergo mild outbursts which apparently do no damage. SS. Cygni rises from a minimum of twelfth magnitude to a maximum of magnitude 8, increasing its luminosity about 40 times.

Off to the east, about 3° below SS Cygni, is another notable star. This is again a faint star—a double whose components are of magnitudes 5.3 and 5.9. This star—61 Cygni—has a tremendous proper motion of about 50 miles per second. This quality made the great German astronomer, Bessel, suspect that it might be a near neighbour to the Sun. The nearer to us a star is, the greater will be the effect upon the observer of any motion that the star might have in space. Bessel, in 1838, attempted to determine the distance of 61 Cygni by triangulation and came up with the figure of 10.3 light years. This would make it one of the very close stars. Later on, Bessel's figures were corrected, by means of more accurate instruments, to 11.1 light years, but even so, 61 Cygni is eleventh in order of distance—if doubles be counted as one. Only four of the stars nearer to the Sun than this one are bright enough to be seen without optical aid.

The long axis of the Northern Cross lies right along the path of the Milky Way which, in this region, is at its brightest. The background for this constellation is magnificent and the awe and wonder inspired by sweeping through this part of the sky with a small telescope can be experienced nowhere else. There is only one dark place. Just below and to the east of Deneb is another of those enormous clouds of cosmic dust, second only to the great cloud in Crux. This is the Northern Coalsack, appropriately associated with the Northern Cross, as the original Coalsack is with the Southern Cross. Aside from this, the entire sky is crowded with stars. When this region passes the field of a telescope, it is very easy to understand how our Galaxy can have a population of 100 thousand million stars.

There are two open clusters that are bright enough to be included in Messier's list. M 39—N.G.C. 7092—lies to the north and east of Deneb and M 29—N.G.C. 6913—is about in the centre of the constellation, not far from Chi (χ) Cygni. Near Deneb, to the east, is another famous but optically invisible nebula that makes a notable impression on photographic plates. This is a combination of gas and dust whose outline bears a weird resemblance to the outline of the continent of North America, and it has been named the North America Nebula.

Far to the southeast there are several of the most beautiful, wispy clouds of celestial gases that exist. They make up the Veil Nebula or the Filamentary Nebula. They are trailing streamers of light and, if the field of the camera is large enough, the entire nebula is seen to be the remnants of what may once have been a circle. This may be the tatters of a gas-cloud, flung out perhaps billions of years ago, by some unseen supernova.

661. Where did Delphinus get its name? The constellation of Delphinus, the Dolphin, has been called that by the Western world since the earliest astronomical records. There are many legends which have grown around the figure of a dolphin. The best known of these is that of Arion, the poet, who was threatened with death by the crew of a ship upon which he was travelling. Arion begged his captors to allow him to sing one last song. They agreed, and the music attracted a dolphin who swam by and rescued Arion when he was at last jettisoned by the crew. To the Arabs, Delphinus was a camel. The Chinese saw a gourd in these same stars. Delphinus is also called "Job's Coffin." The reason for the coffin is obvious from the lozenge shape outlined by the brighter stars, but no one seems to know why Job was designated as the owner of the coffin. Delphinus is in a part of the sky which contains many other marine constellations. Near it are Aquarius, Capricornus, Piscis Austrinus, and so on.

662. Where is Delphinus? Delphinus extends from the twelfth to the twentieth degree of north declination and, except for one small portion, lies between the twentieth and twenty-first hours of right ascension. It culminates at 9 P.M. on September 14. Sagitta and Vulpecula border it on the north; Pegasus and Equuleus on the east; Aquarius and Aquila on the south, and Aquila and Sagitta on the west. Its brighter stars form a fairly symmetrical lozenge-shaped figure which has an appendage of stars falling from its western end. Delphinus can be found on Map 4.

663. What are the notable objects in Delphinus? The two stars that form the western side of the lozenge of Delphinus are interesting more for their names than for any other reason. The upper of the two is Alpha (α) Delphini (Sualocin) and the lower is Beta (β) Delphini (Rotanev). Alpha is a double star whose components are of magnitudes 4.0 and 9.5, respectively. Beta is also a double made up of two stars of magnitudes 4.0 and 9.5. One particularly interesting object in the constellation is Nova or HR Delphini, discovered in July 1967 by the English amateur astronomer G. E. D. Alcock. It was bright enough to be seen with the naked eye, and a year later, in July 1968, it was still of about the sixth magnitude, so that it was an exceptionally "slow" nova. By 1974 it had declined to the 11th magnitude. Alcock's discovery was not due to pure chance. He had spent years in learning his way around the sky, and had been carrying out a systematic search for novae, using specially-mounted binoculars. As he also has four comets to his credit, his record is most distinguished.

664. Why is Dorado so called? *Dorado,* "The Swordfish," is one of the constellations named in modern times by Bayer. The name *Dorado* is still given to a type of dolphin, but in astronomy the word is translated usually as "Swordfish" or simply "Fish" in order to avoid confusion with Delphinus. It is impossible to see, in the figure outlined by the faint stars of Dorado, any suggestion of a fish, much less of any specific genus of fish.

665. Where is Dorado? Dorado extends, in a steplike formation, from a trifle north of 50° south declination to 70° south and, irregularly, from west of the fourth hour of right ascension nearly to the seventh hour line. It is bounded on the east by Pictor and Volans; on the south by Mensa; on the west by Hydrus, Reticulum and Horologium; and on the north by Caelum. It culminates on January 31 at 9 P.M. Dorado may be found on Map 5.

666. What are the notable objects in Dorado? There is only one star in Dorado brighter than fourth magnitude. This is Alpha (α) Doradûs, magnitude 3.3, but Dorado does contain the major part of the Larger Magellanic Cloud—the Nubecula Major—which lies across the southern boundary of Dorado and runs over into Mensa. This is one of the two neighbour galaxies which are near enough and bright enough to be seen, on clear nights, by the casual observer without the aid of a telescope. The presence of the two Magellanic Clouds was first reported by members of the Magellan Expedition who completed the first voyage around the world in 1519, almost 100 years before telescopes were developed.

The Nubecula Major is easily visible to the naked eye as a lovely glowing patch of light. Its appearance has beguiled individuals into believing it a local cloud. It is a galaxy, lying outside our Milky Way system and is, as a whole, almost 200,000 light years distant. It is classified as a barred spiral galaxy because it has a brighter core, or axis, from the ends of which wispy arms seem to extend. The diameter of the main body of the Larger Cloud is about 10,000 light years. It is probably about one fifth as large as the Milky Way Galaxy, if its outlying regions are taken into account. The theory has been advanced that both this cloud and the Smaller Cloud might be satellite galaxies of our own system, but this has yet to be proven.

In the Nubecula Major there have been found many of the phenomena that are present in the Milky Way Galaxy, such as galactic clusters, nebulae, both light and dark, and stars of all types, including Cepheid variables. More Cepheid variables have been located in the Larger Cloud than in our own galaxy, although this does not necessarily mean that there are actually more of these stars in the cloud than there are in the local Galaxy. It means that we can see more of the cloud than we can of our Galaxy. It was in the Magellanic Clouds that Miss Henrietta Leavitt discovered the relation between the period and the absolute magnitude of the Cepheid variables that led to the tremendously valuable period-luminosity law for the determination of distance. (See question 559.)

The stars that can be seen as individual stars in the Magellanic Cloud must all be of tremendous brightness, for at that distance our own Sun would not be visible under any circumstances. These stars are all intensely hot, bright supergiants. There are many bright nebulae in the Clouds. The Tarantula Nebula, also known as the Great Looped Nebula—30 Doradûs—is the brightest nebula known. If it were in the position of the Great Nebula in Orion, it would be bright enough to cause objects upon the surface of the Earth to cast shadows. Within the Larger Cloud is one great star which seems to be the brightest object yet discovered. This is S Doradûs, whose visual magnitude ranges from 8.2 to 9.4. The absolute magnitude of this blazing star must be brighter than —8 magnitude or considerably more than 600,000 times as bright as the Sun. S Doradûs is a double star, both of whose components are probably larger than the entire orbit of the Earth around the Sun.

667. Where did Draco get its name? Draco, "The Dragon," seems to have been identified with some form of serpent or dragon in most civilizations from their beginnings. In the early days of astronomy, about 3,000 B.C., one of the stars in Draco was the nearest bright star to the north celestial pole, the position now held by *Polaris*. Because Draco, as seen from the northern latitudes, never sets, it has always been considered a sort of reptilian guardian of the pole. The Greeks applied to this constellation two different legends relating to dragons. One was that Draco represented the dragon that guarded the Gardens of the Hesperides, where grew the Golden Apples. This dragon was slain by Hecules, whose constellation is adjacent to one portion of Draco. Draco was also pictured as the dragon overcome by Cadmus and whose teeth, when sown, produced an army. The Arabs and Egyptians concurred with their contemporaries, although the Arabian names for many of the stars in Draco refer rather to the ubiquitous sheep and goats which are represented in so many of their stars.

668. Where is Draco? Draco is a circumpolar constellation, and while it may be seen in northern skies all the year round, the centre of the constellation culminates about July 8. Draco is very irregular in shape and winds around one side of Polaris, between Ursa Major and Ursa Minor. Its southern limit is 48° north declination, and it runs north from there until one small segment of it touches about 87° north. It extends slightly less than halfway around the celestial sphere, from halfway between the twentieth and twenty-first hours of right ascension to east of the eleventh hour. Ursa Minor and Camelopardalis cover its irregular northern boundary. Ursa Major touches it on the east and on much of the south. Boötes, Hercules, Lyra and Cygnus cover much of its rambling southern and western sides and Cepheus completes its borders. Draco may be found on Map 6.

669. What are the notable objects in Draco? For all its rambling area, Draco contains only eight stars brighter than fourth magnitude. Alpha (α) Draconis, Thuban, is not the constellation's brightest star, although it was given the prime designation by Bayer and recorded by him as second magnitude early in the seventeenth century. Thuban is probably the best-known star in Draco. Five thousand years ago, Thuban was the nearest bright star to the celestial north pole, occupying the position held today by Polaris. The precession of the equinoxes (see question 59) has swung the poles of the Earth in a great, slow circle until the north pole now points temporarily almost at Polaris. Thuban lies about halfway between the central star in the handle of the Big Dipper and the brightest star in the front of the bowl of the Little Dipper. It is now of magnitude 3.6 and is a spectroscopic binary. *Thuban* is part of the Arabic name for the entire constellation and means "Dragon."

The brightest star in Draco is Gamma (γ) Draconis, Eltanin. It is part of a tiny asterism of four stars which forms what is known as the Head of the Dragon, very near to the border of Hercules. It was a study of this star that led Bradley to the discovery of the aberration of light. (See question 998.) *Eltanin* is Arabic for "The Dragon's

Head," and a neighbouring star, Beta (β) Draconis (Rastaban), has a name which is a synonym for "The Dragon's Head." Rastaban is a double star whose components are 2.7 and 11.5, while Eltanin is of magnitude 2.2. The north pole of the ecliptic lies within the confines of Draco, and 15° north of Gamma (γ) Draconis. The point marks the centre of the great circle of the ecliptic, that path which the Sun appears to follow through the year, but which is really the path of the Earth as seen from the Sun, in the opposite sense.

670. Why is Equuleus so called? *Equuleus* means "The Colt" or "The Foal," and is, in spite of its small size and general insignificance, one of the ancient constellations, having been included in the Babylonian star groups. Since Equuleus lies in that region of the skies where the constellations begin to take on a marine significance, it is generally pictured as a hippocampus or sea horse.

671. Where is Equuleus? Equuleus lies cheek by jowl with Delphinus and is, next to Crux, the constellation having the smallest area. It culminates at 9 P.M. on September 22. South of it is Aquarius, to the east is Pegasus, and Pegasus with Delphinus makes up its northern boundary. Delphinus also borders it to the west. Equuleus extends about 10° north and south, from 3° north declination to 13°, and from slightly west of the twenty-first hour to about 25 minutes east of it. There are three stars of fourth magnitude in Equuleus. The constellation can be found on Map 1.

672. Where did the name Eridanus come from? In every ancient culture, the constellation of Eridanus has been identified with a river. The long, curving shape of the constellation certainly lends itself to the image of a river better than to anything else. The Egyptians called this stream of stars the Nile; the Babylonians named it the Euphrates. The name *Eridanus* is the name of a real river in Turkey, known now as the Strong River. Modern Greek mythology gives it the identity of the River Po in Italy.

673. Where is Eridanus? Eridanus culminates on December 25 at 9 P.M.—at least, one of its more prominent stars does—but it actually meanders all over the southern sky. The stream of stars that marks the river begins just above Rigel in Orion and wanders thence south and west for almost 60°. Part of its northern boundary is the celestial equator and it zigzags along the first and fifth hour lines between the equator and 58° south declination. Taurus, Cetus and Orion border it on the north; Cetus, Fornax and Phoenix on the west. Orion, Lepus, Caelum and Horologium mark its eastern boundary and Hydrus its southern. Parts of it may be found on Maps 1, 2 and 5.

674. What are the notable objects in Eridanus? In all the vast area of Eridanus there are exactly six stars brighter than fourth magnitude. The brightest star in Eridanus lies far to the south, out of range of observers in northern latitudes. This is Alpha (α) Eridani (Achernar). Achernar ranks ninth among all stars in visual brightness, with a magnitude of 0.53. It is a giant at a distance of 118 light years and has an absolute magnitude of —2.3, which makes it about 800 times as bright as the Sun. The name *Achernar* is a part of the Arabic name for the star which means "The End of the River." The brightest of the northern stars of Eridanus is Beta (β) Eridani, Cursa, not far from Rigel in Orion. *Cursa* is Arabic, too, and refers to the position of the star. It means "The Front Part of Orion's Chair." Beta is of magnitude 2.9.

675. Where did Fornax get its name? *Fornax* is Latin for "Furnace." Fornax was one of the inventions of Lacaille, who, after he had called a neighbouring group of stars The Sculptor, provided his sculptor with a furnace for casting his bronzes.

676. Where is Fornax? Fornax culminates at 9 P.M. on December 17. It is almost rectangular, except for a slight irregularity on its eastern edge, and lies across the second and third hour lines, extending about 12 minutes west of the second hour and 50 minutes east of the third hour. It runs from 25° south declination to 40° south. It is a rather blank space lying alongside the main stream of Eridanus, which curves around it, making its eastern and part of its southern and western borders. Cetus takes care of about half of its northern line. It may be found on Plate 1. In spite of its name—The Furnace—there are no stars in Fornax brighter than fourth magnitude.

677. Where did Gemini get its name? Gemini are The Twins, Castor and Pollux, sons of Leda by Jupiter, and the constellation is one of the V.I.P.'s of the sky. In ancient times, the two bright stars were considered to be twins, but no one bothered to identify them particularly until the imaginative Greeks came up with the Castor-Pollux legend.

In the most ancient of all existing records—the boundary stones marking the division of property in Babylon—there is a symbol which is repeated again and again. It is known as "The Triad of Stars"—two stars, and near them a crescent with its points upward. This figure is now believed to refer to the two bright stars Castor and Pollux which, because of their position thousands of years ago near the solsticial colure which then ran just east of them, were always near the first new Moon of the year. The "Triad of Stars" signifies what the men of those remote times saw in the skies at the beginning of their year.

To the Arabs, Castor and Pollux were two kids, in line with the pastoral culture of those people. The name *Gemini* is Latin and is fairly recent. Castor and Pollux were famous for two sporting accomplishments. Pollux was the originator of the manly art of self-defence and Castor was the first man ever to bestride a horse. As a pair the two were also the patrons of sailors. Legend says that the mild expletive "By Jiminy!" is a descendant of the ancient invocation of naval men to their protectors—"By Gemini!"

678. Where is Gemini? Gemini culminates at 9 P.M. on February 19. It lies generally between the sixth and eighth hour lines of right ascension and extends from 10° to 35° north declination. Auriga, Taurus and Orion border it on the west; Auriga and Lynx on the north; Cancer on the east and Canis Minor and Monoceros on the south. It may be found on Map 2.

679. What are the notable features in Gemini? The two almost equally bright stars Castor and Pollux easily identify the constellation of Gemini, The Twins. They lie directly above Procyon in Canis Minor and a bit to the east and above Orion. Alpha (α) Geminorum (Castor) is the upper of the two stars and is not the brighter of the two in spite of its designation. When Bayer assigned the Greek letters to the stars in 1603, he apparently saw Castor as the brighter. Whether he was mistaken in his judgment or whether Castor has become dimmer or Pollux brighter, or both, is not known. Castor was called by Herschel the finest double star in the northern hemisphere. Its two principal components are of magnitudes 1.97 and 2.95, and each of these again is a spectroscopic binary. In addition, there is a third star in the Castor system, magnitude 9.08, which is also a spectroscopic binary. There are, then, six stars in all, each pair revolving around the other pairs and the two stars in each pair pursuing mutual orbits about each other. The system is about 45 light years distant.

Beta (β) Geminorum (Pollux) of magnitude 1.16, is about ten light years nearer to us than its twin. Eta (η) Geminorum (Propus) lies to the west, tucked into the corner of the constellation near Orion. *Propus* means "The Forward Foot" and refers to the locational name of the star in the foot of one of the twins, as indicated in Ptolemy's catalogue. Propus is a variable whose mean magnitude is 3.33 and it is also a double star whose component has a magnitude of 6.7. West of Propus is Mu (μ) Geminorum, another variable with a very small range of brightness whose mean magnitude is 2.92. Below and to the east of Mu is the third brightest star in Gemini, Gamma (γ) Geminorum (Alhena), with a magnitude of 1.93. *Alhena* is Arabic for "The Brand Mark," and refers to the Arabic conception of the constellation as two kids. Below Alhena is Zeta (ζ) Geminorum (Mekbuda), magnitude 3.38, and some distance above Mekbuda is Epsilon (ϵ) Geminorum (Mebsuta), an even magnitude 3. *Mekbuda* and *Mebsuta* are names given to these stars by the Arabs when this part of the constellation was separate from the rest of Gemini. They pictured this region as occupied by a lion; *Mekbuda* means "The Folded Paw of the Lion" and *Mebsuta,* "The Outstretched Paw of the Lion."

Gemini holds one Messier object, M 35—N.G.C. 2168—an open cluster of bright and faint stars. M 35 lies just above Eta (η) Geminorum. There is a fine planetary nebula lying a bit east of the centre of the constellation, which bears the designation N.G.C. 2392.

Just inside the western edge of Gemini runs the sixth hour line of right ascension—the solsticial colure. When the Sun reaches this point in its apparent journey through the skies, summer begins. The apparent journey of the Sun is caused by the Earth's revolution about the Sun and the inclination of the Earth's axis. At this point in the skies, the Sun stands farthest north above the surface of the Earth, on the imaginary line called the Tropic of Cancer. Cancer lies to the east of Gemini, and the westward shift of this most northerly aspect of the Sun from before the stars of Cancer, where it was thousands of years ago when men first began to record the changes in the Sun's position, was caused by the precession of the equinoxes. The imaginary line, 23½° north of the equator, might now more fittingly be called the Tropic of Gemini.

The position of the planet Uranus when it was discovered by Sir William Herschel in 1782, was not far from Eta (η) Geminorum. It was also in this region that Pluto stood when, in 1930, Clyde Tombaugh found the tiny dot that appeared to move among the many faint stars scattered in the sky here. Both these discoveries were complicated by the fact that the edge of the star-strewn stream of the Milky Way crosses this corner of Gemini, and makes the proof of the discovery, particularly of an object as faint as Pluto, a project of tremendous difficulty.

680. Where did Grus get its name? *Grus* is Latin for "The Crane," and was contributed by Bayer in his catalogue of 1603.

681. Where is Grus? Grus is far to the south, below Aquarius and Piscis Austrinis. It culminates at 9 P.M. on October 12, and part of it should be just visible from the latitude of New York, given a perfectly clean southern horizon. Grus extends from west of the twenty-first hour of right ascension to about halfway between the twenty-third and zero hours, and from 37° south to 57° south. Piscis Austrinus and Sculptor border Grus on the north, and Sculptor and Phoenix on the east. Tucana and Indus cover its southern edge, and Indus and Microscopium complete the western boundary. Grus may be found on Map 1.

682. What are the notable objects in Grus? There are only two bright stars in Grus. Alpha (α) Gruis, magnitude 1.76, and Beta (β) Gruis, magnitude 2.17, are almost on a line with each other, east and west, about 8° apart. Beta (β) is slightly variable. These two stars make an equilateral triangle with Delta (δ) Gruis, which is much fainter, but which is a naked-eye double star whose components are Delta^1 and Delta^2.

683. Where did Hercules get its name? Hercules was probably the best known of all the heroes of Greek mythology. The very human figure of the strong man, with his weaknesses and his triumphs, has always captured the imagination of all ages. As a constellation, Hercules has been in its present form as far back as history goes. The Arabs pictured this constellation as a kneeling giant without a specific name. It is not difficult to see how the Greek culture overlaid the gigantic anonymous figure of the Arabs with that of their muscular hero.

684. Where is Hercules? Hercules is a tremendous constellation, ranking fifth in size. Its most southern limit is about 4° north of the celestial equator, and it extends from that point to slightly north of 50° north declination. Its east-and-west dimension carries it from just west of the sixteenth hour line of right ascension almost to the nineteenth hour. It is bounded on the west by Boötes, Corona Borealis and Serpens Caput; on the north by Draco; on the east by Lyra, Vulpecula, Sagitta and Aquila, and on the south by Ophiuchus. Hercules is found on Map 4. None of the stars of Hercules is brilliant, but the pattern outlined by its brightest stars is distinctive and easy to identify. There are six stars there which make a figure somewhat like a great butterfly, flying westward. The upper portion of this figure is often called the Keystone. The butterfly can be found just east of the curved line of stars that makes up Corona Borealis.

685. What are the noteworthy objects in Hercules? Alpha (α) Herculis (Ras Algethi) is not the brightest star in the constellation, but it is one of the notable stars in all the skies. It is a variable star of mean visual magnitude 3.1 and is a double with a companion of magnitude 5.4. This is one of the largest of stars. Its diameter has been estimated as anywhere from 400 to the almost incredible figure of 6,500 times the diameter of the Sun—from 360 million miles to 65 thousand million miles in diameter. There is not much doubt that it is a tremendous object, a cool supergiant, extremely tenuous, with a density that must be measured as a minute fraction of that of the Sun. If the lower figure for the diameter of Ras Algethi is correct, the density of the star is 0.0000003—3 ten-millionths of the density of the Sun. If the higher figure is correct, Ras Algethi is still more tenuous—considerably more of a vacuum than can be created mechanically on the Earth. *Ras algethi* is Arabic for "The Kneeler's Head," and comes from the Arabic conception of a kneeling giant. Since Ras Algethi is in the far southern part of the constellation, the Arab's kneeler was upside down. In all early star charts, Hercules was shown thus inverted, treading the figure of the Dragon as represented by Draco, to the north.

Only seven of the many stars in Hercules are brighter than fourth magnitude, and three of these are doubles. In addition to Alpha (α), Zeta (ζ) and Mu (μ) Herculis are multiple stars. Zeta (ζ) is the central star of the top of the butterfly figure. We see it as one star of magnitude 2.81, but there are two stars there, one of magnitude 2.91 and the other of 5.46. Mu lies considerably to the east of and a bit to the south of Zeta, and is made up of two stars, of magnitudes 3.42 and 9.78. In the northern section of the butterfly, a little more than a third of the way between Eta (η) and Zeta, is one of the wonders of the skies. This is the globular cluster M 13—N.G.C. 6205. Under ideal conditions, the Great Cluster in Hercules would be visible to the naked eye as a vaguely seen star on the threshold of vision. Actually, it is a great sphere of giant stars covering an area of the sky which has about two thirds the diameter of the full moon. The real diameter of the cluster must be in the neighbourhood of 100 light years and the central portion, about 30 light years in diameter, is so crowded with stars that no individuals can be resolved. They must be spaced as closely as the planets in the solar system. This cluster is 34,000 light years distant, and is one of the nearest of the slightly more than 100 such objects that are known to us. (See question 533.) In addition to the Great Cluster in Hercules, there is another globular cluster lying well to the north of the centre of the upper wing of the butterfly. This one is M 92—N.G.C. 6341—similar to M 13, but about twice as far away and hence only half the apparent diameter and brightness.

In the eastern region of Hercules—that part of the constellation that lies below the small neighbouring constellation of Lyra—is the Apex of the Sun's way. (See question 6.) Somewhere near here—and various astronomers have pinpointed the precise spot in slightly different locations—is the part of space toward which the Sun is moving at about 12 miles per second, carrying with it all its entourage of planets, satellites, comets, meteors and so on.

686. How did Horologium get its name? Horologium is another of Lacaille's fancies. *Horologium* is Latin for "The Clock." Lacaille is supposed to have named this undistinguished region for that essential instrument so that the voyagers in the good ship *Argo* might not lack for the time of day.

687. Where is Horologium? Horologium is far to the south and culminates at 9 P.M. on December 25. Horologium begins at 40° south declination and goes thence in a series of steps to 67° south. It slants from about four hours of right ascension westward to past the second hour line. It is just a place in the sky. There is, in all its area, one star of fourth magnitude. Horologium may be found on Map 5, with a bit of its northern area on Map 2.

688. Where did Hydra get its name? In Greek mythology, Hydra was a horrid serpentine monster that had a hundred heads. One of the tasks given to Hercules was to kill the Hydra. The job was complicated by the fact that as fast as one of the heads was cut off, two others grew in its place. Hercules, willing but not too bright, was having quite a time of it when a more clever friend suggested burning off the heads. This seems to have inhibited the duplication factor, and the mission was finally accomplished by this method. The Greeks gave the name of Hydra to a constellation which had been variously called a serpent and a river by more ancient cultures.

689. Where is Hydra? Hydra is all over the sky. It starts at the eighth hour of right ascension and goes eastward to the fifteenth hour, traversing seven out of the 24 hours, or almost a third of the celestial circle. The only part of Hydra north of the celestial equator is its very beginning, which touches about 7° north declination. From that point, Hydra winds south and east, averaging slightly more than 5° in width, with its most southern point touching 35° south declination. Along its northern boundary from west to east lie Cancer, Leo, Sextans, Crater, Corvus, Virgo, and Libra. Its southern side, in the same order, is bounded by Puppis, Pyxis, Antlia and Centaurus, while Canis Minor touches it on the west and Libra closes its career on the east. At its western end, a bit to the east of and on the same line with Procyon in Canis Minor, is a group of six faint stars which form an asterism known as the Head of Hydra. Hydra begins on Map 2 and extends all the way across Map 3.

690. What are the notable objects in Hydra? Only two of the six stars that form the small asterism known as the Head of Hydra are brighter than fourth magnitude. They bear the designations Epsilon (ϵ) and Zeta (ζ) Hydrae. Epsilon is a complicated affair. It is made up of four stars whose magnitudes are 3.7, 5.2, 6.8, and 12. To the east and considerably south of the Head of Hydra is Alpha (α) Hydrae (Alphard). *Alphard* is Arabic for "The Solitary One." This is an apt name, for there is nothing in the vicinity anything like as bright as Alphard to keep it company. Alphard is of magnitude 1.98 and is a perceptively red star about 94 light years away.

On the border of Hydra and Canis Major, Messier listed object number 48 in his catalogue. There is absolutely nothing there to warrant any listing of any sort. This is one of several such items catalogued by Messier of which no trace could ever be found. It is just possible that Messier, whose primary purpose was the discovery of comets, may have found comets without realizing it, and noted them as permanent clusters in his catalogue. He lists M 48 as an open cluster, and the listing may be an ironic memorial to a comet whose identity will never be known.

Much farther east, south of Corvus and in line with the two eastern stars of that constellation, is M 68—N.G.C. 4590. This is still there: a globular cluster, similar to other globular clusters but extremely remote, as it appears to be only about $\frac{1}{10}$ of the diameter of M 13, the Great Cluster in Hercules. Still farther east is M 83—N.G.C. 5235—an external galaxy.

691. Where did Hydrus get its name? Hydrus is the masculine of *Hydra,* so it must be a male water serpent. Bayer included Hydrus in his catalogue of 1603, feeling perhaps that Hydra needed a mate.

692. Where is Hydrus? Hydrus is in the far south. Its most southern limit lies about 7° from the south celestial pole—at 83° south declination. From there it runs north to 70° south for the most part although it has one bay that goes as far north as 62°

south. Its most eastern edge touches the zero hour line of right ascension and it extends west from there to about 4 hours and 30 minutes. It is bordered by a number of constellations as undistinguished as itself, such as Octans, Tucana, Horologium, Reticulum and Mensa. The only neighbours of Hydrus with any claim to distinction are Eridanus, which touches it on the north for a brief distance, and Dorado, which is on the west. Hydrus culminates at 9 P.M. on December 10, and may be found, for what it is worth, on Map 5.

693. Where did Indus get its name? Indus is another of Bayer's contributions. Bayer attempted, it is said, to honour the American Indian in this far southern constellation.

694. Where is Indus? Indus lies entirely south of 45° south declination. Its southern border is along the sixtieth parallel of south declination for the most part with one extension that stretches to 75° south. It lies mainly between 20 hours, 25 minutes and 22 hours, 5 minutes of right ascension, although its southern extension is slightly farther west, reaching 23 hours, 25 minutes. Indus culminates, for those far enough south to see it, at 9 P.M. on September 26. It may be found on Map 5.

695. What are the notable objects in Indus? The only star brighter than fourth magnitude in Indus is Alpha (α) Indi, which is of magnitude 3.1. There is one other star of some interest within the constellation boundaries, but the interest in this star lies in its location rather than in anything remarkable about the make-up of the star. Epsilon (ϵ) Indi, a fourth-magnitude star near the western border of the constellation, is one of our near neighbours in space. You will find it in question 470 on the list of the nearest stars. It is one of the very few on that list that is visible to the naked eye. Epsilon Indi is 11.3 light years distant. Apart from these two objects, Indus is quite unremarkable.

696. Where did Lacerta get its name? *Lacerta,* "The Lizard," is one of the rare regions in the northern sky that were not assigned to any of the ancient constellations. It might well have been included either in Cygnus or in Andromeda, which border it on the east and west, but it was neglected. Hevelius seized upon it in the seventeenth century and claimed it. He named it Lacerta because he said that only a lizard could be sinuous enough to wriggle into that tiny space.

697. Where is Lacerta? Lacerta is well placed for observers in the northern hemisphere. It reaches the zenith of the latitude of 40° N. every October 12 at 9 P.M. Lacerta extends from 35° north declination to 57° north and lies between the twenty-second and twenty-third hours of right ascension. Cygnus touches it on the west; Cepheus on the north; Cassiopeia and Andromeda on the east, and Pegasus on the south. There are several interesting variable stars within the boundaries of Lacerta, but it contains no visible stars brighter than fourth magnitude. Lacerta may be found on Map 1.

698. How did Leo get its name? Leo is among the oldest of all the constellations. All known astronomical records show this group of stars in substantially the same boundaries that it has today. Leo is also the most famous of all the zodiacal constellations. The reason for this distinction is that when man first began to mark the stars rising and setting with the Sun, it was Leo which accompanied the Sun at the time of its greatest northern elevation—the summer Solstice. The constellation was considered then as marking the supreme place of the Sun in the sky, of its return to full power and to the ability to give life to the Earth. The lion was traditionally the king of beasts and perhaps of all living things. In ancient Egypt, the lions were made bold by the overwhelming heat of the Sun and ventured into the inhabited parts of the Nile valley at this season. It was the time of the Lion. The early Babylonians also pictured this constellation as a lion, as did the Arabs. The Greeks accepted the figure and wove about it the legend of the Nemean lion that Hercules was compelled to slay as one of his twelve tasks.

699. Where is Leo? Leo culminates at 9 P.M. on April 15 and extends generally from about 8° north declination to 30° north, with a southern extension that reaches 6° south declination. Its western edge is at 9 hours, 20 minutes of right ascension and it runs east almost to the twelfth hour. Cancer forms its western boundary; Leo Minor and Ursa Major border it on the north; Coma Berenices and Virgo on the west, and Crater and Sextans on the south.

The figure of the constellation can be easily recognized by an asterism called The Sickle, a curved line of stars open to the west with one bright star where the handle of the sickle ends. Somewhat to the east of the Sickle is a right-angle triangle of stars with its hypotenuse uppermost. This triangle marks the eastern end of Leo. Leo may be found on Map 3.

700. What are the notable objects in Leo? The brightest star in Leo is the star at the base of the Sickle. This is Alpha (α) Leonis (Regulus). This is a double star whose components are of magnitude 1.36 and 10.8. Regulus is a main-sequence star (see question 494) about 84 light years distant and is almost 100 times as luminous as the Sun. Regulus has always been a famous star, more from its position than from its brilliance. It lies almost on the ecliptic and has held names in various cultures that were all indicative of royalty. *Regulus* means "The Prince" and Copernicus himself christened it so. The Arabs gave it a name which could be translated as "The Heart of the Lion" and it has often been called *Cor Leonis,* which has the same meaning.

Probably because it was the brightest star in the constellation of the Lion as well as because of its position near the path of the Sun, Regulus has held a commanding position in the dream world of astrology. It was considered an unfortunate star which, nevertheless, ruled all the affairs of heaven. Leo, as a whole, was given jurisdiction over the hearts and the backs of men and is responsible for the affairs of Italy, France, Bohemia, Sicily, Rome, Bristol, Bath, Taunton and Philadelphia. Obviously, if the fates of all these nations and cities do not coincide, Leo and Regulus are not tending to business.

Epsilon (ϵ) Leonis is the star at the outer extremity of the Sickle. It is of magnitude 2.99 and is about 340 light years distant. Mu (μ) Leonis, also known as Rasselas, is a fourth-magnitude star that marks the top of the curve of the Sickle. *Rasselas* is an Arabic translation of Ptolemy's locational name for this star. It means "The Northern Part of the Lion's Head." Zeta (ζ) Leonis continues the curve of the Sickle. It is of magnitude 3.46. Gamma (γ) Leonis, below Zeta, is called *Algieba,* Arabic for "The Forehead," and is a double which has been called one of the finest in the sky. The combined light of the two stars reaches us as magnitude 1.99. One deep orange-coloured star of magnitude 2.29 and a yellow-green star of magnitude 3.54 contribute to this light.

The most northern of the stars in the triangle that marks the eastern portion of Leo is Delta (δ) Leonis, Zosma, a star of magnitude 2.57. *Zosma* is a Greek word and means "The Loin Cloth"—possibly an overdressed lion? The companion of Zosma, about 5° south, is Chertan, Theta (θ) Leonis, magnitude 3.34. *Chertan* is Arabic for "The Two Small Ribs." Last of the bright stars in Leo is Beta (β) Leonis, Denebola. We meet another Deneb in Cygnus and two more in Cetus with a fourth in Capricornus. *Denebola* is part of the Arabic name which means "The Lion's Tail." Denebola is magnitude 2.14. It has a distinction, too, because of its position. It is the westernmost star in a vast figure of four bright stars that form the Diamond of Virgo. Topmost and faintest of the four is Cor Caroli in Canes Venatici. The eastern corner of the Diamond is Arcturus, in Boötes and the lowest point is Spica in Virgo. The Diamond of Virgo can be seen on Map 3.

Leo lies at the western edge of the vast field of galaxies that swarm in this region of the sky, lying principally in Coma Berenices, to the east, and Virgo, to the south. Four of these incredibly distant systems in Leo were bright enough to attract the attention of Messier, who gave them the numbers M 65, M 66, M 95 and M 96. Their N.G.C. numbers are 3623, 3627, 3351 and 3368 respectively. Since Messier's day, better instruments have revealed scores of similar objects in the same field and charts of this region are thickly dotted with symbols, each of which stands for a galaxy at an awesome distance from our own. Gerard de Vaucouleurs has built a hypothetical metagalaxy—a galaxy of galaxies—which includes our own Milky Way Galaxy. He constructs it in somewhat the same disk formation as the conventional spiral, but with galaxies taking the places of stars. The Milky Way Galaxy is placed in this tremendous formation in a place analogous to the place of the Sun in our galaxy. This great Coma-Virgo Cluster of galaxies, says de Vaucouleurs, is the centre—the hub—of his metagalaxy, taking a place similar to that taken by the star-clouds of Sagittarius in our Galaxy.

701. How did Leo Minor come to be called that? The region which contains Leo Minor was unclaimed by any of the large and important constellations surrounding it. Hevelius, abhoring any vacuum in the celestial sphere, filled up this space with a name which he obviously took from the great constellation to the south. *Leo Minor* means "The Lesser Lion."

702. Where is Leo Minor? Leo Minor adjoins Leo to the north, and is bordered on the west by Lynx, on the north by Ursa Major and on the east by Ursa Major and by Leo. The constellation lies mainly between 30° and 40° north declination with small bays that extend over the parallels in both directions. It extends from 9 hours, 20 minutes of right ascension eastward to beyond the eleventh hour line.

In spite of Hevelius, Leo Minor is still empty to the casual observer. There are four stars of the fourth magnitude in Leo Minor, and one of these bears the designation Beta (β). There does not appear to be an Alpha (α) Leonis Minoris in any charts of this constellation. Leo Minor culminates at 9 P.M. on April 9, and may be found on Map 3.

703. Where did Lepus get its name? The origin of the name *Lepus* is hidden in doubt and speculation. *Lepus* is Latin for "The Hare," but none of the more ancient civilizations pictured a hare in this part of the sky. Lepus, which lies directly below Orion, was variously known as a serpent, as a boat in which Orion was riding, as a chair for Orion and as four camels drinking from a stream—the River Eridanus which runs past Lepus to the west. The Greeks, liking none of these interpretations, may have supplied this region with a hare to provide quarry for the two dogs, Canis Major and Canis Minor, which are to the east.

704. Where is Lepus? Lepus is a very small, almost rectangular constellation directly below Orion. It lies between the fifth and sixth hours of right ascension, overlapping each of them a trifle, and runs from south of 10° south declination to just short of 30° south. Orion bounds it on the north; Eridanus, on the west; Columba on the south, and Canis Major on the east. Monoceros touches its upper eastern corner. Lepus culminates at 9 P.M. on January 28 and may be found on Map 2.

705. What are the notable objects in Lepus? There are four fairly bright stars in Lepus, although the whole constellation is overshadowed by the tremendous authority of Orion, its neighbour to the north. Alpha (α) Leporis (Arneb) is almost in the centre of the constellation. It is magnitude 2.58. *Arneb* is Arabic for "The Hare," and is part of the relatively modern Arabic name for the entire constellation. Beta (β) Leporis, almost directly south of Arneb, is a double made up of two stars of magnitudes 2.81 and 9.4. Beta is called *Nihal,* the older Arabic name taken again from the constellation name of "The Camels Quenching Their Thirst." To the east and above Arneb is Mu (μ) Leporis, magnitude 3.29, and below it and farther west is Epsilon (ε) Leporis, magnitude 3.21. The figure suggested by these four stars is that of a bow tie put together with studied carelessness.

On the western border of Lepus, far too faint to be seen without a telescope, is R Leporis, known also as Hind's Crimson Star. This star is, as its designation indicates, a variable star. At its brightest, it is of sixth magnitude, at the theoretical limit of naked-eye visibility. It diminishes to magnitude 10.4 and returns to its maximum in 430 days. Its deep red colour indicates that it is fairly cool and spectroscopic analysis indicates that it is inside a dense cloud of gas through which it sends its own brilliance with difficulty.

706. Where did Libra get its name? Libra is the only constellation in the zodiac which represents an inanimate object. *Libra* is the Latin word for "The Scales" and the constellation is generally indicated as a pair of weighing scales. In Arabian drawings of ancient times, Libra was also pictured as a pair of scales, but the more modern Arabian pictures show it as the claws of Scorpius, The Scorpion, which lies east of it. This was a Greek conception, and must have been reborrowed from the Greeks by the more modern Arabians.

707. Where is Libra? Libra lies entirely south of the celestial equator. Part of its northern edge runs along the equator and one extension of the constellation reaches as far south as 30° south declination. The western edge of Libra is at about 14 hours, 20 minutes of right ascension, and it goes eastward from there to the sixteenth hour.

Virgo and Serpens Caput border it on the north; Ophiuchus and Scorpius on the east; Lupus and Hydra on the south, and Virgo again on the west. Libra culminates at 9 P.M. on June 23. It can be found on Plates 3 and 4.

708. What are the notable objects in Libra? The four brighter stars of Libra form a rather distorted rectangle whose longest dimension runs northwest-southeast. Beta (β) Librae (Zubeneschamali) is actually the brightest of the stars of Libra, and again we have a situation in which the eye of Bayer, who assigned the Greek letters to the stars, was at fault or the brightness of the star has changed. Beta is magnitude 2.61. Its awe-inspiring name is from the Arabic, and comes from the more modern Arabic conception, taken from the Greek, in which the two brightest stars of Libra were considered to be the claws of the Scorpion, which lies just to the east. *Zubeneschamali* is a corruption of *Zuben el Gemal,* "The Northern Claw." Its opposite number, *Zuben el Genubi,* the Southern Claw, is Alpha (α) Librae, below it and to the west. *Zuben el Genubi* is a double star whose components are 2.76 and 5.15 magnitude. It would pay to go out some fine evening in the early summer and look at these two stars steadily and carefully. Beta (β) Librae, *Zubeneschamali,* the Northern Claw, is said to be the only definitely green naked-eye star. Star colours are not always vivid, but they do exist and can usually be seen when the eye has been long enough in the dark to be fully accommodated. Try it out on this one. The third star of the rectangle is Sigma (σ) Librae, magnitude 3.31 and the fourth is Gamma (γ) Librae, above it and to the east. Gamma is fourth magnitude.

709. Where did Lupus get its name? *Lupus,* which is Latin for "The Wolf," is one of the 48 Ptolemaic constellations. It was probably named for the wolf that suckled Romulus and Remus, but there seems to be a strange lack of references about it.

710. Where is Lupus? Lupus is well to the south, but the northern reaches of the constellation are theoretically visible from the latitude of 40° north. It extends from 30° south declination south to 55°, and from 14 hours 15 minutes of right ascension eastward to beyond the sixteenth hour. Centaurus borders it on the west, Libra on the north, Scorpius and Norma on the east, and Circinus and a part of Centaurus on the south. Lupus culminates at 9 P.M. on June 23 and may be found on Maps 3 and 4, with a bit of it on Map 5.

711. What are the notable objects in Lupus? Lupus has the general appearance of a curved line of stars superimposed upon a straight line. The southern end of the curve is Alpha (α) Lupi, magnitude 2.32. Incidentally, by sheer coincidence of position, a line drawn from the south celestial pole northward just about here would pass through four stars with Alpha (α) designations: Alpha (α) Apodis, Alpha (α) Circini, Alpha (α) Centauri, Alpha (α) Lupi and, with a little cheating, Alpha (α) Librae, too.

Zeta (ζ) Lupi, far to the southeast, is a double star whose components are magnitude 3.42 and 7.8. Gamma (γ) Lupi, almost in the middle of the constellation, is another double, 3.5 and 3.7 magnitudes, as close to being twins in brightness as the two members of a star team ever are. Eta (η) Lupi, on the eastern edge of the constellation, is another double, magnitudes 3.47 and 7.70.

712. How did Lynx get its name? Lynx is a modern constellation, formed, like so many other small star families in the northern sky, in vacant space between the larger ancient constellations. Hevelius is responsible for Lynx. In naming it, he perpetrated one of the minor witticisms that the science of astronomy has permitted itself. He looked at the dark and empty spaces that make up the area covered by this new constellation and declared that anyone would need the eyes of a lynx in order to see anything at all there—and Lynx it was.

713. Where is Lynx? Lynx has a most irregular outline. It extends from 33° north declination to 60° north and angles from 4 hours, 15 minutes to 9 hours, 15 minutes of right ascension, but nowhere does it fill the space in which it lies. It is shaped like a crooked stairway. Auriga borders it on the west; Ursa Major and Leo Minor on the east; Camelopardalis and Ursa Major on the north, and Cancer and Gemini on the south. It culminates on March 5 at 9 P.M., and may be found on Map 2.

Just for the record, there is one star in Lynx that is brighter than fourth magnitude. That is Alpha (α) Lyncis, 3.17 magnitude and about 180 light years distant.

714. Why is Lyra so called? Lyra is the Lyre, the ancient Greek musical instrument resembling the harp. The lyre is conventionally pictured as a series of strings held between two beautifully curved side pieces whose shape gave the lyrebird its name. There is a bit more to it than that. The bright stars of Lyra appeared to the ancients as a tortoise. The transition from tortoise to lyre is not nearly so difficult as it sounds, since the original instrument that developed into the lyre was actually the empty shell of a tortoise. When strings were strung across the opening of the shell, the hollow curve formed a natural sounding board, and the first such instruments were made in this way.

To the Arabs, Lyra was a bird. They called it The Swooping Eagle to differentiate between it and Aquila, not too far away, which was the Flying Eagle. Many of the ancient cultures pictured Lyra as some sort of a bird.

When the Greeks came across the conception of a lyre, which was originally Phoenician, they could not let it rest. They immediately wove around it the story that here was the very lyre of Orpheus himself, given to him by Apollo, or Hermes, and used by him in his attempt to woo Eurydice out of Hades.

715. Where is Lyra? Lyra is a summer constellation. It culminates on August 18 at 9 P.M. and its brightest star, Vega, is circumpolar to those who live north of the latitude of London. The constellation extends north and south from about 25° to 47° north declination and runs from east of 18 hours eastward to about halfway between the nineteenth and twentieth hours of right ascension. Draco borders it on the north; Cygnus on the east; Vulpecula and Hercules on the south and Hercules on the west. Small as it is, Lyra is a fascinating congregation of stars and other objects. It may be found on Map 4.

716. What are the noteworthy objects in Lyra? The brightest star in Lyra is Alpha (α) Lyrae (Vega). This star holds the distinction of being fifth in brilliance of all the stars, and is third in order of brightness of the stars visible from the northern hemisphere, exceeded only by Sirius and Arcturus. Vega is of visual magnitude 0.05. It is a main-sequence star of spectral classification AO, very hot and blue-white, and it looks the part as it pours forth its radiance on a clear summer night. Vega is just short of that critical distance from which absolute magnitudes are measured. It is 26.5 light years distant as against 32.5 light years to determine absolute magnitude, so there is only a slight difference between its visual and absolute magnitudes. Vega is 0.5 absolute magnitude. At Vega's distance, the splendid Sun would be a dim object of magnitude 4.86. Vega has a companion of magnitude 10.

Vega is near to the point in space that the Sun, with all its court, is approaching at the rate of a little more than 12 miles a second—the Apex of the Sun's way. Herschel's first estimate of the location of this point placed it somewhat southeast of Vega, but modern astronomy, with better and more accurate means of measurement available, keeps bringing it nearer and nearer to Vega with each cast. The last two locations calculated are very close together, just a little south of Vega. Vega is approaching us at about 8½ miles a second and this, combined with our speed of 12 miles a second, assures a close passage in about 325,000 years. It almost seems worth waiting for.

The name *Vega* is Arabic and goes back to the old Arabic conception of the constellation as a bird. It means "The Falling Eagle."

Just below Vega, four stars form a parallelogram whose longer dimension runs almost directly north and south. The two lower stars of this figure are Beta (β) Lyrae to the east and Gamma (γ) Lyrae to the west. Beta (β) Lyrae is *Shelyak,* which is the Arabic word for "The Tortoise" and was the later Arabic name for the constellation. Beta is a remarkable star. It is a double—an eclipsing variable star whose light drops almost one full magnitude in a little less than 13 days, going from magnitude 3.38 to 4.36. The brighter component of Beta is 3.38 magnitude and the fainter is 7.8. These two stars are so near to each other that there are only about three million miles of space between them, although they are both much larger than the Sun. One is 31 million miles in diameter and the other about 24 million. They must be pear-shaped because of the tremendous gravitational tug between them and there is very likely a transfer of gases from one star to the other because of their proximity to each other.

The other bright star in the parallelogram, Gamma (γ) Lyrae, is called *Sulafat,* which is an Arabic synonym for "Tortoise." Sulafat is of magnitude 3.25.

A little above and to the east of Vega is an entrancing star. It is Epsilon (ϵ) Lyrae, a famous double-double. A good field glass will show here two stars of magnitudes 4.6 and 4.9. One reason why this first separation is so easy is that the two stars are so evenly matched in brilliance. Then, if a stronger telescope is used, each of these two stars may again be separated. With the brighter of the original pair is another star of magnitude 6.3 and with the fainter is one of magnitude 5.2. Each of these pairs is in itself a binary and the whole system is gravitationally connected. Below Epsilon is Zeta (ζ) Lyrae, another fine double whose two components are of magnitudes 4.2 and 5.5.

About one third of the way between Gamma and Beta is M 57—N.G.C. 6720—the Ring Nebula. Photographs of long exposure by great telescopes show this to be an apparent ring of hazy light surrounding a bright central star. This is one of the most famous of the planetary nebulae. It is possible that the central star is emitting gases slowly in an effort to maintain a balance of atomic processes. The debris of this slow leakage surrounds the central star like a shell and looks like a ring because we are looking through a greater depth of the surrounding gases at the edge of the shell than at the centre. The central star is of about fifteenth magnitude.

To the southwest and almost on a line with Beta and Gamma is M 56—N.G.C. 6779. This is a distant globular cluster, very faint and hazy. Among the faint stars in Lyra is a variable star, RR Lyrae. This is an intrinsic variable star, pulsating with a period of less than one day between maximum and maximum. It is the prototype of untold millions of stars scattered through the farthest reaches of our Galaxy, in the great globular clusters that lie like sentinels along the boundary of the Milky Way system. RR Lyrae stars also occur in external galaxies. Together with the Cepheid variables of longer period, the RR Lyrae stars have given us the most accurate measurements to date of our own Galaxy.

717. Where did Mensa get its name? Mensa is a creation of Lacaille, who is responsible for 14 of the southern constellations. *Mensa* is Latin for "Table," and was named for Table Mountain in South Africa, where Lacaille stayed for some time making a detailed study of the southern skies.

718. Where is Mensa? Mensa is a small keystone of a constellation lying far to the south and culminating, for those who live in the southern hemisphere, at 9 P.M. on January 28. It runs from 70° to 85° south declination. Its southern portion lies between 3 hours, 20 minutes and 7 hours, 40 minutes of right ascension. Its narrower northern end extends from 5 hours, 20 minutes east to 6 hours, 40 minutes. It is bordered on the north of its key by Dorado, and by Hydrus and Volans, which touch the north edges of its wider part. Octans surrounds its southernmost section south, east and west; Hydrus covers most of its western side and Chamaeleon touches it to the east. The sole claim to distinction possessed by Mensa is that a small part of the Larger Magellanic Cloud overruns the border of Dorado and lies in the extreme northern part of this tiny constellation. Mensa may be found on Map 5.

719. Why is a constellation called Microscopium? This is another of the 14 southern constellations named by Lacaille, who, on the evidence, was a most unimaginative and unpoetic astronomer. Microscopium, it hardly needs be said, means a Microscope.

720. Where is Microscopium? Microscopium is a perfectly rectangular and very small area in the sky containing one star of the fourth and one of the fifth magnitude and an unknown number of stars visible only through telescopes. Microscopium lies athwart the twenty-first hour line and is about one hour in width. It extends from 30°15′ to 40°30′ south declination. Sagittarius borders it on the west; Capricornus on the north; Piscis Austrinis and Grus on the east, and Indus on the south. It culminates on September 18 at 9 P.M. and can be found on Maps 1 and 2.

721. Why was Monoceros so called? Monoceros is a relatively modern constellation made up in what was unclaimed territory surrounded by several of the mightiest of the ancient constellations. It may have been created by Bertschius, who staked two other astronomical claims in the vacant regions of the sky, naming Camelopardalis and Columba, but some astronomers say that Bertschius revived a much earlier name that had somehow become lost or overlooked. Monoceros means "The Unicorn."

722. Where is Monoceros? Monoceros lies across the celestial equator east of Orion, south of Gemini and Canis Minor, west of Hydra, and north of Puppis and Canis Major. At its widest point, it extends from east of the eighth hour to west of the sixth hour of right ascension, and runs from a little above 10° north declination to just south of the equator. It culminates at 9 P.M. February 19 and can be found on Map 2.

There are no stars brighter than fourth magnitude within its borders. Monoceros does contain several faint variable stars and Messier found one rather undistinguished open cluster about in the centre of the southern edge of Monoceros. This is M 50—N.G.C. 2323.

723. Where did Musca get its name? *Musca* means "The Fly" and is the only insect among the constellations. It was included by Bayer, in 1603, in his catalogue.

724. Where is Musca? Musca lies south of Crux, the Southern Cross. It runs south about 10° from 65° south declination to 75° south, and from 11 hours, 20 minutes eastward to 13 hours, 40 minutes of right ascension. Centaurus and Crux form its southern border; Circinus lies to the east, Chamaeleon to the south and Carina to the west. There are two stars in Musca brighter than fourth magnitude. Alpha (α) Muscae is a variable star whose brightness ranges from 2.66 to 2.73, and Beta (β) Muscae is a double star whose components are magnitudes 3.7 and 4.1. Musca culminates at 9 P.M. on May 14 and can be found on Map 5.

725. How did Norma get its name? *Norma* means a "straightedge," a "ruler" or "square" and is one of the navigational instruments that Lacaille provided for the crew of the ship *Argo*.

726. Where is Norma? Norma lies far to the south and runs from declination 42° south down to 60° and from 15 hours, 20 minutes eastward to 16 hours, 30 minutes of right ascension. Lupus and Circinus border it on the west; Lupus and Scorpius on the north; Scorpius and Ara on the east, and Triangulum Australe on the south. Norma culminates at 9 P.M. on July 3 and may be found partly on Map 4 and partly on Map 5. There are no bright stars in Norma.

727. Why was Octans so named? Octans, which means "The Octant," was named by Lacaille, along with a number of other instruments such as the clock, the microscope, the telescope and so on, and placed in the southern sky. An octant is a navigating instrument upon the same plan as a sextant. A sextant is built upon an angle of 60°, which is the sixth part of the circumference of a circle, and gives the sextant its name. An octant is built upon an angle of 45°, the eighth part of the circumference of a circle, and is hence called an octant.

728. Where is Octans? Octans is at the south celestial pole. The constellation is shaped somewhat like a conventional keyhole and the south pole of the skies is in the circular portion of the keyhole. This circular section follows the line of 82°30′ south declination except for one small segment bitten out of it by Musca. The wider bay of the keyhole extends as far north as 75° south declination. This wide bay runs from the eighteenth hour to the twenty-fourth or zero hour of right ascension, while the circular portion goes completely around the full hour circle. Octans may be found on Map 5.

729. What is noteworthy about Octans? The south celestial pole is in one of the most poverty-stricken regions of the sky, so far as bright stars are concerned. There are only two stars of the fourth magnitude within a circle whose centre is the Pole and whose diameter is 10 degrees. The nearest bright stars to the south celestial pole are 15° away and are only third magnitude at that. Among the nearest first-magnitude star to this important point in the skies are Alpha (α) and Beta (β) Centauri, both of which lie about on the 60° line of south declination, a good 40° from the south pole. Alpha (α) Crucis (Acrux) is three degrees nearer.

730. Who was Ophiuchus? Ophiuchus has been identified with Aesculapius, who was the first physician. The name *Ophiuchus* is formed from two Greek words which mean "The Serpent Bearer." The Latin name for the constellation is "Serpentarius," which means the same thing. Ophiuchus has also been thought of as Laocoön, the man who protested against the admittance of the Wooden Horse into Troy and was crushed to death by great serpents for his pains. As a constellation, Ophiuchus is at least 5,000 years old. Many cultures have pictured these stars as representing a giant either in conflict with a serpent or carrying a serpent. Tradition says that most of the medicines of those far-off days could be obtained by distilling the essential oils from serpents. The constellation of Ophiuchus is crossed by the constellation of the Serpent, so that it may very well be considered to be the classical representation of the physician and his ubiquitous little black bag.

731. Where is Ophiuchus? Ophiuchus culminates at 9 P.M. on July 24 and covers one of the largest areas of any constellation. It is most irregular in outline and runs, roughly, from 30° south declination to 13° north, and from 16 hours of right ascension almost to 19 hours. It is bordered on the west by Scorpius, Libra and Serpens Caput; on the north by Hercules; on the east by Aquila, Serpens Cauda and Aquarius, and on the south by Scorpius. For all its tremendous size, Ophiuchus is not conspicuous. Its general shape is that of a great barn with a peaked roof. It may be found on Map 2.

732. What are the notable objects in Ophiuchus? There are very few bright stars within the borders of Ophiuchus. At the peak of the roof of the big barn which the general outline of Ophiuchus resembles, is Alpha (α) Ophiuchi (Rasalhague). The word *Rasalhague* is part of the Arabic name for the entire constellation and means, approximately, "The Head of the Serpent Gatherer." The magnitude of Alpha is 2.09, which is a pretty good indication of the brilliance of the rest of the constellation. Eta (η) Ophiuchi (Sabik) is a double star, far to the south, and is made up of two stars, one of magnitude 3.0 and the other of magnitude 3.4. The name *Sabik* is common to two stars in Ophiuchus, this one and Zeta (ζ) Ophiuchi. *Sabik* means "The Conqueror," and the rest of the names of the two stars indicate that Zeta (ζ) is "The First Conqueror" and Eta (η) is "The Second Conqueror." These names refer to the more ancient Arabic conception of the constellation as depicting a man struggling with a serpent.

Two stars on the western border of Ophiuchus have curious names. Their designations are Delta (δ) and Epsilon (ϵ) Ophiuchi. They are one degree apart in declination. Delta is of magnitude 2.72 and Epsilon of magnitude 3.22. They are known as "The Yeds." Delta, the uppermost of the two, is Yed Prior and Epsilon is Yed Posterior. These names are a strange combination of Latin and Arabic, for *Yed* means "Hand"—*Yed Prior* is "The Leading Star in the Hand" and *Yed Posterior* is the "Following Star in the Hand." They represent the hands of Ophiuchus gripping the Serpent.

As if to make up for its lack of virtuosity in the matter of stars, Ophiuchus holds within its boundary a large number of clusters and nebulae. Two bays of the Milky Way come into Ophiuchus, one from the south and one from the north, bringing with them several of these prizes. Messier lists no less than seven globular clusters in this constellation. They are all faint and obviously lie at tremendous distances from us. M 9—N.G.C. 6333—is far to the south, below Eta (η) Ophiuchi. M 10—N.G.C. 6245 and M 12—N.G.C. 6218—lie farther north, below the celestial equator about in the middle of the constellation. M 14—N.G.C. 6402—is some distance east of M 10, and M 19—N.G.C. 6273—is farther south, on a line with and east of Alpha (α) Scorpii (Antares). Below it, exactly on the southern border of Ophiuchus, is M 62—N.G.C. 6266. M 107—N.G.C. 6171—is west of Eta Ophiuchi.

Ophiuchus has been the locale of four bright novae, most notable of which was Kepler's Star in 1604. In the vicinity of the faint star Rho (ρ) Ophiuchi, just above Antares, there is a famous nebulous region in which great clouds of gas combine with even greater clouds of cosmic dust to form a fascinating neighbourhood. E. E. Barnard lists many dark nebulae here in which vast agglomerations of dust hide the light of stars beyond them.

Although Ophiuchus is not listed as one of the zodiacal constellations, a southern bay belonging to it juts south far enough to intercept the ecliptic so that the Sun and the planets occasionally find themselves "in" Ophiuchus. In fact, the length of that portion of the ecliptic which crosses Ophiuchus is about four times as long as the section in Scorpius. The planets, which have the privilege of wandering some distance to either side of the ecliptic, often slide below the border, but the Sun inevitably spends some weeks of the winter before the stars of this strange constellation.

733. Who was Orion? Orion, in Greek mythology, seems to have been the prototype of big, good-natured, well-meaning and clumsy youth. He was a giant and extremely powerful. He fell in love with Merope, a daughter of King Oenopion of the island of Chios. Oenopion did not, evidently, hold a very high opinion of his daughter's suitor for, although Orion made himself useful around the place, ridding Chios of all its wild beasts, Oenopion kept putting off the date of the wedding. Finally Orion tired of these tactics and, being a fairly simple and direct sort of fellow, decided to take matters, including Merope, into his own hands. Oenopion, employing very subtle methods, managed to get Orion into a state of extreme intoxication and threw him into the ocean. Then he persuaded Diana to try her marksmanship upon the head of Orion, which she could see only as an unidentified dark object in the sea. She scored a bull's eye and, with proper remorse, placed Orion among the stars.

Most of the ancient cultures picture this constellation as a giant, a king, a mighty man. To the Egyptians, he was Osiris; to the Arabians, an anonymous giant. The Chinese called Orion The Three Kings, from the three bright stars that make up what we now call the Belt of Orion. The Greeks picture Orion as a great hunter, striding among the stars with a shield made of the hide of a lion held before him—to the west—on his left arm. In his right hand, held above his head, he is flourishing a club.

734. Where is Orion? Orion dominates the winter sky. It culminates at 9 P.M. on January 27 and is the finest of all the constellations. When it is rising, it is tipped over to the left above the eastern horizon. It is standing upright at culmination and slanting to the right at setting. It is easily identified by the general outline made by four bright stars of a somewhat irregular rectangle with a conspicuous line of three stars crossing the centre of this great rectangle from northwest to southeast. The uppermost of these three belt stars lies almost exactly on the celestial equator. If the night is exceptionally clear, a bow of faint stars, opening to the east, may be seen just west of the main figure of Orion, and three faint stars are visible above a line formed by the upper two stars of the great rectangle. If the line of the Belt of Orion is followed to the southeast, it will come very close to Sirius, and if taken to the northwest, it will pass just below Aldebaran in Taurus.

Orion's greatest north and south extent carries it from 23° north declination to 11° south and it runs from 4 hours, 40 minutes to 6 hours, 15 minutes of right ascension. Taurus borders Orion on part of the west and part of the north; Gemini completes its irregular northern boundary and turns the corner to the east. Monoceros covers the rest of the eastern side of Orion and Lepus completes its boundary on the south. Orion may be found on Map 2.

735. What are the notable objects in Orion? Alpha (α) Orionis (Betelgeuse), has been fully discussed in question 565. The name *Betelgeuse* is difficult to interpret, but it appears to be a corruption of an Arabic phrase which means "The Armpit of the White-Belted Sheep." This strange name may be a combination of the locational name from Ptolemy—the armpit—with the more ancient Arabic name for the constellation. Betelgeuse has always been considered the right shoulder of Orion, or of any of the large figures identified with the constellation.

Across from Betelgeuse at the top of the figure is Gamma (γ) Orionis (Bellatrix). This is a star of magnitude 1.64 and, in any other constellation, would make much more of an impression than it does here.

The star in the southwest corner of the great rectangle of Orion is Beta (β) Orionis (Rigel), and Rigel too has been described in detail in question 465. East and a little south of *Rigel,* which means "The Foot" or "The Knee," is Saiph, Kappa (κ) Orionis. *Saiph* is part of the Arabic phrase which means "The Sword of the Giant." Saiph is not a particularly distinguished star. It is of magnitude 2.20, but is tremendously distant, lying at more than 2,000 light years from us, and is extremely luminous.

The three belt stars are all of them remarkable. The uppermost of these, at the northwestern end of the row, is Delta (δ) Orionis (Mintaka). The word *Mintaka* is part of the Arabic phrase which means "The Girdle of Orion." It is a complicated star, an eclipsing variable. We see it as a star whose stated magnitude of 2.20 varies to magnitude 2.35 because there are two mutually eclipsing stars revolving about each other in our line of sight. The period of the star is 5.7 days and the magnitude of the companion star is 6.47. The whole arrangement is about 1,500 light years distant and, if seen from the standard distance of 32.6 light years, the magnitude of the system would be —6.1.

The middle star of the belt, Epsilon (ϵ) Orionis (Alnilam) is another tremendously bright star. At its distance of about 1,600 light years, it appears to be of magnitude 1.7, but assumes an absolute magnitude of —6.8 at the standard distance. Most of the stars in Orion are really tremendously hot, bright supergiants. Why all these great atomic torches should be located in this particular part of the sky is not known. *Alnilam* comes from an Arabic term meaning "A String of Pearls."

The most eastern of the three belt stars is *Alnitak,* another Arabic synonym for "belt." Its astronomical designation is Zeta (ζ) Orionis, and it is another luminous supergiant at a tremendous distance—about 1,600 light years. We see it as magnitude 1.79, but it is really a double with an absolute magnitude of —6.6. The visual magnitudes of the two stars involved are 1.91 and 4.05.

Directly below Zeta is a short line of fainter stars and other objects which forms what is popularly known as Orion's Sword. The uppermost of this short row of luminaries is a very faint star involved in a mild nebulosity. Just below it is one of the most magnificent nebulae in our galaxy and one of the very few that can be perceived at all without a telescope. This is the Great Nebula in Orion. It is a cloud of gas surrounding a complicated star that has the designation of Theta (θ) Orionis. Theta (θ) is made up of four stars in one system, so placed as to form a trapezoid. These four stars are of visual magnitudes 6.0, 7.0, 7.5, and 8.0. Surrounding them is the nebula itself, M 42 and M 43, for Messier divided the nebula into two sections. These bear the N.G.C. numbers 1976 and 1982 respectively.

To the eye alone, the Great Nebula appears like a rather faint star. If you have excellent eyes and know what you are looking at, it may have a slightly hazy appearance, but that will probably be wishful thinking. In a field glass or small telescope, you will see a definitely fuzzy and irregular patch of light. As more powerful telescopes are used, the detail of this marvel increases. It has a slightly greenish colour and resembles the flame of a blowlamp. It is only in photographs taken in great observatories that all its intricate modelling and varying tone and colour become clear.

The Great Nebula is a cloud of gas. It has been described by one astronomer as being as tenuous as the few molecules of air remaining in a good, workable electric light bulb would be if they were expanded to fill a space the size of the Grand Central Station in New York City. The Great Nebula is about 16 light years in diameter and lies about 1,000 light years from us. It is also the centre of a vast but infinitely fainter nebulous field that apparently involves the entire area of the constellation. The gases in this and in similar nebulae are excited into brilliance by the radiation from bright, hot stars that are situated either within the clouds or near them. The high-frequency radiation of these stars causes the gases in the clouds to become fluorescent and to glow and shine because of the ionization of their atoms. In the case of cooler stars involved in nebulae, the light of the nebulae is largely reflected light—the same light as that of the stars. There is considerable difference, however, between the kind of light produced by the very hot stars inside nebulae such as the Orion nebula and the light of the nebula itself. The radiation of the star is absorbed by the nebula and then re-emitted. This is evidence of the radiation effect of such hot stars on the gases of the nebulae. The greenish tone of the Orion Nebula is caused by atoms of oxygen which have been ionized to the extent of losing two of their electrons.

Eta (η) Orionis, lying below the western end of the belt, is another complicated and interesting double. It is an eclipsing binary with a range of about ⅕ of a magnitude and is made up of two stars of magnitudes 3.59 and 4.98 which revolve mutually about each other in eight days. At the very top of Orion, above Betelgeuse and Bellatrix, is Meissa, Lambda (λ) Orionis, a triple system with a visual magnitude of 3.40. The three stars that form the system are of magnitudes 3.56, 5.54 and 10.92. *Meissa* is Arabic and means "The Sparkling Star." It is usually considered to be the beard of Orion and, although there is nothing there to indicate it, Orion's head must be somewhere in the vicinity.

Below Zeta (ζ), the easternmost star in the belt, is the famous Horsehead Nebula which, next to the Coalsack, is probably the best known of the dark nebulae. It is a vast cloud of opaque, inert cosmic dust which obscures the light from stars that lie beyond it. In photographs, it looks somewhat like the head of a horse with a proudly arched neck, seen as though looking out of the upper leaf of a double stable door. This particular cloud was first seen by the third Earl of Rosse, the great Irish astronomer, and named by him because of its resemblance to a horse's head. It is also number 33 in Barnard's great catalogue of dark objects.

736. Why is Pavo so named? *Pavo* means "The Peacock," and is one of the constellations in the far south which were not seen at all from the part of the world in which the ancient civilizations flourished. It was named by Bayer in 1603. It is not possible to find in the outline of the stars of Pavo any reason for its being named as it was.

737. Where is Pavo? Pavo runs from 57° to 75° south declination and extends generally from 17 hours, 20 minutes of right ascension to 21 hours, 25 minutes. Octans covers its southern border; Indus is on the east, Telescopium on the north and Ara and Apus on the west. Pavo culminates at 9 P.M. on August 29 and may be found on Map 5.

738. What is noteworthy in Pavo? By contrast to its dim and undistinguished neighbours, Pavo has one second-magnitude star and two stars of the third magnitude. Its brightest star, Alpha (α) Pavonis, is of magnitude 1.95. This star is in the extreme northern part of Pavo on the boundary of the constellation where Indus and Telescopium join. South of Alpha, almost in a straight line, are the two third-magnitude stars.

Kappa (κ) Pavonis, on the same line with the two fainter stars but farther west, is a Cepheid variable with a period of slightly less than 10 days. Its range of magnitudes is from 4.0 to 5.5. There are no other objects of interest in Pavo.

739. Where did Pegasus get its name? Pegasus was an actor in the Andromeda Story. (See question 582.) Judging from the names of many of the individual stars in the constellation, the Arabs also regarded this constellation as a horse, but the equine picture is a bit blurred by the eternal herds and flocks which form so many of the Arabian constellation figures.

740. Where is Pegasus? Pegasus is an autumn constellation and culminates at 9 P.M. on October 19. It is one of the easier constellations to identify because of the striking formation taken by its brighter stars. Three of these stars combine with Alpha (α) Andromedae (Alpheratz) to make a distinct square of stars. This Great Square in Pegasus is almost 10° on a side and is not quite directly overhead when seen from the latitude of 40° north. A bright star a short distance out from the upper right-hand corner of the Square extends the Square's diagonal in that direction and a strikingly similar formation is at the lower right-hand corner, except that here two stars prolong the diagonal. The Square occupies the eastern end of Pegasus and takes up a little more than half the area of the constellation. In its greatest north-south extent, Pegasus reaches from about 3° to 36° north declination. It stretches from a little east of the twenty-first hour to beyond the twenty-fourth or zero hour of right ascension.

Pegasus is bordered on the north by Cygnus, Lacerta and Andromeda; on the east by Andromeda and Pisces; on the south by Pisces and Aquarius, and on the west by Equuleus, Delphinus, Vulpecula and Cygnus. Pegasus is one of the larger constellations, ranking seventh in area. It may be found on Map 1.

741. What are the notable objects in Pegasus? The northeastern star of the Great Square in Pegasus is Alpha (α) Andromedae (Alpheratz). (See question 584.) The matching star at the top of the Square, in the northwestern corner, is Beta (β) Pegasi (Scheat). This name is a faulty rendering of the Arabic word for "leg." Scheat is a giant, about 100 times the diameter of the Sun. It is a great red variable star whose magnitude ranges from 2.4 to 2.7.

Alpha (α) Pegasi (Markab) marks the lower western corner of the Square. It is of magnitude 2.5 and is white. *Markab* is Arabic and can mean either "riding" or some kind of vehicle, or an animal to ride on.

The fourth star of the Square, in the southeast corner, is Gamma (γ) Pegasi (Algenib). *Algenib* is the Arabic word for "The Side." It is a variable star with an extremely short period. Its light ranges through $\frac{2}{100}$ of a magnitude—from 2.83 to 2.85 and back again in $3\frac{1}{2}$ hours.

Inside the Square, in an area of slightly more than 100 square degrees, a great German astronomer, Argelander, counted 30 naked-eye objects. This record is bested by that of one Schmidt, who claimed to be able to see over 100 faint—and they must have been faint—stars inside the Square. One of the best star charts, one that shows objects down to the limit of naked-eye visibility, indicates 32 objects in that area. A chart on a still larger scale, giving objects as faint as seventh magnitude, gives 86 objects. There are just two there that can be depended upon—these are two fourth-magnitude stars lying almost on the northwest-southeast diagonal about half way between Beta (β) Pegasi and the centre of the Square. Outside the Square, northwest of Beta, is Eta (η) Pegasi (Matar) of magnitude 2.95. *Matar* is Arabic for "Rain" and is only a part of the Arabic name which was "The Lucky Star of the Rain." Southwest of Alpha is Zeta (ζ) Pegasi (Homam), "The Great King." Homam is magnitude 3.40. Continuing along the same line from the corner of the Square out beyond Zeta (ζ) lies Theta (θ) Pegasi (Biham). This star is a modest 3.70 magnitude. In its name is one of the frequent references in Arabic names to the pastoral culture of those people. *Biham* crams into its few letters the meaning of "A Flock of Lambs, Kids or Camel Colts."

If we make a 90° turn to the northwest here, we find Epsilon (ϵ) Pegasi, (Enif), which is from the Arabic words that mean "The Horse's Nose." Enif is a double star, seen as magnitude 2.31, with an eleventh-magnitude companion.

On to the northwest beyond Enif is M 15—N.G.C. 7087—a beautiful globular cluster, very bright, nestling right alongside a faint fifth-magnitude star.

742. How did Perseus get its name? The legend involving Perseus and Andromeda, together with several other neighbouring constellations, may be found in question 582, on Andromeda. The legend of Perseus was current in Greek literature by the fifth century B.C., and it is not certain whether the legend was given to the stars or whether the stars produced the legend.

743. Where is Perseus? Perseus culminates at 9 P.M. on December 22, when it will be directly overhead for observers along the 40th parallel of latitude, north. Perseus is a straggling, irregular constellation which extends northward from above 30° north declination to 58° north, and runs from 1 hour, 30 minutes east to 4 hours, 50 minutes of right ascension. Andromeda and Cassiopeia bound it on the west, Cassiopeia and Camelopardalis on the north, Auriga on the east; and Taurus, Aries and Triangulum on the south. Perseus can be found on Maps 1 and 2.

744. What are the notable objects in Perseus? A brilliant section of the Milky Way runs from southwest to northeast through the centre of Perseus, and the constellation is crowded with bright stars. To the north, a curve of stars is known as the Segment of Perseus, and from the lower, or southwest end of the Segment, two more curved lines of stars drop away to the south. The easternmost of these two star lines seems to point directly at the Pleiades, and the great giants of Taurus and Orion spread out to the south and east.

The brightest of Perseus' stars is Mirfak, Alpha (α) Persei, well to the north of the centre of the constellation. Mirfak is a supergiant of magnitude 1.80 and is surrounded by several much fainter stars so that, in a small telescope, it appears to be the centre of an open cluster. *Mirfak* is a name that bears no relation whatever to Perseus. It is much older than the Perseus legend and refers to the former Arab conception of this part of the heavens as being associated with the Pleiades, to the south. *Mirfak* is part of the name of the star in those days and means "The Elbow" (of the Pleiades).

Beta (β) Persei is some distance below and to the west of Alpha. This is the famous variable star, Algol. (See question 541.) *Algol* is part of the Arabic name *Ras al Ghul,* which means the "Demon's Head." This name was considered by many to refer to the variability of Algol. It is most doubtful, however, that the Arabs, who named the star, were aware of the peculiarity of Algol, whose light changes were confirmed scientifically in 1669. The Arabs did know of the legend of the Medusa and of the horror in which that particular evil thing was held. They had, in their own legends, an equally terrible being, the Ghul, a nightmare that haunted men and destroyed them. Not to be outdone by the Greeks, they gave to the same star which, to the Greeks, represented the head of the Medusa, the name of the head of the Ghul. It must be coincidence which arranged things so that this star with the fearful name is the one whose mysterious change in brightness must have made men wonder for years after demons were discarded.

There are two other stars with Mirfak that make up the Segment of Perseus. These are Gamma (γ) Persei to the northwest and Delta (δ) Persei to the southeast. Gamma is magnitude 2.91 and Delta is 3.03.

Near the centre of the eastern curve of stars below the Segment is Epsilon (ϵ) Persei, a double star whose components are 2.99 and 7.99. At the bottom of this line of stars is Zeta (ζ) Persei, another double made up of two stars of magnitudes 2.83 and 9.36. Below Algol is Rho (ρ) Persei, a variable, with a light range of a little more than half a magnitude. At irregular intervals, Rho changes from magnitude 3.2 to 3.8 and back again.

Far to the northwest, at the upper end of the Segment of Perseus, is a beautiful and amazing double cluster of stars. Herschel first publicized this pair and the mystery is that Messier did not include it in his catalogues, for these clusters can be seen as glowing blurs in the stream of the Milky Way. They are open clusters, and in a small telescope they are near enough together to be in the same field. About halfway down the western edge of Perseus is M 34—N.G.C. 1039. This is a small, faint planetary nebula. Almost precisely in the centre of the constellation is the point where a bright nova flamed up in 1901 and became, at that time, the brightest nova since Kepler's Star in 1604. This nova was discovered by a Scottish physician and amateur astronomer, Dr. Anderson, who took nightly walks and observed the stars. During one of his walks, without instruments, he found and identified this nova.

745. Where did Phoenix get its name? Phoenix was named for the mythical bird who was burned in a ceremonial fire every hundred years and who rose renewed from the flames to begin another cycle. Phoenix is not one of the older constellations, but it was another of Bayer's contributions. It lies east of Grus, The Crane and Tucana, the Toucan, and was probably named to keep Bayer's aviary intact, since he contributed all three of them.

746. Where is Phoenix? Phoenix lies below the horizon of an observer at latitude 40° north. Its northern boundary is just above 40° south declination and it extends south to above 60° south. Phoenix reaches eastward from zero hours, 20 minutes to 2 hours, 20 minutes of right ascension. Grus and Tucana, two more celestial birds, border Phoenix on the west; Sculptor and Fornax touch it on the north and Tucana on the south. Phoenix can be found on Maps 1 and 5.

747. What are the notable objects in Phoenix? There are only three stars in Phoenix that are brighter than fourth magnitude. Alpha (α) Phoenicis is of magnitude 2.39. With Beta (β) and Gamma (γ), it forms a right-angle triangle with its hypotenuse uppermost. Beta (β) Phoenicis is a perfectly matched double star whose components are both magnitude 4.1, and Gamma is magnitude 3.44.

748. Why is Pictor so named? Pictor, the Painter, is another vacant southern constellation. The unimaginative quality of the name points to Lacaille, who christened this region.

749. Where is Pictor? Pictor culminates on January 30 at 9 P.M. It extends irregularly from 42° south declination to 64° south and from 4 hours, 40 minutes of right ascension to 6 hours, 50 minutes. Carina borders it on the east; Dorado on the south and much of the west; while Caelum finishes off the western boundary. Columba touches its northern border, and Puppis covers a very small stretch of the eastern edge. Pictor has just one bright star, Alpha (α) Pictoris, magnitude 3.27. One of the bright novae of recent times flared up near Alpha in 1925. Pictor can be found on Maps 2 and 5.

750. Where did Pisces get its name? The part of the sky in which Pisces is located has been regarded as the sea of the heavens by many ancient civilizations. The neighbouring star groups also take on a marine flavour, for Cetus, the Whale, is one of the bordering constellations and Capricornus, whose name has been corrupted to mean a Sea Goat, is not far away. The Persians, Babylonians, Turks and Greeks all pictured this area as containing two fishes. One possible source for this duplicity is that when the Sun was standing before Pisces, it was the time of the year in which the ancient Babylonians inserted into the year their intercalary month, causing a doubling of the months in certain years. The fish conception may have come from the fact that this was the season when, in ancient Egypt, the fish were fattest and easiest to catch.

Conventional drawings of the constellation show two fishes, one to the north and the other to the west, tied together by a long piece of ribbon attached to their tails. Because of the precession of the equinoxes, Pisces now contains the point at which the Sun seems to cross the equator on its way north to bring spring to the northern hemisphere. This point, where the twenty-fourth or zero hour line of right ascension crosses the celestial equator, was once located in Aries and is still known as The First Point of Aries. (See question 336.)

751. Where is Pisces? Pisces culminates on November 11 at 9 P.M., and can be most easily located by using the Great Square in Pegasus as a reference point. A western extension of Pisces lies directly south of the Square, and Pisces runs east and then north from there. The constellation is not at all conspicuous. In east and west direction, Pisces runs from 22 hours 50 minutes of right ascension eastward to 2 hours, 10 minutes. Its southernmost border is about 6° south of the celestial equator; and from there it goes irregularly north as far as 33° north declination. Aquarius lies at its southwestern corner and Pegasus touches its western and much of its northern side. Andromeda finishes off the north and Triangulum, Aries and Cetus cover the east, while Cetus also takes care of most of the south. The constellation covers a large area but in all the region there is just one star as bright as third magnitude. Pisces may be found on Map 1.

752. What are the notable objects in Pisces? The brightest star in Pisces, by a fraction of a magnitude, is Eta (η) Piscium. A close second is Alpha (α) Piscium (Al Rischa), which is of magnitude 3.94. *Al Rischa* means "The Rope" or "The Cord," and is sometimes considered to represent the cord or ribbon by which the two fishes are joined. It is more likely, however, that this name comes from an older and more obscure conception held by the Arabs, in which this part of the constellation was considered to be a well, and the rope was attached to a bucket. Al Rischa is a double star whose components are of magnitude 4.3 and 5.2.

Messier has located only one object in Pisces. This is a very faint galaxy east of Eta. Messier numbered it 74, and its N.G.C. number is 628.

The point at which the zero hour line crosses the equator—the vernal equinox—is found somewhat south of the star Omega (ω) Piscium, in the part of the constellation known as the Western Fish. A very faint circle of stars here, below the Great Square, locates one of the two fishes, while the other is to the east and north, east of the Square.

753. Where did Piscis Austrinus get its name? *Piscis Austrinus* means "The Southern Fish." This appears to be a relatively recent name, given to the constellation because of its place in the sea of the heavens and because it is situated so far to the south. The Arabs pictured both a frog and an ostrich in these stars at different epochs in their astronomical history.

754. Where is Piscis Austrinus? Piscis Austrinus culminates on October 9 at 9 P.M. It lies just above the southern horizon for observers at latitude 40° north and is bisected by the thirtieth parallel of declination, since it extends to about 5° both north and south of that line. Its eastern edge is at 23 hours, 10 minutes of right ascension and it extends westward to 21 hours, 20 minutes. Piscis Austrinis may be found on Map 1.

755. What are the notable objects in Piscis Austrinus? The only bright star in Piscis Austrinus is Fomalhaut, Alpha (α) Piscis Austrini. *Fomalhaut* is from the Arabic and is the rather recent acceptance by the Arabs of Ptolemy's place name for the star. The name comes from the Arab phrase which means "The Mouth of the Southern Fish." If the eye follows the western side of the Great Square in Pegasus, continuing the line laid out by those two stars down to the horizon below the Square, Fomalhaut will loom up far to the south. Fomalhaut is of magnitude 1.16 and is very much alone among the fainter stars near it. It is one of the relatively near stars to our system, for it lies only 21 light years from us. Its isolated position lent it considerable authority among astrologers in ancient days and, with Regulus, Antares and Aldebaran, it made the astrological team of the Four Royal Stars. None of the other stars in Piscis Austrinus is as bright as fourth magnitude.

756. Why is Puppis so called? *Puppis* means "The Deck" and the stars of this constellation were once members of the tremendous constellation of Argo Navis, The Ship *Argo*. Argo Navis covered so much of the celestial sphere that when astronomers began to use letters and numbers to identify the stars, they found that Argo Navis was much too cumbersome. By general agreement, the great constellation was broken up into three of the important parts of a ship. There are now Carina, the Keel; Vela, the Sails; and Puppis, the Deck.

757. Where is Puppis? Puppis cuts across the southern horizon, culminating at 9 P.M. on the night of February 22. Its greatest overall extent is north and south, for it runs from 12° south declination to 52° south. Its greatest east and west extent is from 6 hours, 5 minutes of right ascension to 8 hours, 35 minutes. Canis Major lies to the west of most of it, with Columba and Pictor completing the southern part of the western boundary. Monoceros and Hydra touch it on the north, with Hydra, Pyxis and Vela forming its eastern boundary. Carina bounds it on the south. The general location of the constellation is directly south of Canis Major and directly east of the great stars in Canis Major. Puppis may be found on Map 2.

758. What are the notable objects in Puppis? The brighter stars of Puppis lie in the southern part of the constellation. There is no Alpha (α) Puppis, for the Alpha of Argo Navis was assumed by Canopus, now in Carina. The brightest star of Puppis is Zeta (ζ) Puppis, lying very near to the 40° line of south declination. Zeta (ζ) is of magnitude 2.23, and is one of the truly bright stars since, at its distance of 2,400 light years, it must have an absolute magnitude of —7.1 to throw as much light as it does across all that awful distance.

There are several notable variables in Puppis. L_2 Puppis is a famous long-period variable star whose light ranges from magnitude 3.4 to 6.2 in 141 days. This involves a change in brilliance of almost three magnitudes or about 12 times in actual luminosity. Rho (ρ) Puppis is a variable star with a very small change in brightness, from magnitude 2.72 to 2.87.

The Milky Way cuts across the eastern side of Puppis and there are two of Messier's clusters within the constellation boundaries. M 46—N.G.C. 2437—is an open cluster lying southeast of Sirius. About 5° due south of it is M 93—N.G.C. 2447—another open cluster without particular distinction. A nova appeared in the eastern part of Puppis in 1942.

759. How did Pyxis get its name? Pyxis is the Greek word for "Box," but in this special case, it is a ship's compass and the box that encloses it. The name was given to the constellation by Lacaille, who named many of the southern constellations. Pyxis is intended to be the compass of the Ship *Argo,* for Pyxis is adjacent to the site of the old Argo Navis, lying between Puppis and Antlia, the Deck and the Air Pump.

760. Where is Pyxis? Pyxis extends irregularly from 17° south declination south to 37° south, and from 8 hours, 20 minutes east to 9 hours, 20 minutes right ascension. Puppis borders it on the west; Hydra on the north; Antlia on the east; and Vela on the south. Pyxis culminates at 9 P.M. on March 21 and can be found on Map 3, with a small section on Plate 4. There are only two stars brighter than fourth magnitude in Pyxis, and no other objects worthy of special note lie within its borders.

761. Why is Reticulum so called? Reticulum, "The Net," is a contribution to the welfare of the Argonauts. The constellation has generally been attributed to Lacaille, although the name of an otherwise unknown German is often given as the originator of this group. The man's name was Habrecht, but nothing further is known of him.

762. Where is Reticulum? Reticulum is far to the south, lying entirely below the fiftieth parallel of declination. It extends from 53° south to about 57° south and goes from 3 hours, 10 minutes of right ascension to 4 hours, 20 minutes. Dorado borders it on the east; Hydrus on the south; and Horologium on the west and north. It culminates at 9 P.M. on January 3 and may be found on Map 5. The brightest star in Reticulum, Alpha (α) Reticuli, is a double in which one component is of magnitude 3.33 and the other of magnitude 12.

763. How did Sagitta get its name? Sagitta is "The Arrow," and from its general appearance, it could hardly avoid being pictured as an arrow. In this tiny constellation, there are two stars almost in line with a pair of stars set at right angles to the line between the first two and at about the same distance beyond the second star. This arrangement gives a very definite arrow shape to Sagitta. All of the ancient civilizations pictured it as an arrow.

764. Where is Sagitta? Sagitta lies between Cygnus and Aquila, with the Arrow flying to the northeast toward one wing of Cygnus, passing through Vulpecula to reach its goal. Sagitta is a rather flat constellation lying, for the most part, below the parallel of 20° north declination, although two small extensions at either end bring it to 22° north. It runs from 18 hours, 50 minutes of right ascension eastward to 20 hours, 20 minutes. Hercules touches its western end; Vulpecula borders it on the north; Delphinus on the east; and Aquila on the south. Sagitta culminates at 9 P.M. on August 30, and can be found on Map 4.

765. What are the notable features of Sagitta? Alpha (α) and Beta (β) Sagittae are the two stars that mark the western feathered end of the arrow and, in spite of their designations, neither of them is as bright as Delta (δ) Sagittae, the centre of the arrow, or Gamma (γ), the arrow's point. The Milky Way sweeps across this tiny constellation and the whole region in which it lies is magnificent. Just about halfway between Gamma (γ) and Delta (δ), the point and centre of the arrow, is M 71—N.G.C. 6838—an open cluster which, in spite of the richness of its surroundings, is outstanding. It is a cluster in a heaven of clusters! M 71 is very compact, almost a globular cluster, but it is not quite symmetrical enough to be in that class. It is also very much nearer than the globular clusters, for its outlying stars can be resolved with a 3-inch telescope, which is not possible in the case of any object at the distance of the globular clusters.

766. Why is Sagittarius so called? Sagittarius is one of the old constellations. It is the zodiacal constellation before which the Sun now pauses at the time of the winter solstice—that point in the Sun's apparent journey around the Earth when it is overhead above the most southern portion of the Earth's surface. Thousands of years ago, this point was located in Capricornus, to the east, but the precession of the equinoxes has moved it westward almost to the western edge of Sagittarius. In Greek mythology, Sagittarius represents the great centaur, Chiron, the tutor and trainer of most of the heroes of

Greek lore. Chiron is supposed to be shooting his arrow across the entire width of the sky at Taurus, the Bull, whose stars are setting in the west as Sagittarius rises in the east. Most of the ancient cultures pictured this constellation either as an archer or as a bow and arrow alone. In one conception the Arabs called it the bow, but in another these stars represented a herd of ostriches with their keeper. In Egypt, and earlier, in Babylonia, the figure of a bowman is carved to represent Sagittarius among the other constellations.

767. Where is Sagittarius? Sagittarius culminates at 9 P.M. on August 21 and is brilliant along the southern horizon for much of the late summer. It reaches west as far as 17 hours, 45 minutes of right ascension and goes eastward to 20 hours, 20 minutes. Its farthest southern extent takes it to 45° south declination and it comes north to 8° south declination. Scorpius and Ophiuchus border it on the west; Serpens Cauda, Aquila and Scutum on the north; Capricornus and Microscopium on the east; and Telescopium and Corona Australis on the south.

The constellation is striking and is easy to identify. Its best known asterism is the Milk Dipper, a figure of four stars whose arrangement is remarkably like those that form the bowl of the Big Dipper, with a handle of two more stars extending away from the bowl toward the northwest. The Milk Dipper is upside down. The first star in the handle of the Milk Dipper plays a double role as the uppermost star in the great bow of Sagittarius, which is bowed to the west. The tip of the arrow being drawn is plainly marked by a star to the west of the convex side of the bow. The entire combination of the Milk Dipper and Bow is often likened to a teapot, tipped slightly to the west with the handle of the teapot made of the stars of the bowl of the Milk Dipper and with the tip of the arrow marking the spout of the teapot. The teapot has a pointed lid.

The Milky Way pours down through Sagittarius. The point in space at which Harlow Shapley and others have placed the hub of our Galaxy lies in the direction of Sagittarius.

Sagittarius can be found on Map 4.

768. What are the notable objects in Sagittarius? Sagittarius is crowded with wonders. In spite of this, one of the faintest of the stars in the entire constellation is given the designation of Alpha (α)

Sagittarii. Alpha lies far to the south, a poor thing of third magnitude. Why this should be so, it is impossible to say, for it is inconceivable that Bayer, who first assigned the Greek letters to the stars, could have been so mistaken in his estimate of their brightness, nor is there any apparent order of position which might account for the choice of designations. Beta (β) Sagittarii, a naked-eye double, lies down in the corner next to Alpha (α) and is no brighter.

Farther north, just above the horizon of 40° north latitude, Sagittarius is glorious. In this region are 11 stars brighter than magnitude 4, most of them much brighter. The lowest star in the bowl of the Milk Dipper is Zeta (ζ) Sagittarii, Ascella. *Ascella* is a corruption of the Latin word *Axilla,* "The Armpit," for it represents the upper end of the arm of the bowman as he draws back on the arrow. Zeta is a double star of visual magnitude 2.61, made up of two components of magnitudes 3.3 and 3.5. In the Arab conception, Zeta was an ostrich.

Above and slightly to the west of Zeta is Tau (τ) Sagittarii, forming the corner of the bowl of the Milk Dipper farthest from the handle or, in the toxophilite conception, the elbow of the archer. Tau is magnitude 3.30.

Above and to the west of Tau is Sigma (σ) Sagittarii, Nunki, magnitude 2.12, about as bright as *Polaris.* The name *Nunki* is given by Davis as coming from the Sumerian name of the star, which meant "The Star of the Holy City." The holy city was Eridu, on the Persian Gulf, and has nothing to do with archers or ostriches. Phi (ϕ) Sagittarii, magnitude 3.20, completes the bowl of the Milk Dipper. Lambda (λ) Sagittarii, Kaus Borealis, is the first star in the handle of the Milk Dipper and the uppermost star in the Bow. Its name, taken from a Latin-Arabic combination, means "The Northern Part of the Bow." It is of magnitude 2.80. Below it is Delta (δ) Sagittarii, Kaus Media, "The Middle of the Bow." Delta is magnitude 2.71. The Bow is completed to the south by Epsilon (ϵ) Sagittarii, Kaus Australis, "The Southern Part of the Bow." Epsilon is magnitude 1.81. The point of the arrow, to the west of the curve of the bow, is Gamma (γ) Sagittarii, Al Nasl. The words *Al Nasl* are Arabic for "The Point of the Arrow," and, in older interpretations, the star was another of the ostriches which the Arabs earlier saw in the constellation. Al Nasl is of magnitude 2.97.

Eta (η) Sagittarii, southwest of Epsilon, is a double whose brighter component is magnitude 3.17, with a faint companion of tenth magnitude. Directly north of the Milk Dipper are two stars taken as a unit, although they are not an optical double. They are Xi[1] and Xi[2] (ξ^1 ξ^2) Sagittarii. Xi[1] is very faint, but Xi[2] is magnitude 3.51. East of these two is Pi (π) Sagittarii, a triple star. Its three components are 3.7, 3.8, and 6.0 magnitudes. To the northwest, forming the extension of the handle of the Milk Dipper, is Mu (μ) Sagittarii, third magnitude, mentioned here because it is the marker for the magnificent region of nebulae and clusters which lies about it.

Above Mu (μ), at the border of the constellation, is M 17—N.G.C. 6618. This is the Omega Nebula, sometimes called the Horseshoe Nebula. It is a gigantic, diffuse nebula, a great cloud of tenuous gas shaped like a horseshoe or like the Greek capital omega (Ω). It is not easy to see the detail in this cloud, but a dark streak separating the upper portion from the lower is easily discernible. Below this is M 18—N.G.C. 6613—an open cluster. A little farther south again is M 24—N.G.C. 6603—another open cluster. This one can be seen on fine dark nights as a slight thickening of the field of stars in the Milky Way wherein it lies. East of this is M 25—N.G.C. 4725—another open cluster which is large and rather scattered. Well to the west is M 23—N.G.C. 6494—another open cluster, again widely scattered.

A little below and to the west of Mu (μ) is M 21—N.G.C. 6531—another open cluster. M 20 is again close below this and is N.G.C. 6514, the famous Trifid Nebula. This is another diffuse nebula, a cloud of gas which has three distinct segments when seen in photographs. It is tremendously complicated and intricate, with a fascinating pattern of shading and obscuring cosmic dust. In a small telescope it is a hazy glow involving a few faint stars.

Not far below the Trifid Nebula and again to the west is M 8—N.G.C. 6523. This is the Lagoon Nebula, another diffuse galactic nebula. In a small telescope, this is a patch of bright, luminous cloud involving several bright stars. The Lagoon Nebula can be seen as a faint blotch of light against the Milky Way even without a telescope.

Off to the east and a little way from Lambda (λ) is M 28—N.G.C. 6626. This is a globular cluster, small but bright against the glowing field of the Milky Way.

Not far to the west and about on a line with it is M 22—N.G.C. 6656. This is possibly the finest globular cluster visible in the northern sky. It is not so well known as the Great Cluster in Hercules, M 13, because of its less favourable location, but it is amazingly beautiful. It does not seem to be as closely packed with stars as does the Hercules cluster, and it appears to be slightly elliptical with its long axis running northwest and southeast. It is one of the nearer of these amazing objects, about 22,000 light years away, and contains a possible 100,000 stars, all tremendously bigger and brighter than the Sun.

Two more globular clusters lie in the neighbourhood. They are M 69 and M 70, far to the south, somewhat ahead of the opening of the bowl of the Milk Dipper. Their N.G.C. numbers are 6637 and 6681 respectively. They are both very faint and far away.

To the west of them, south and east of Zeta (ζ), is M 54, still another globular cluster—N.G.C. 6715. This too is very remote and very faint. There is another globular cluster quite a way to the west and about on the same east and west line. This is M 55—N.G.C. 6809. It is more diffuse than most globulars and seems to be involved in some nebulosity. One more, M 75—N.G.C. 6864—is farther west and higher, on the border of Capricornus. This is again a very faint and distant globular cluster.

In 1898, a nova flared up in the far northwest corner of Sagittarius, and in 1936, another blazed out just below Al Nasl. Here, in this region, the Milky Way seems almost to move. It does not require too much imagination to feel, on a clear and moonless night, the tremendous, great circling of the galaxy.

769. How did Scorpius get its name? From the very beginning, this striking group of stars has been known as a Scorpion. The Egyptians, the Arabs and their predecessors all show the image of a scorpion in their early pictures of this part of the sky. The sharply curved formation of stars that marks the tail of The Scorpion does look more like a scorpion's tail than anything else. It is easier to see a scorpion in the constellation than it is to see in most of the constellations the things or animals for which they were named.

The Greeks, with their lively imaginations, have linked the Scorpion with Orion, saying that it was the sting of the Scorpion that really killed Orion, not the arrow of Diana. Orion sets as Scorpius rises. Scorpius is also associated with Ophiuchus, which is directly above Scorpius. In this conception, Ophiuchus is treading on the Scorpion, and above Ophiuchus, the other giant, Hercules, is stamping on Draco, the Dragon. This arrangement would make Hercules upside down; and the figure of the Greek hero has been pictured in that topsy-turvy manner for many hundreds of years.

Long ago, the western boundary of Scorpius was extended to include the constellation of Libra. (See question 708.) The two brighter stars in Libra bear names which mean the Northern and the Southern Claw (of The Scorpion).

770. Where is Scorpius? Scorpius is about as far south as a constellation can be and still be visible from 40° north latitude. The southern border of Scorpius is about 45° south declination and one very narrow extension of it reaches as far north as 8° south. It is a most irregularly shaped constellation. Its westernmost region is at 15 hours, 50 minutes of right ascension and it reaches as far east as 17 hours, 55 minutes. On the west, it is touched by Centaurus and Libra; Ophiuchus covers most of its northern edge, with Sagittarius just turning the corner and taking care of part of its western side as well, although Ophiuchus too comes down to cover the upper western edge. Corona Australis finishes up the western side; and Ara and Norma border it on the south.

Scorpius is striking, and should be identified without too much difficulty. It culminates at 9 P.M. on July 8. The great red supergiant Antares, glowing like a furnace on the southern horizon, should mark it beyond question. Antares is accompanied by two lesser lights, one ahead of it and one behind, and a spray of three fairly bright stars fans out to the west of Antares. From here, the Scorpion drops almost straight south and then the wicked looking tail curls up to the east—a tail made up of bright stars with a brilliant, hot-looking double as the stinger. Scorpius can be found on Map 4.

771. What is remarkable about Scorpius? Alpha (α) Scorpii (Antares) is the leader of this noble family. Antares is a supergiant, a tremendous red star. This is a big one. For many years Antares was the largest star known. Its diameter of 480 times the Sun's diameter is only now exceeded by a few freak stars. If the Sun were placed at the centre of Antares, all the planets almost to Jupiter would be buried inside this vast star. For all its size, Antares has a mass which is only about 30 times the mass of the Sun. It is a red-hot vacuum. Antares is a variable star. Its magnitude is given as 0.98, but it ranges from

0.86 to 1.02, and it has a tiny, green companion of magnitude 5.46. This star is so close, visually, to Antares, that it is most often hidden in the glare of its gigantic mate, and with small telescopes can be seen only under the most perfect atmospheric conditions.

Antares means "The Rival of Mars." Both the star and the planet are unmistakably red, and since Antares lies very near the ecliptic and the path of the planets, Mars is occasionally found in its neighbourhood. Antares has two faint attendants. To the northwest is Sigma (σ) Scorpii, and Tau (τ) Scorpii is to the southeast. Sigma is a double. The brighter component is a variable whose light changes from magnitude 2.82 to 2.90 and back again in six hours. Its companion is magnitude 8.49. Tau (τ) is magnitude 2.85.

Topmost of the spray of stars ahead of Antares is Beta (β) Scorpii (Graffias). Here is another complicated luminary. Graffias is a triple that we see as one star of magnitude 2.65, but actually there are three stars of magnitudes 2.78, 5.04 and 4.93, all contributing to this light. *Graffias* is a word of doubtful meaning, but it is probably a corruption of a Greek word whose transliteration is *Grapsias,* and which could mean either "a crab" or "a scorpion."

Below Beta is Detla (δ) Scorpii (Dschubba), magnitude 2.34. *Dschubba* means "The Forehead," presumably of the Scorpion. Below Delta is Pi (π) Scorpii, magnitude 2.92.

South of Antares and his escorts is Epsilon (ϵ) Scorpii, which with Mu (μ) Scorpii, still farther south, outlines the body of the scorpion. Epsilon is magnitude 2.28. Mu is a bit complicated. First of all, there are two stars here, not gravitationally connected but visually so closely associated that they bear the same Greek letter and are distinguished by superscripts 1 and 2. Mu^1 (μ^1) is an eclipsing variable whose components, by passing between one another and our line of sight, make its light vary from magnitude 2.99 to 3.09 in a little less than 36 hours. Mu^2 (μ^2) is a faint star of fourth magnitude.

Below this point the stars fall into a line which marks a sharp curve to the east, then north and a trifle west to make the upraised tail of The Scorpion. First of these stars is Eta (η) Scorpii, magnitude 3.33. Almost due east is Theta (θ) Scorpii, much brighter, with a magnitude of 1.86. To the east and somewhat above Theta is another visual pair which have the same Greek letter designation with superscripts. Iota^1 (ι^1) Scorpii is magnitude 2.99, while Iota^2 (ι^2) Scorpii is fifth magnitude. Kappa (κ) Scorpii, magnitude 2.39, starts the final curve to the west of the tail, which ends in a magnificent star, Lambda (λ) Scorpii (Shaula). *Shaula,* which is part of the Arabic name meaning "The Cocked-up Part of the Scorpion's Tail," is magnitude 1.60, and is very close, visually, to Nu (ν) Scorpii (Lesath), magnitude 2.71. *Lesath* is a corruption of the Arabic word for "The Scorpion's Sting," and though it appears to be alongside Shaula, it is really some 200 light years more distant. To the west again is G Scorpii. This is the first example of a star bright enough and far enough along in the hierarchy of a constellation to be beyond the use of Greek letters or numbers. G is magnitude 3.21.

The main stream of the Milky Way, just below the hub of the local Galaxy, runs wide through practically all of Scorpius. It brings with it a great number of bright stars and many clusters and nebulae. Above and a bit to the west of Antares is M 80—N.G.C. 6093. This is a globular cluster, very small and condensed. About one degree and 15 minutes directly west of Antares is M 4—N.G.C. 6121—another globular cluster and again very small and condensed, probably because its light is quenched by the nearer and therefore stronger light of Antares through which it must pass in order to reach us. Somewhat to the south and west is M 62—N.G.C. 6266—still another globular cluster, again very small and condensed, as we see it.

Farther to the west, just above the cocked-up part of the scorpion's tail, is M 6—N.G.C. 6405. This is an open cluster, fairly widespread and almost rectangular in outline. It is beautiful and looks a little like a butterfly with its wings open. Below it, directly above G. Scorpii, is M 7—N.G.C. 6475—another open cluster, even larger. This one is on the edge of the Milky Way and is a bit difficult to define exactly. Almost all of Scorpius is rich in stars and it is a rewarding experience to wander telescopically through the crowded suns of this region.

772. Why is Sculptor so named? Sculptor, which does not need translating, is a modern constellation, added by Lacaille, with so many of the southern constellations. Its name is really The Sculptor's Workshop, but even though it now has a name, it is still a large, blank region.

773. Where is Sculptor? Sculptor is far to the south, but the entire constellation should be visible from the latitude of 40° north. The southern side of Sculptor runs straight just 1° above and parallelling the 40° of south declination, and its northern boundary lies exactly on 25° south. Sculptor extends from 23 hours, 10 minutes eastward to 1 hour, 40 minutes of right ascension. Aquarius and Cetus bound Sculptor on the north; Fornax on the east; Phoenix and Grus on the south and Piscis Austrinis on its west. Sculptor culminates at 9 P.M. on November 10. It can be found on Maps 1 and 5.

The only claim to distinction that Sculptor can possess is that within its boundaries is located the south galactic pole—the southern end of the apparent axis of an imaginary sphere whose equator is the Milky Way.

774. Why is Scutum so called? The Danzig astronomer, Hewelcke, whose name is generally Latinized as Hevelius, introduced this tiny constellation in a chart of the heavens that he published in 1690. He called the constellation *Scutum Sobieskii*, "The Shield of Sobieski." The Sobieski was John Sobieski—King John III of Poland.

775. Where is Scutum? Scutum lies west of Aquila and above the Milk Dipper in Sagittarius. It is rectangular and extends from 4° to 16° south declination, and from 18 hours, 20 minutes to 18 hours, 50 minutes of right ascension. Serpens Cauda borders it on the west; Serpens Cauda and Aquila on the north; Aquila and Sagittarius on the east; and Sagittarius on the south. Scutum culminates at 9 P.M. on August 15 and may be found on Map 4.

776. What are the notable features of Scutum? For all its tiny area and its complete lack of bright stars, Scutum is crossed by the Milky Way and in the awesome depths of space seen within the constellation boundaries there are some of the densest concentrations of distant stars in the northern sky. The star clouds of Scutum are distinctly visible on a fine, dark summer night, their light glowing even after a journey of thousands of light years. In photographs taken of this part of the sky, the prints show immense numbers of stars—stars beyond stars and still stars. It is not possible to look at this part of the sky without feeling much of the mystery and wonder of space.

Placed against this awesome backdrop are two of Messier's objects. M 11—N.G.C. 6705—is an open cluster of stars. It is one of the first such objects that I ever saw through a telescope and notes written on that night tell the story: "A very compact and wedge-shaped cluster of stars with one star considerably brighter and more prominent than the others, at its apex. I did not know it was wedge-shaped as seen in a small telescope and was amazed to find it so. It is not easily resolvable with the exception of its bright top." Then, several years later: "7/24/49: Glorious! Close double just below it. Seen now many times and easily found. On clear nights, what are evidently the brighter stars seem to stand out against the main body of the cluster. Running across the sky below it is a very dark, narrow streak which is one of Barnard's Dark Objects, B 318." B 318 could not be seen in the telescope.

About in the middle of Scutum is M 26—N.G.C. 6694. This is—from the same note-book—"a rather condensed coarse cluster in a very rich field."

777. How did Serpens get its name? *Serpens,* The Serpent, is unique in that it is the only constellation that crosses another. For convenience, Serpens is divided into two sections, *Serpens Caput,* The Serpent's Head; and *Serpens Cauda,* The Serpent's Tail. In Greek mythology, the Serpent is being firmly held by Ophiuchus. (See question 730.) The Serpent's Head lies to the west of Ophiuchus and the Tail to the east.

778. Where is Serpens Caput? The Serpent's Head is very nearly rectangular with its greatest dimension north and south. Its base lies 3° south of the celestial equator and it extends north from there to 25° of north declination. The western edge of Serpens Caput runs precisely parallel to the fifteenth hour line about 10 minutes east; and its greatest western width brings it to 16 hours, 20 minutes of right ascension. Hydra takes up most of its southern border and starts the western side. Virgo and Boötes complete the western edge and Corona Borealis covers the north. Hercules and Ophiuchus take care of the eastern border and Ophiuchus also finishes off its southern side. Serpens culminates at 9 P.M. on July 21. It may be found on Map 4.

779. What are the notable features of Serpens Caput? A triangle of faint stars below the curve of the Northern Crown marks the head of the serpent, whose body begins with a sharp curve to the southeast, indicated by a line of three stars. The first of these three stars is faint, the centre one is bright and the last is of medium brilliance. The brightest star is Alpha (α) Serpentis (Unuk al Hay). This name means "The Neck of the Serpent." Alpha is magnitude 2.65. The other stars are all of fourth magnitude or fainter.

Along the western edge of Serpens Caput, well to the south, is M 5—N.G.C. 5904. This is a bright globular cluster, showing up like a brilliantly glowing bit of fluff in a small telescope. Almost in line of sight with it is a faint star, 5 Serpentis, which appears in the same field with the cluster seen with a low power.

780. Where is Serpens Cauda? *Serpens Cauda,* "The Serpent's Tail," lies to the east of Ophiuchus and, in theory, the entire constellation of the Serpent crosses Ophiuchus, going in at the west and coming out on the east. Actually, all charts show that the Serpent is broken into two distinct constellations, one on either side of Ophiuchus. Serpens Cauda begins in a bay which juts into the eastern side of Ophiuchus, rather far to the south. From this point, the constellation extends in an irregular series of steps east and north. Its most northern point is at 7° north declination from which it drops to 16° south declination. Its easternmost point is at 18 hours, 55 minutes of right ascension, and its western extremity is 17 hours, 10 minutes. There are two stars of about fourth magnitude in Serpens Cauda so placed that a line connecting them will run northeast, carrying out the general shape of the tail of a serpent. Serpens Cauda may be found on Map 4.

781. Where did Sextans get its name? Sextans—The Sextant—is again the responsibility of Hevelius. The story, possibly apochryphal, is that his house burned down in 1679. In the fire, Hevelius lost several astronomical instruments, among them a sextant which he prized. He was, at the time, engaged in gathering data for his new star catalogue which was published in 1690. In it, among many other new constellations which he named and placed in previously unclaimed areas, was this one, named in memory of the lost sextant.

782. Where is Sextans? Sextans culminates at 9 P.M. on April 10. It is a perfectly rectangular region lying across the celestial equator and the tenth hour of right ascension, devoid of any bright objects. Hydra borders it on the west and south; Leo on the north and most of its eastern side; and Crater finishes off its eastern boundary. Alpha (α) Sextantis, above and to the west of the centre of the constellation, is about fourth magnitude, and is far and away the brightest object in this tiny waste of the sky. Sextans may be found on Map 3.

783. Where did Taurus get its name? When the first primitive artists began to decorate caves and temples, the bull, the first domesticated animal, was an invariable subject. The bull has been worshipped in many ancient religions. Apis, in Egypt, the sacred white bull, was one of the forms assumed by Osiris, the chief of the Egyptian hierarchy. In Assyria, the crumbling sculptures show sacred figures of a human form with the head of a bull. In the religion of Mithra, from which many of the forms and symbols of Christianity were taken, the bull was an important figure. The constellation of Taurus, the Bull, is the oldest and one of the most splendid of all the constellations. About 5,000 years ago, when man first began to connect the position of the Sun with the seasons, the vernal equinox probably stood before the stars of Taurus. This most important constellation was given the identity of the most important animal in the rites of ancient peoples.

Around this figure of the celestial bull, the Greeks wove the legend of Jupiter and Europa, in which the chief deity assumed the form of a bull in order to abduct the beauty. From this legend, the constellation and the continent were named. In Rome, when the year began with the month of March, Taurus was rising and setting with the Sun, and Virgil mentions Taurus in his bucolic verses in this connection. Even in China, where this area of the sky was originally known as the White Tiger, its character was later altered and it became the Golden Ox.

784. Where is Taurus? Taurus culminates at 9 P.M. on January 14, and extends from the celestial equator north to a little above 30° north declination. Its western boundary lies mainly on 3 hours, 20 minutes of right ascension and its eastern edge does not quite reach the sixth hour. Cetus and Aries border it on the west; Perseus and

Auriga on the north; Gemini and Orion on the east; and Eridanus on the south. Taurus may be found on Map 2.

785. What is noteworthy about Taurus? Probably the best-known cluster of stars in the sky is the Pleiades, which are in the upper western quadrant of Taurus. This lovely formation is easy to find and looks like a tiny edition of the Great Bear, with six visible stars and perhaps more under exceptional conditions. The old name for the Pleiades was "The Seven Sisters," and it is possible that one of the stars has become dimmer in the thousands of years that have passed since man began to watch them. It is also possible that man has polluted the air to an extent that has obscured the faintest of these visible stars, which may have been easily seen in the darker, clearer skies of long ago.

In mythology, the Pleiades were the daughters of Atlas, and were, as a group, the objects of the attention of Orion. They were a bit frightened at the boisterous wooing of the rather crude giant and ran away from him. To help them, Jupiter changed them into a flock of doves, and they are still flying across the skies with Orion still in pursuit. The names of these stars are lovely: Alcyone, Celaeno, Electra, Taygeta, Maia, Asterope, Merope, Pleione. With them is Atlas, their father. Brightest is Alcyone, whose designation is Eta (η) Tauri, of magnitude 2.96.

The Pleiades bears a number in Messier's catalogue—M 45—but it has not been given a number in the New General Catalogue. It is difficult to understand why Messier gave it a number, for the nature of the cluster is so obvious, both as to location and character, that it scarcely needs further identification. There are actually between 250 and 300 stars which are considered as belonging to the Pleiades, and most of them are hot, blue giants, ranging from tremendous size and temperature down to about the size and temperature of the Sun. The cluster is involved in a bright nebulosity which is apparent only in photographs of long exposure. This nebula shows the same characteristic spectral lines as do the stars themselves, and is therefore known as a reflecting nebula, as opposed to other nebulae, excited by stars even hotter than the Pleiades, which absorb and re-emit the energy from such stars, in addition to reflecting their light. In either case, it is the energy and radiation from stars in or near the nebula that gives it light. In a small telescope, the Pleiades is one of the loveliest sights to be seen. The whole field of the telescope is blazing with stars in glorious profusion, in streams and loops and in all possible combinations, and all this wonder is set against a background of fainter, far-off stars which obviously do not belong to the cluster. Pleione is slightly variable—sufficiently so to have received a variable-star designation: BU Tauri. On more than one occasion, Pleione appears to have undergone a slight outburst, like a nova.

The brightest star in Taurus, Alpha (α) Tauri, is Aldebaran. Its visual magnitude is 0.85. Aldebaran is a red giant, about 35 times the diameter of the Sun, shining with a luminosity about 120 times that of the Sun. It lies within 5° of the ecliptic and is, therefore, periodically occulted by the Moon. *Aldebaran* is Arabic, and means "The Follower," from the position of the star east and south of the Pleiades, so that it seems to follow them up from the horizon and across the sky. Aldebaran is 65 light years distant. Beyond it lies another star cluster, not quite so well known as the Pleiades. This is the Hyades.

The Hyades are much nearer to us than are the Pleiades—about 130 light years as against the 350 light years which separate us from the Pleiades. The stars of the Hyades are larger but cooler and therefore more red than those of the Pleiades, and their proximity makes them appear to be more widely separated. The most notable star in this cluster is Theta (θ) Tauri, at the western end of the great arrowhead formed by the Hyades. Theta (θ) is a wide double star whose components are 3.6 and 7.0 magnitudes. It is purely by accident of position that Aldebaran appears among the stars of the Hyades. There is no possible connection between them, for the Hyades are at least twice as far from us as is Aldebaran.

Far to the northeast in Tauris is Beta (β) Tauri (El Nath), which lies on the boundary between Taurus and Auriga. Until the rearrangement of the constellations in 1930, El Nath was considered common to both constellations, and bore the designations of both Gamma (γ) Aurigae as well as Beta (β) Tauri. As a consequence of the new constellation alignment, the star was assigned to Taurus, and Auriga now has no Gamma. *El Nath,* which means "The One Butting with Horns," is pictured as the point of one of the horns of the bull, whose eye is Aldebaran and whose face is outlined by the Hyades. Beta Tauri is magnitude 1.65.

Below Beta and a bit to the east is Zeta (ζ) Tauri, representing the tip of the other horn. Zeta is magnitude 3.07. A little to the west of Zeta and slightly above it is M 1, the first object in Messier's catalogue. This is N.G.C. 1952, the Crab Nebula. (See question 512.)

786. How did Telescopium get its name? *Telescopium,* The Telescope, is the gift of Lacaille to the crew of Argo Navis.

787. Where is Telescopium? Telescopium culminates at 9 P.M. on August 24. It lies well to the south across the 50th parallel of south declination, running from about 46° south to 57° south, and from 18 hours, 5 minutes right ascension to 20 hours, 20 minutes. It is evenly rectangular and is bordered on the west by Ara; on the north by Corona Australis and Sagittarius; on the east by Indus; and on the south by Pavo. It can be found mainly on Map 5 with its northern regions on Map 4. The three brightest stars of Telescopium, all fainter than third magnitude, make an inconspicuous, small triangle high in the northwest corner of the constellation and there is nothing else of note within its boundaries.

788. How did Triangulum get its name? The Triangle is one of the older constellations, apparently given a place of its own from ancient times in spite of its lack of bright stars. The shape outlined by the positions of three fairly bright stars is a long, narrow isosceles triangle with its apex to the southwest. The Greeks called the constellation Deltoton, and it was so known to the early astronomers. The delta is the Greek letter *D,* which is a triangle.

789. Where is Triangulum? Triangulum is in the northern sky, extending from 25° north to 37° north declination and from 1 hour, 30 minutes to 2 hours, 50 minutes of right ascension. It is bordered on the west by Pisces and Andromeda; on the north by Andromeda and Perseus; and by Perseus and Aries on the east. Aries and Pisces also cover its southern border. Triangulum culminates at 9 P.M. on December 7. It may be found on Map 1.

790. What are the notable objects in Triangulum? Alpha (α) Trianguli, the apex of the Triangle, is magnitude 3.45. Beta (β), the uppermost of the two stars that mark the base of the Triangle, is slightly brighter, magnitude 3. The other corner of the Triangle is Gamma (γ) Trianguli, magnitude 4.1, with Delta (δ) even fainter, close to Gamma.

Almost on the eastern edge of Triangulum is M 33—N.G.C. 598—the Great Spiral Nebula in Triangulum. This is an external galaxy at about the same distance from our Galaxy as is M 31, the Andromeda Galaxy—over 2,000,000 light years. M 33 is smaller, however, and much fainter. It is diffuse and without definite arm structure and without nearly so concentrated a nucleus as has M 31. M 33 cannot be seen at all without a telescope, and in a small telescope it requires considerable luck and study to find it, although its visual magnitude is given as 7.

791. How did Triangulum Australe get its name? Triangulum Australe, "The Southern Triangle," was introduced by Bayer in 1603. It is in the far southern region which was visually out of reach of the early astronomers. If they had been able to see it clearly, it would undoubtedly have been named very much as Bayer named it, for the constellation is a well defined and conspicuous triangle of second- and third-magnitude stars. Like its northern counterpart, Triangulum, it is almost an isosceles triangle, but tilted to the southeast.

792. Where is Triangulum Australe? Triangulum Australe culminates at 9 P.M. on July 7 and lies neatly between 60° and 70° south declination, touching each of the parallels. Its east-to-west extent is from west of the fifteenth hour to east of the seventeenth hour of right ascension. Ara and Apus touch its eastern side and Apus forms its entire southern boundary. Circinus is on its west and Norma to its north. Triangulum Australe may be found on Map 5.

793. What are the notable objects in Triangulum Australe? Alpha (α) Trianguli Australis, the apex of the Southern Triangle, is magnitude 1.93. Beta (β) Trianguli Australis, the most northern of the three stars, is 2.87 and Gamma (γ) Trianguli Australis, the third corner, is 2.94. Two fainter stars lie on either side of Beta (β) and the Milky Way cuts through the northeastern corner of the constellation.

794. Where did the name *Tucana* come from? *Tucana,* "The Toucan," is another introduction into the constellation hierarchy by Bayer. It may have been so named to keep it in the same category as its apiarian neighbours, Phoenix and Grus.

795. Where is Tucana? Tucana culminates at 9 P.M. on November 1 and runs from 57° to 76° south declination. Its eastern edge is almost at 1 hour, 20 minutes of right ascension while its western edge just fails to coincide with the twenty-second hour line. Eridanus and Hydrus border it on the east; Hydrus and Octans on the south; Indus on the west; and Grus and Phoenix on the north. Tucana may be found on Map 5.

796. What is noteworthy about Tucana? Tucana's only claim to distinction is that, within its boundaries, far in its southeastern corner, lies the Nubecula Minor, the Smaller Magellanic Cloud.

The Smaller Cloud, like its larger counterpart in Dorado (see question 666), is a tremendous concentration of stars and nebulae lying at a distance of about 180,000 light years from the Sun. It is a small, irregular galaxy; one of the nearest to our own Milky Way Galaxy. The precise distance of this tiny universe and, indeed, of all remote objects in space, must be considered as still open, for telescopes of increasing efficiency and more precise methods of the determination of distances have made many changes in the last few years in the original estimates of such distances. The best estimate for the distance of both the Magellanic Clouds is about 180,000 light years. This figure, recently accepted, doubles the distance originally estimated for the Clouds. This means that the stars in them that are visible to us range from tremendous supergiants with an absolute magnitude of about —10 down to those with an absolute magnitude of about zero.

There are many nebulae discernible in the Larger Cloud, but very few in the Smaller Cloud, and the Smaller Cloud seems to contain much less dust and obscuring matter. The manner of determining this is based on the varying ability to see more distant galaxies through the two clouds. Many more can be seen through the Smaller Cloud, for unit of area, than through the Larger Cloud. Hence, there is believed to be less dust in the Smaller Cloud. It was the presence of Cepheid variable stars in great numbers in the Clouds that led to the determination, by Henrietta Leavitt and Harlow Shapley, of the famous period-luminosity law, which is a good yardstick for the measurement of extragalactic distances.

797. Where did Ursa Major get its name? *Ursa Major,* "The Greater Bear," is probably the best known constellation, at least in the northern sky. Its position near the north pole of the heavens and the fact that, for most of the regions of the Earth that have been inhabited for thousands of years, this constellation never sets, combine to make Ursa Major one of the oldest, if not the oldest of all the constellations. The arrangement of the stars in Ursa Major does not in any way suggest the picture of a bear, and yet this constellation has been known as "The Bear" by civilizations which are so widely separated as to preclude any possibility of communication. The Greeks called it *Arktos Megale,* "The Greater Bear." The Finnish tribes called it a bear; the American Indians called it a bear, but they modified the picture slightly. Instead of placing the three stars which form the handle of the Big Dipper where the tail of the bear would be, they said that these three stars represented three hunters pursuing the bear. The Arabs originally depicted this constellation as a bier with three mourners following it in melancholy procession, but later they fell in with the Greeks and accepted the Bear figure. In ancient Egypt, carvings of the star figures on monuments show Ursa Major as the thigh of an animal, probably a hippopotamus, and the Egyptians identified Ursa Major with a hippopotamus.

The origin of the bear idea is obscure, but one learned authority theorizes that it may have come from the Sanskrit. The word for "bright star" in Sanskrit is *riksha.* The same word also means "bear." It is possible that a punning form of the Sanskrit name for the constellation gave rise to the bear picture. It is from the Greek name of *Arktos* that we get our word for the northern end of the Earth; the Arctic—the country of the bear.

798. Where is Ursa Major? Ursa Major is irregular in shape with extensions and indentations in every direction. Its southern boundary touches the 28° line of north declination and it goes as far north as 73°. Its most western part is at about 8 hours, 5 minutes of right ascension and one bay goes as far east as 14 hours, 25 minutes. On the extreme north, it is bordered by Draco and Camelopardalis.

Boötes touches it to the east, in part, and Canes Venatici juts into the field to make a sort of peninsula out of part of its eastern side. Coma Berenices brushes the southeastern corner; and Leo thrusts up one bay to touch it on the south, where Leo Minor carries on for several steps. Lynx completes most of its western edge. Ursa Major culminates at 9 P.M. on April 25, when it is directly overhead for people living along the fortieth parallel of north latitude. Ursa Major may be found mainly on Map 6, with small bits of it running down into Maps 2 and 3.

799. What is noteworthy about Ursa Major? The most prominent feature of Ursa Major is, of course, the group of seven bright stars that form the Great Bear. This is the eastern end of the constellation and, in the ancient conception, it represents the haunches and tail of the bear. West of the Great Bear, the stars of Ursa Major are less bright, but their arrangement does provide, for those with vivid imagination, a couple of legs for the bear in the form of a sort of Plough. The western end of the constellation seems to come to a point at Omicron (o) Ursae Majoris (Muscida) far to the west. *Muscida* is a corruption of the Latin word *musum*, which means the "muzzle" of an animal. Omicron is a double star whose principal component is magnitude 3.57. It has a very faint companion of magnitude 15.

The bright stars of the Plough are designated in order of position rather than by brightness. The first star is Alpha (α) Ursae Majoris (Dubhe). *Dubhe* is relatively recent Arabic and means "The Back of the Bear." This name superseded the Arabic conception of the bowl as a coffin and replaced the Babylonian conception of the constellation as "The Long Chariot."A strange throwback to the Long Chariot idea can be found in one of the English names for the Plough, "Charles's Wain"—the Charles being Charles I of England. Even today, the Japanese call the Plough "The Emperor's Carriage." Dubhe is a double star whose components are magnitudes 1.88 and 4.82, producing, to the unaided eye, one star of magnitude 1.81.

The lower front corner of the Plough is Beta (β) Ursae Majoris, *Merak,* part of the Arabic name which means "The Loins of the Greater Bear." Merak is magnitude 2.37. Next comes Gamma (γ) Ursae Majoris (Phecda). *Phecda* is from the Arabic phrase which means "The Thigh of the Greater Bear."A transliteration of the Arabic words which are included in the names of all these stars is *Dubb al Akbar*, "The Greater Bear." *Phecda*, for example, is rendered completely as *Fakhidh ad Dubb al Akbar*. Phecda is magnitude 2.44.

In the upper eastern corner of the quadrilateral to the handle, is Delta (δ)Ursae Majoris (Megrez), "The Root of the Tail of the Greater Bear." Megrez is the faintest of the Dipper stars, magnitude 3.30. The first tail-star is Epsilon (ε) Ursae Majoris (Alioth), the brightest of the Plough stars. Alioth is magnitude 1.79. The name *Alioth* is something of a mystery. It appears to be a corruption of the Arabic name for the very bright star Capella, in Auriga, but the name the Arabs actually used for Alioth was *Al Hawar*, "The Very Bright One." The central star in the Bear's tail is Zeta (ζ) Ursae Majoris (Mizar), from the Arabic word for "The Veil," which has nothing whatever to do with a bear. Mizar is a famous star. It is a double whose components are magnitudes 2.26 and 3.94. In addition, the brighter of these two stars is a double whose two components are so far from us and so close to each other that the star's duplicity can be found only through the periodic doubling and closing of the darker lines in its spectrum. This was one of the first spectroscopic binaries to be discovered, about 70 years ago. Zeta has a period of mutual revolution of 20½ days.

Very near Mizar is 80 Ursae Majoris (Alcor), making a visual double with Mizar, but a double only in that sense, since the two stars are not at all physically associated. *Alcor* is a Persian name and means "The Friendless One," but the pair is popularly known as "The Horse and Rider." Alcor is fifth magnitude, and the legend is that the combination once served as an optical test. The recruit who wished to serve in the army of the Sultan was required to observe Mizar and Alcor. If he saw two stars, he was optically eligible. It is no trick now to see both stars. It may be that a change in position has brought the fainter star farther from the brighter and so made their separation easier—or perhaps the recruits were not too enthusiastic to pass this test.

The last star in the tail of the Bear is Eta (η) Ursae Majoris (Alkaid), magnitude 1.87. It is also one of the most distant of the Plough stars which are, in general, somewhat less than 100 light years distant. Alkaid is about 210 light years away. Alkaid is also known as Benetnasch; both names are from the Arabic phrase *Qaid al Banat ad*

Nash, "The Chief of the Daughters of the Greater Bier"—*not* the Greater Bear. This name comes from the more ancient Arabian conception of the Plough as a coffin and mourners.

Down from the lower rear corner of the quadrilateral there extends a formation of stars that might be called the Bear's hind leg. The first joint of the leg is Psi (ψ) Ursae Majoris, magnitude 3.0. Just to the west of and a little below Psi is the hind foot of the bear, made up of two stars, Lambda (λ) Ursae Majoris and Mu (μ) Ursae Majoris. Lambda is *Tania Borealis,* from a strange combination of Latin and Arabic which refers to an Arabic conception of this part of the constellation as two gazelles leaping about in play. The name of Lambda means "The Northern Star of the Second Leap." Mu is *Tania Australis,* the "Southern Star of the Second Leap." Lambda is magnitude 3.45 and Mu is 3.05.

The front leg of the bear is made up of Theta (θ) Ursae Majoris, as the joint of the leg, and Iota (ι) Ursae Majoris and Kappa (κ) Ursae Majoris as the foot. Iota bears the name of *Talitha,* which the Arabs applied to both Iota and Kappa, and which means "The Third Leap" of the gazelles. Theta (θ) is a double whose brighter component is magnitude 3.19 with a very faint fourteenth-magnitude companion. Iota (ι) is another double made up of one star of magnitude 3.12 and one of 10.8. Kappa (κ) is still another double whose components are magnitudes 4.0 and 4.2.

There are six Messier objects in Ursa Major, but there are literally hundreds of galaxies there that become apparent on long-exposure photographs of the constellation. M 97—N.G.C. 3587—is one of the best-known planetary nebulae. This is the Owl Nebula. (See question 513.) The Owl Nebula is named from two conspicuous dark spots in the general pale, circular plane of the planetary which make it look like a staring owl. The Owl Nebula is close to Merak, the lower of the two Pointers. It is large, as planetaries go, but it is very faint. It looks somewhat like two circles not quite in alignment.

West of the Owl is M 108—N.G.C. 3556—a faint external galaxy. Below Gamma is M 109—N.G.C. 3992—another external galaxy, even fainter than M 108. M 101—N.G.C. 5457—is still a third galaxy, much larger and brighter than either of the other two. This can be found about 3° east of Mizar, the central star in the handle of the Plough.

Above and to the west of the bear's nose—Muscida—is one of the prizes of the sky. Here are two galaxies, M 81 and M 82—N.G.C. numbers 3031 and 3034 respectively. These two distant systems are so nearly in the same line of sight that they can be brought into the same field with a low-power eyepiece. One of them is presented edge on to view and the other is canted at an angle of about 30°. The two are fascinating to find and see.

Dubhe and Merak are famous as a team. They are the Pointers, for they are so situated that a line drawn from one to the other and then extended to the north will strike so close to Polaris that they serve as a sure guide to that famous star. The distance between Dubhe and Merak, too, is a sort of celestial yardstick, for it is just five degrees. An astronomer usually knows which of his knuckles, when his hand is held at arm's length, will span the apparent distance between these two stars. This knowledge gives him a quick and fairly accurate measure of distance on the celestial sphere.

The tail of the Bear strikes out a curve of stars against the sky; it makes the arc of a circle. From this circumstance has been built the mnemonic "Follow the arc to Arcturus and speed to Spica." *Arcturus* is Alpha (α) Boötis (the name *Arcturus* means "The Bear Driver"), while Spica is Alpha (α) Virginis. A line continuing the curve of the Bear's tail to the south will fall very near to Arcturus and, with just a little cheating, come close to Spica.

800. How did Ursa Minor get its name? Ursa Minor goes naturally with Ursa Major, for *Ursa Minor* means "The Lesser Bear." Ursa Minor was once, long ago, known as Cynosura, which means "The Dog's Tail," because the older picture of the constellation was that of a dog with a long and prominent tail. The word *cynosure* has been drafted into the English language to mean something that is the focus of attention. Cynosura was at the focal point of the skies and was the place toward which, from time to time, all eyes were turned.

801. Where is Ursa Minor? Ursa Minor covers the north celestial pole and extends south from that point in the region between 13 hours and 18 hours of right ascension. Its most southern bay goes as far south as 65° north declination. It culminates at 9 P.M. on June 27. Ursa Minor is almost surrounded by Draco, which wraps it up on the east, the south and much of the west. The portion of Ursa Minor nearest the pole is touched by Cepheus and Camelopardalis. The familiar figure of stars—familiar, that is, verbally—of the Little Bear occupies most of the constellation. In practice, the Little Bear is decidedly faint. Ursa Minor is on Map 6.

802. What are the notable objects in Ursa Minor? At the end of the tail of the Little Bear is Polaris, the North Star, or the Pole Star. At this time Polaris is at north declination 89° 05′, 55′ from the north celestial pole. The precession of the equinoxes will bring the Earth's pole nearer to Polaris during the next 100 years, although the star and the pole will never coincide. They are near enough, though, for all practical purposes. Polaris thus swings in a tight little circle almost 2° in diameter about the pole every 24 hours. There are more than 200 stars inside that circle, stars which are apparent on photographs and which are, therefore, nearer to the extension of the pole than is Polaris.

Polaris is a Cepheid variable star whose light changes from magnitude 1.99 to 2.10 in four days. It is also a double whose fainter component is magnitude 8.9. Polaris is a supergiant more than 100 times the diameter of the Sun and is extremely tenuous. There seems to be some difficulty in calculating the precise distance of Polaris from us. Estimates have ranged from as near as 400 light years to as far as 1,035 light years. A sound estimate would probably lie somewhere in between these extremes, and one excellent authority gives 600 light years. In any event, it is not one of our near neighbours.

The stars that form the tail of the Little Bear are faint, and the next bright stars are the two which mark the outer edge of the quadrilateral of the Little Bear. These are called, as a pair, "The Guardians of the Pole," which is a fine, romantic name. They are Beta (β) Ursae Minoris (Kochab), and Gamma (γ) Ursae Minoris (Pherkad). *Kochab* comes from the Arabic phrase *Al Kaukab ash shamali,* "The North Star," which position Kochab occupied between 1,500 B.C. and 300 A.D. It is magnitude 2.04, sometimes slightly outshining Polaris. *Pherkad* means "The Calf" and comes from the original Arabic conception of the constellation as two calves. It is magnitude 3.08.

803. Where did Vela get its name? Vela, "The Sails," was once a part of the rather overwhelming constellation, Argo Navis, "The Ship *Argo*." This vast arrangement of stars was broken up in relatively modern times into sails, keel and deck—Vela, Carina and Puppis. Vela is the more northerly portion of the old constellation.

804. Where is Vela? Vela is far to the south. Only its northern edge can be seen from latitude 40° north, and a perfectly clear southern horizon is needed to make that possible. Its northern boundary is, in the main, 40° south declination, and Vela reaches south as far as 57° south. Its western extremity almost touches the eighth hour of right ascension and it extends eastward to beyond the eleventh hour. Centaurus borders it on the east; Carina on the south; Puppis on the west and Pyxis and Antlia on the north. Vela culminates at 9 P.M. on March 30, and may be found largely on Map 5, with snatches on Maps 2 and 3.

805. What are the notable objects in Vela? The Milky Way angles through the western end of Vela and the bright stars of the constellation are concentrated there. The Alpha (α) of the old constellation Argo Navis, Canopus, is the only star in the three constellations which used to form Argo Navis with a leading designation, and is now Alpha (α) Carinae. (See question 626.) Vela, therefore, has no Alpha. Gamma (γ) Velorum, on its western border, is its brightest star. Gamma is a double of magnitudes 1.88 and 4.31. Somewhat above and to the east of Gamma is Lambda (λ) Velorum, magnitude 2.24. Along the southern edge of this part of Vela are Delta (δ) and Kappa (κ) Velorum. Delta is a complicated star. It has four components. The first of these is magnitude 1.95 and the second 5.1. The third and fourth compose a spectroscopic binary which is seen as one star of magnitude 10. Kappa is magnitude 2.45.

By pure accident of position, Kappa and Delta Velorum, together with Iota (ι) and Epsilon (ϵ) Carinae, form a figure which is almost a duplicate of the Southern Cross, both in size and in arrangement of stars. This figure is called the False Cross, and it behooves mariners to beware of it. The long axis of the Southern Cross points almost due south. The long axis of the False Cross points some 45° west of south. A mistake could lead and, in fact, it has led to complications of navigation. The Southern Cross can be recognized by two distinguishing characteristics. First, in the southwestern sector of the real Cross, there is a star which spoils the symmetry of the Cross; second, the two bright stars Alpha (α) and Beta (β) Centauri form a line which, if continued, would almost touch the northern tip of the Southern Cross.

806. How did Virgo come to be so called? In almost every culture, from the beginning, this constellation has been identified with a woman, usually a maiden. Its name, *Virgo,* means "The Virgin." In most of the ancient sky maps, a woman is shown in this part of the heavens, usually with a sheaf of wheat in her arms. The Sun lies before the stars of Virgo in August, the month in which growing things reach their peak of ripeness. Virgo has been Ceres and Proserpine to the Greeks; Isis to the Egyptians; Ishtar to the Babylonians. Even the Chinese named the constellation "The Frigid Woman." The Peruvians called it "The Earth's Mother." To the Romans she was Rhea, daughter of Sky and Earth. The early Arabs, who refused to picture a human form in the skies, called Virgo "The Wheat Ear," but later even they called the constellation "The Innocent Maiden."

One part of Virgo was named, by the Arabs, "The Retreat of the Howling Dogs." This is the upper western corner of Virgo, just below the constellation of Coma Berenices. Although the Arabian astronomers could not possibly have known it, this region is one in which countless distant galaxies are now known to be clustered—the great field of Coma-Virgo.

807. Where is Virgo? Virgo culminates at 9 P.M. on May 26. Next to Hydra, Virgo is the largest constellation in the heavens, taking in almost 1,300 square degrees. Virgo is much more impressive than Hydra because the area of Virgo is concentrated while that of Hydra is wound in a rather narrow belt around a large part of the celestial sphere. The greatest dimension of Virgo is east and west, extending from 11 hours, 40 minutes of right ascension eastward to 15 hours, 10 minutes. The constellation is somewhat irregular in shape, but its most southern bay is at 22° south declination and its northernmost point at 14° north. Sextans and Leo bound it on the west; Coma Berenices and Boötes border it on the north; Serpens Caput and Libra touch its western edge and Hydra its southern side. Virgo can be found on Map 3.

808. What is noteworthy about Virgo? The brightest star in Virgo is Alpha (α) Virginis (Spica). The word *Spica* is Latin for "The Ear of Wheat," or any kind of grain.

Spica can easily be located by using the curve of the tail of the Great Bear. (See question 799.) Spica is an important star in many ways. It is an eclipsing variable whose brightness is considered to be almost exactly first magnitude. Its two components are magnitudes 0.91 and 1.01, so the visual brightness of the star must be very close to magnitude 1. There is evidently not much difference in the star's appearance during eclipses, which occur every four days. Spica is, however, of absolute magnitude —3.3 and is almost 220 light years distant. It is a hot, white star. Spica lies very near the ecliptic.

Somewhat to the northwest is Gamma (γ) Virginis (Porrima). It is a binary star whose almost equal components may be seen individually with a small telescope. Porrima is magnitude 2.86. The most western bright star in Virgo is Beta (β) Virginis (Zavijava) which lies just above the ecliptic and the equator, which are very close together at this point. Beta is magnitude 3.80. The name *Zavijava*, one of the most exotic among star names, is part of an Arabic phrase which may be transliterated as *Zawiyat al Awwa*, and which means, "The Corner of the Barking Dog." This name may have some connection with the adjacent constellation of Leo, the Lion, and with the dogs which were traditionally understood always to follow the lion.

Far to the north, well above Spica, is another Roman lady, Epsilon (ϵ) Virginis, *Vindemiatrix*, "The Grape Gatherer" or "The Mistress of the Vineyards." The legend says that when the Sun rose at the same time as Vindemiatrix, the time had come to harvest the grapes. Epsilon is magnitude 2.86. Except for Spica and Vindemiatrix, Virgo is not visually conspicuous. Within its boundaries at present, however, is located the autumnal equinox, where the twelfth hour line of right ascension, the equator and the ecliptic, meet and where the Sun is located at the instant of its passage from northern to southern skies about September 23, when summer ends and fall begins. This point is about 2° east of and 2° south of Beta (β) Virginis.

There are eight Messier objects in Virgo, all of them galaxies and most of them faint and tremendously distant. The brightest of these is M 104—N.G.C. 4594—below Gamma (γ) and in line with and to the west of Spica. This is a fairly bright galaxy seen edge on, looking so much like a wide western hat that it is called the Sombrero Nebula.

To the north, in the great field of galaxies that lies across the border of Virgo and Coma Berenices, Herschel located and catalogued more than 300 galaxies, and about that many more have since been found in this field by improved observing methods. The great French astronomer, Gérard de Vaucouleurs, has formulated a theory that we live in a metagalaxy—a sort of galaxy of galaxies—in a great organization of galaxies made up of thousands of these star cities. This collection of galaxies appears to be shaped very much like the average galaxy, many of which are supposed to be in the metagalaxy. It is rather flat and disk-like, with a definite hub at its centre. Our view of the centre of this overall organization—the point of the greatest concentration of these galaxies—is the Coma-Virgo Cluster. Our own Milky Way Galaxy is placed fairly near to one edge of the metagalaxy and very nearly in the plane of its greatest diameter.

809. Where did Volans get its name? Volans is due to Bayer. *Volans* means "The Flying Fish," and is snuggled alongside of Carina, the Keel of the *Argo,* and is beside Dorado, the Swordfish.

810. Where is Volans? Volans is in the far south. It is a tiny, square constellation running from the ninth hour of right ascension westward to about 6 hours, 40 minutes, and from 75° south declination north to 64° south. Carina touches it on the east; Chamaeleon and Mensa on the south; Mensa, again, and Dorado cover its western border; while Carina also covers its northern side. There are seven stars faintly visible in Volans, but none of them is as bright as third magnitude. Volans culminates, for watchers south of the equator, at 9 P.M. on March 4, and may be found on Map 5.

811. Why is Vulpecula so called? Vulpecula was introduced by Hevelius in his catalogue of 1690. It had been unclaimed territory until that time. Hevelius called it *Vulpecula et Anser*, "The Fox and the Goose," because it lay, as he explained, between Cygnus, the Swan, and Aquila, the Eagle, in a sort of barnyard region of the sky.

812. Where is Vulpecula? Vulpecula is a long, narrow and most irregular constellation. It extends westward from 18 hours, 55 minutes of right ascension to 21 hours, 30 minutes, and from 19° to 29°

north declination. Lyra and Cygnus border it on the north; Pegasus on the east; Pegasus, Delphinus and Sagitta on the south; and Hercules on the west. Vulpecula culminates at 9 P.M. on September 8 and may be found on Map 3.

813. What is noteworthy about Vulpecula? There are no notable stars in Vulpecula, but the Milky Way runs right through the middle of the constellation and some of the glory of that great river of stars has washed to its shores in Vulpecula. It has one very famous planetary nebula within its borders. This is M 27—N.G.C. 6853—the Dumbbell Nebula. This nebula is superficially like the Owl Nebula in Ursa Major. It is a definite disk with dark markings that are so arranged as to make the object look a little like a dumbbell—if you have that kind of an imagination.

In 1968 the British amateur astronomer G. E. D. Alcock discovered a nova in Vulpecula, just visible to the naked eye and easily seen in binoculars. This was Alcock's second nova discovery in a year (he had found Nova Delphini in 1967). The nova in Vulpecula was less exceptional than his first, since it declined below naked-eye visibility in a few weeks instead of remaining near maximum for a long time as with Nova Delphini.

X. GALAXIES

814. What is a galaxy? A galaxy is the overall aggregation of stars, singly and in groups and clusters; of nebulae, dust and gas; of possibly every kind of astronomical object.

815. How many different kinds of galaxies are there? There is no very definite boundary to the various kinds of galaxies, and the variation is one of form. Galaxies are loosely grouped into three different classifications—elliptical, spiral, and irregular galaxies, with a great range of difference within each class.

816. What is an elliptical galaxy? An elliptical galaxy is one in which no arm structure is found. The nuclei of conventional spiral galaxies resemble elliptical galaxies very closely. The overall shapes of elliptical galaxies may range from almost spherical—M 87—through various degrees of elongation to one like N.G.C. 3115. The small companion of the Great Spiral in Andromeda—N.G.C. 204—is a typical elliptical galaxy.

817. What is a spiral galaxy? A spiral galaxy is one in which the formation of arms exterior to the nucleus is discernible. Here again there is a wide variety of arm production, ranging from those in which the nucleus is at a minimum and the arms widely spread, like M 101, to those in which the arm structure is barely apparent, like M 31.

818. What is a barred spiral galaxy? A barred spiral galaxy is one which appears to have, for a nucleus, a central, bright bar, from the ends of which the arms, usually two and fairly well defined, are produced. An example of a barred spiral is N.G.C. 1300.

819. What is an irregular galaxy? A galaxy which cannot be classified under the rather loose categories of elliptical or spiral is an irregular galaxy. The irregular galaxies make up a very small percentage of the total known galaxies.

820. Are there subdivisions in the classification of galaxies? Edwin Hubble, who classified galaxies roughly into the three groups according to their general outlines, put those galaxies which were of smooth outline, without arms, into his first class. Because these galaxies were elliptical, ranging from almost spherical to rather long ellipses, Hubble called this first class "elliptical." This group contained about 17% of the known galaxies. The second class consisted of galaxies which showed a definite nucleus which was surrounded by a disk in which there was some arm development. This class comprised about 80% of the known galaxies. Hubble designated the elliptical galaxies by the letter E, followed by a number from 0 to 7. This number indicated the degree of departure from the circular cross section, which was given the classification EO. The flattest type of elliptical galaxy was E7.

The spiral galaxies, Hubble designated by the letter S, followed by the small letter a, b or c. Sa was the most tightly wound spiral with the least arm development. Sb and Sc indicate successive stages of arm development. The same division held for the barred spirals, which are designated SBa, and SBb and Sbc, depending upon the amount of arm development.

In Hubble's third group were the galaxies which did not conform to either of the two preceding classifications. These do not have any standard of appearance except that most of them are somewhat flattened—longer in one dimension than in the other. These were the irregular spirals, and accounted for the remaining 3%. Hubble's designation for these was the letter I, for irregular. Recently, Gérard de Vaucouleurs has built a classification containing a much more detailed breakdown upon Hubble's foundation. De Vaucouleurs uses the "elliptical" classification for those galaxies which are very nearly spherical in cross section without any arm development, and gives them Hubble's designation E. Then there is a classification called "lenticular" or bean-shaped, rather long ellipses without arms, which bear the classification SO. Then come the Spirals which are divided into normal spirals, SA, barred spirals, SB and intermediate objects, SAB. Finally there are the irregular galaxies, designated by the letter I. De Vaucouleurs has also made a new enumeration of the percentage of spirals in each group. The apparent frequency of E-type galaxy is 23.4%; SO, 21.0%; SA, 24.4%; SB, 26.3% and I, 4.9%.

821. How many different parts are there to a galaxy? Galaxies are generally divided into the nucleus, which is the concentration of stars and gas at the centre of the galaxy; the disk, which is the flatter portion of the galaxy extending out around the nucleus; and the arms, which extend outward from the disk to the greatest diameter of the galaxy and which may be considered as belonging to the disk.

822. Why do some galaxies have arms and others do not? The reason for this is not fully known. It is possible that the arms came into being in those galaxies where there was a rather slow initial rotation of the material out of which the galaxy was formed. The more rapid the rotation, the smaller likelihood would there be of the existence of arms. In galaxies of slower rotation and more gas and dust, the arms would be formed by the drag of the gas and dust. One of the puzzles of galactic construction is how and why the arms persist against the shearing effect of the uneven rotation of a galaxy. It would seem that with the nucleus rotating much more rapidly than the outer regions, the arms would be cut off from the galactic centre before very many revolutions. It has also been suggested that the elliptical galaxies, which are comparatively free from gas and dust and have no spiral arm structure, are galaxies which have collided with other galaxies. Such collisions would not disturb the stars in either of the two colliding galaxies. There is sufficient space between the stars to permit two galaxies to collide and to have the stars of each pass between the stars of the other without any possibility of stellar contact. The dust and gas in each galaxy, however, would be expelled by such a collision. Without the drag of the dust and gas, the rotation of the galaxies would increase, the arms would be drawn in and, after several rotations, all the stars would be concentrated into what was originally the nucleus, but which had, by then, become the entire galaxy. Nowadays, astronomers have rejected this theory.

823. Is there any difference in the materials found in various parts of a galaxy? The nucleus is composed mainly of stars and does not contain very much dust or gas. The disk has the greatest variety of material within it—stars, dust, gas and so on, and it is in the disk, as a consequence, that the greatest changes, perhaps amounting to evolution, are taking place. The arms contain most of the dust and gas of a galaxy—perhaps 50% of the mass of the arm structure of a galaxy is dust and gas, but only about 10% of the total mass of a galaxy is dust and gas.

824. What is meant by stars of Population I and Population II? The stars of Population I are found mainly in the arms and disks of galaxies. The regions in which Population I stars abound are rich in dust and gas. The brightest stars of Population I are blue and are about —7 in absolute magnitude or brighter. They are generally in the spectral classifications from O through F and are, consequently, the hotter stars.

Population II stars are found in the nuclei of galaxies where there is little or no dust and gas. They are red and relatively cool stars, with absolute magnitudes on the order of —3.

825. Are the nuclei of galaxies made of stars or gas alone? The nuclei of some of the nearer and brighter galaxies have been distinctly resolved into stars with the 200-inch Hale Reflector. Some of the elliptical galaxies have been made to disclose their stars through the same agency. For years previous to this, there was some doubt as to the presence of stars in elliptical galaxies or in the nuclei of spiral galaxies because of the difficulty of resolving these bright regions into stars.

826. Is it possible to see the dust in distant galaxies? Yes. In most of the photographs of the brighter galaxies, there are definite dark streaks and spots which are dust. In cases of galaxies that happen to lie so that they are edge on to our line of sight, there is, more often than not, an irregular dark band which bisects the image of the galaxy along its greatest dimension, making it look, sometimes, like a sandwich. This dark streak is dust.

827. How many galaxies are there? There is no way of knowing exactly. The 200-inch Hale Reflector at Mount Palomar, America's largest telescope, has an effective range of about six thousand million light years. Within this range, which represents a tiny bubble about four thousand million light years in diameter in the great expanse of space, there must be several thousands of millions of galaxies. If space, on the average, is populated equally densely throughout, it might be possible to calculate the number of galaxies there were, if we knew precisely the extent of space. To make the situation perfectly clear, remember that the Milky Way Galaxy, in which we live, is an average galaxy, containing about 100 thousand million stars. There are smaller galaxies and there are also larger galaxies. If the average picture is correct—how many stars are there?

828. What are the brightest galaxies? Here is a list of the 20 brightest galaxies, compiled by Dr. Harlow Shapley.

Name	**Type*	*Constellation*	*Magnitude*
Large Cloud	I-SB?	Dorado	1.2
Small Cloud	I	Tucana	2.8
M 31; NGC 224	Sb	Andromeda	4.3
M 33; NGC 598	Sc	Triangulum	6.2
NGC 253	Sc	Sculptor	7.6
NGC 55	Scp	Sculptor	7.8
M 81; NGC 3031	Sb	Ursa Major	7.8
M 83; NGC 5236	Sc	Hydra	8.0
M 101; NGC 5457	Sc	Ursa Major	8.2
M 104; NGC 4594	Sa	Virgo	8.6
M 64; NGC 4826	Sb	Coma Berenices	8.7
Sculptor System	Ep	Sculptor	8.8
NGC 2403	Sc	Camelopardalis	8.8
M 51; NGC 5194	Sc	Canes Venatici	8.9
NGC 205	Ep	Andromeda	8.9
M 94; NGC 4736	Sb	Canes Venatici	9.0
Fornax System	Ep	Fornax	9.1
M 82; NGC 3034	I	Ursa Major	9.2
NGC 4945	Sbp	Centaurus	9.2
M 32; NGC 221	E2p	Andromeda	9.2

* I = Irregular; S = Spiral, with degree of arm development as a, b or c, depending on whether arms are less prominent to more prominent; E = Elliptical, with 1, 2 or 3 giving extent of departure from spherical; p = peculiar, with some variation from standard type.

829. What are the nearest galaxies to our own Milky Way Galaxy? The nearest galaxies belong to what is known as the Local Group. Here is a list developed by Dr. Cecilia Payne-Gaposchkin.

Name	*Type*	*Distance in Light Years*	*Size in Light Years*	*Magnitude*	*Abs. Mag.*
M 31	Sb	2,200,000	150,000	4.3	—19.6
Our Galaxy	Sb-c	——	100,000	——	—18.6
M 33	Sc	2,200,000	50,000	6.2	—17.6

Name	*Type*	*Distance in Light Years*	*Size in Light Years*	*Magnitude*	*Abs. Mag.*
Large Cloud	I-SB?	180,000	30,000	1.2	—17.4
Small Cloud	I	180,000	25,000	2.8	—16.0
NGC 205	Ep *	2,400,000	15,000	8.9	—14.0
NGC 6822	I	2,000,000	6,000	9.2	—13.9
NGC 221	E2	2,400,000	5,500	9.1	—13.8
NGC 185	Ep	2,300,000	5,500	10.2	—13.7
IC 1613	I	2,400,000	12,000	10.0	—13.5
NGC 147	Ep	2,400,000	6,500	10.5	—13.4
Wolf-Lund-mark	I	2,300,000	6,500	11.1	—12.7
Fornax	Ep	700,000	7,000	9.1	—11.9
Sculptor	Ep	350,000	3,500	8.8	—10.6

* p = peculiar

830. What is the red shift? The "red shift" is a shift of the darker lines in a spectrum toward the red end of the spectrum. (See question 489.) It means that some motion or characteristic of the light-emitting object is lengthening the waves of electromagnetic energy that are coming from the object. This may be the result of motion away from the object on the part of the observer, or motion on the part of the object away from the observer, or both. It may also signify an intense magnetic field around the light-emitting object.

831. What does the red shift tell us about other galaxies? Most of the observed galaxies show a shift of spectral lines toward the red end of the spectrum. This tells us that they are moving away from us, or that we are moving away from them. With the increase of our knowledge of the motions of the Sun and the Earth, both conditions appear to exist. The galaxies all seem to be moving away from each other. The amount of the shift of the spectral lines also discloses the speed at which the light-emitting object is moving. The more distant a galaxy is from us, the greater is the shift of its spectral lines toward the red and hence the faster it is moving away. Roughly speaking, the external galaxies are receding at the rate of about 300 miles, on the average, per second, for every megaparsec of distance between us and them. A megaparsec is a million parsecs or 3.26 million light years. Thus, we find that the external galaxies are moving away at a rate which increases by 100 miles a second for each 1,000,000 light years of distance. The most distant normal galaxies we can photograph seem to be receding at velocities which rise to about ½ the speed of light.

832. If we measure back in this expanding universe, using the speeds of the galaxies, when did it all start? If the theory of the red shift holds, it might mean that all the galaxies started to move out from one point about 10 thousand million years ago. Again, if this is so, it gives us the size of the universe, for if the galaxies within the range of the 200-inch reflector are moving at about half the speed of light, and the velocity of these objects increases at the rate of 100 miles a second for every 1 million light years of distance, then those bodies at the border of the observable universe must, theoretically, be moving at almost double the speed of light. This, according to Einstein, is impossible. The speed of light is absolute and nothing can move more rapidly; we can therefore estimate the size of the observable universe. Again referring to the proof of Einstein's theory of the curvature of light because of gravitational attraction (see question 391), the universe must be curved. It may be positively curved, like the surface of a sphere, or the curve may be negative, somewhat like the curve of the surface of a saddle. If the curve is positive, the size of the universe is also finite, but if the curve is negative, the curves of its surface will be parabolas or hyperbolas, and the universe is infinite, without boundaries.

833. What are some of the points against this theory? We are not yet sure whether 10 thousand million years is sufficient time for various elements to have evolved from elementary particles of matter; for stars, nebulae and galaxies to have been formed; for the solar system to have come into being. It is also possible that gravitation, working between the galaxies, may have slowed their motion down, so that they may now be travelling at a mere fraction of their original velocities. In that case, the time interval might be much shorter than 10 thousand million years.

834. Is it possible for a galaxy to be so far away that its speed might equal the speed of light? Yes, with several big "ifs." The biggest of these "ifs" is: if the red shift does mean a movement of the galaxy away from us and not that we are looking at fossil light that is two thousand million or more years old and may have had entirely different characteristics when it was created two thousand million years ago. When we look at the most distant galaxies, we are looking at light which left them thousands of millions of years ago—light which was created then and which has been travelling through space ever since. It may be that energy has changed during that time, in some of its characteristics. It may be that energy is affected by time and that we may be looking at "tired" light, and that the energy waves that reach us, after all these years, from these distant objects may not be of the same character as the energy waves produced recently by objects near us. If none of these rather vague theories happens to be true or sound, then it should be possible for an object to recede from us with the speed of light. It would be rather difficult for us to know about such an object, however, for we should never be able to see it.

835. Is there much chance of galaxies colliding? Possibly. In the average region of space, galaxies occur every three million light years. There are an estimated two thousand million galaxies in the volume of space which extends out to 250 million light years from us. The motion of these galaxies would bring about ten collisions continually, calculating a collision to occur even when the outermost regions of two galaxies are in contact. This is an average. There are regions in space where galaxies are closely clustered. However, we have no proof of collisions between galaxies.

836. Can a galaxy suffer more than one collision? In clusters of galaxies, more than one collision is possible. In the cluster of galaxies lying in the constellations of Coma Berenices and Virgo, known as the Coma-Virgo cluster, it was suggested that some of the galaxies may have undergone between five and thirty different collisions during the estimated life of the universe. However, it now seems that the strange, remote galaxies which are strangely energetic in the radio range are not galaxies in collision.

837. What kind of a galaxy do we live in? The Milky Way Galaxy is certainly a spiral galaxy with arms which are trailing back against the direction of rotation of the galaxy. It is believed to resemble, in appearance, the Andromeda Galaxy, M 31, although it is considered to be somewhat smaller.

838. How big is our Galaxy? The Milky Way Galaxy is estimated to have a diameter of about 100,000 light years and to have a thickness, or depth, at its hub of about 20,000 light years. It has a stellar population of about 10 billion stars and contains sufficient additional dust and gas to make up about 10% of the total mass of the galaxy. Most of the dust and gas is concentrated in the arm structure, where it may amount to about 50% of the mass of the arms.

839. What does our Galaxy look like? The Milky Way Galaxy is probably a spiral galaxy of Hubble's type Sb. If we were to see it from a great distance out in space, it might look very much like the Great Spiral Galaxy in Andromeda.

840. Is our Galaxy in motion? It is rotating about a point at its centre. The rotation of the Galaxy is not uniform and even, like that of a wheel, but, as with the solar system, the portions nearer the centre are rotating more rapidly than those farther out. Where we are, the rotation of the Galaxy carries our Sun—and us—around the centre in about 225 million years. At about half the distance from the centre to the Sun, the rotation is more rapid, carrying the stars around in less than half that time. The rotation is complex, however, and takes a rather different form in regions near the centre.

841. How fast does our Galaxy rotate? The Milky Way Galaxy is rotating at about 150 miles per second at the distance from the centre at which our Sun is located. The rotation of stars nearer the centre will be swifter, while those farther from the centre will be rotating more slowly. At its speed, the Sun will travel completely around the centre of the galaxy in about 225 million years.

842. Just where are we inside our Galaxy? We are in the plane of the disk of our Galaxy and about ⅓ the distance between one edge and the centre of the Galaxy—about 15,000 light years from the edge and about 32,000 light years from the centre.

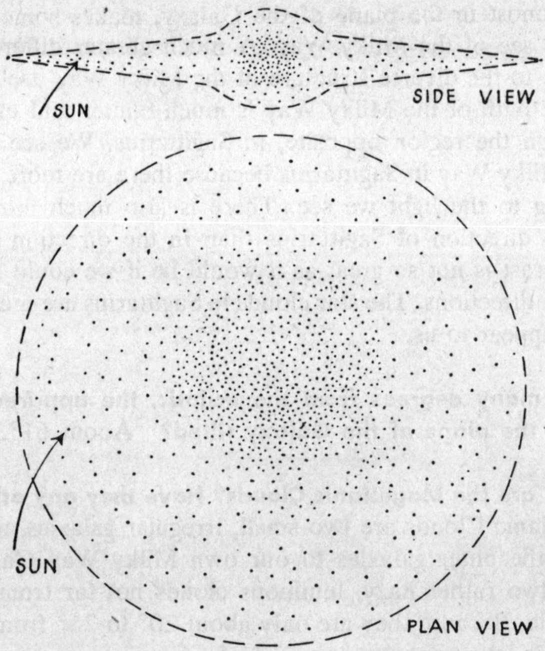

Diagram of the Milky Way Galaxy, showing location of the Sun

843. Can we see from the Earth any trace of the structure of our Galaxy? Yes. The Milky Way, a band of faint luminosity that crosses overhead in summer and in winter, but which is not so high in the spring nor in the autumn, is our view of the disk of our own Galaxy. Telescopes reveal that the Milky Way is made up of countless stars, so far away that they can be resolved into individual stars only by telescopes. This is the edge of our Galaxy, seen from our position inside the Galaxy.

844. Which of the constellations that we can see contains the edge and which the centre of our Galaxy? The edge of our Galaxy nearest us lies in Taurus, and the hub of the Galaxy is in Sagittarius.

845. If we are nearer to one edge of our Galaxy, what difference does our position make in what we can see of the Galaxy? Our location, almost in the plane of the Galaxy, makes some difference in what we see of the Milky Way. A much greater difference, however, is due to the diverse structure of the Milky Way itself. Through Taurus, the path of the Milky Way is much fainter and more diffuse than through the region opposite, in Sagittarius. We see more light from the Milky Way in Sagittarius because there are more stars there contributing to the light we see. There is also much more gas and dust in the direction of Sagittarius than in the direction of Taurus, so the contrast is not so great as it would be if we could see equally well in both directions. The star clouds in Sagittarius are much brighter than they appear to us.

846. How many degrees from the ecliptic, the apparent path of the Sun, is the plane of the Galaxy tilted? About 61°.

847. What are the Magellanic Clouds? Have they any other name? The Magellanic Clouds are two small, irregular galaxies, and are the nearest of the outer galaxies to our own Milky Way Galaxy. They appear as two rather hazy, luminous clouds not far from the south celestial pole. Because they are only about 20° to 25° from the south celestial pole, they can be seen only from points on the surface of the Earth south of 20° north latitude. These clouds were first reported by the survivors of Magellan's voyage around the Earth in 1516, and they have been known ever since as the Magellanic Clouds. Their formal astronomical names are Nubecula Major for the large cloud and Nubecula Minor for the small cloud. *Nubecula* is Latin for "a little cloud."

XI. RADIO ASTRONOMY, ROCKETS AND SATELLITES

848. What is radio astronomy? Objects in space emit or reflect electromagnetic energy in waves of greatly varying lengths. Such waves of a certain very limited range of lengths affect our eyes and we

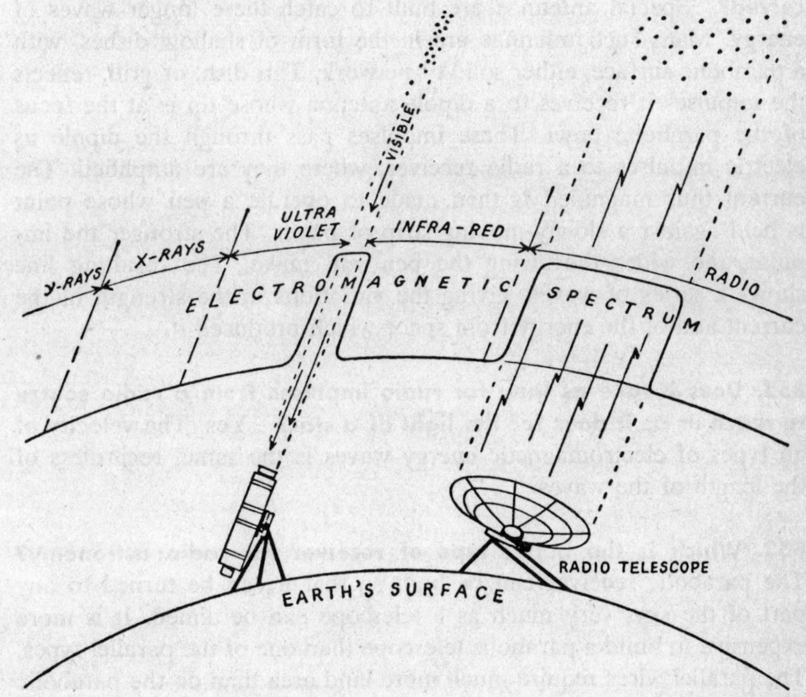

The atmosphere admits short waves of electromagnetic energy which we see—light—and longer waves which we can perceive by means of radio telescopes.

call them "light." Radio astronomy is the detection of electromagnetic energy of waves longer than those of light.

849. How long are the energy waves of which radio astronomy makes use? The energy waves which can be perceived by radio instruments range in length from about 1/10 of an inch to about 100 feet.

850. Why cannot radio astronomy use all wave lengths of energy? The Earth's atmosphere blocks waves whose length falls between the short waves that we can see and those of about 1/10 of an inch in length, and it also stops waves longer than 100 feet.

851. How are the energy waves used in radio astronomy perceived? Special antennae are built to catch these longer waves of energy. Many such antennae are in the form of shallow dishes, with a parabolic surface, either solid or network. This dish, or grill, reflects the impulses it receives to a dipole antenna whose tip is at the focus of the parabolic bowl. These impulses pass through the dipole as electric impulses to a radio receiver, where they are amplified. The current thus magnified is then made to operate a pen whose point is held against a slowly moving strip of paper. The stronger the impulse, the wider the swing the pen will make. The resulting line shows a series of waves, giving the variations in the strength of the current and of the energy from space which produced it.

852. Does it take as long for radio impulses from a radio source to reach us as it does for the light of a star? Yes. The velocity of all types of electromagnetic energy waves is the same, regardless of the length of the waves.

853. Which is the better type of receiver for radio astronomy? The parabolic receiver can be built so that it can be turned to any part of the sky, very much as a telescope can be aimed. It is more expensive to build a parabolic telescope than one of the parallel types. The parallel wires require much more land area than do the parabolic bowls, and they cannot be directed so they must depend upon the passage of various objects over them.

854. Are there other forms of antennae besides the bowl type? Yes. An effective antenna can be made by setting up two parallel wires, supported by as many towers, short or tall, as may be needed for the length of wires to be used. Such antennae may run to 300 feet in length. The energy waves are received by these antennae and

Bowl type radio telescope

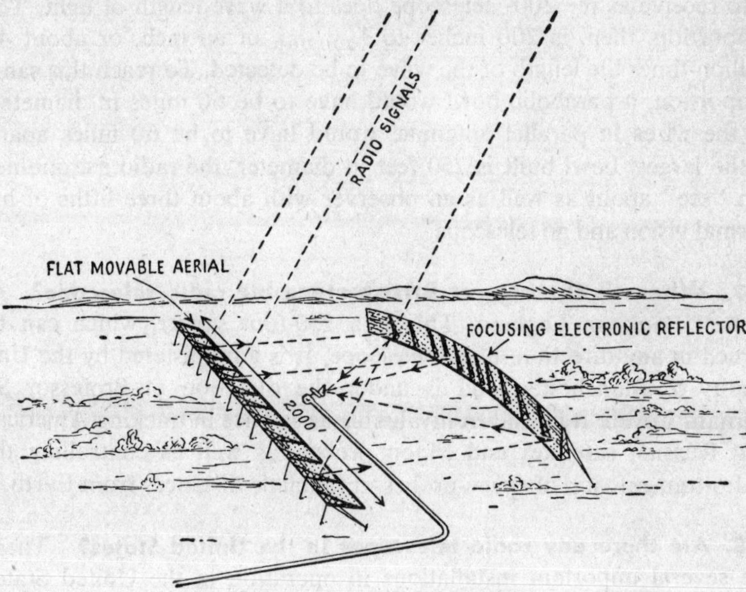

Twin aerial radio telescope

are made to produce wavy lines on strips of paper as in the case of the parabolic antennae.

855. Are there other advantages to the twin-aerial receiver? Yes. As the source of the energy received moves overhead, the change in its position because of the Earth's rotation produces an effect like that made by an interferometer. When the source is overhead, equidistant from each of the lines of receivers, the signal is stronger. As it moves nearer to one of the two lines of wires and farther from the other, the peaks and valleys of its wave formation begin to interfere with each other. The intensity of the signal is diminished and sometimes silenced. By plotting the angle at which the signal strength was greatest, an excellent location of the source can be obtained. Such aerials can be accurate down to about 2′ of arc.

856. How big should a radio telescope aerial be? To get comparable results to those obtained in visual observation by the 200″ Hale Reflector, a radio telescope should have the same bowl diameter or antenna separation in proportion to the wave length of energy it is to receive as the 200″ telescope does to a wave length of light. The proportion, then, is 200 inches to $1/200{,}000$ of an inch, or about 40 million times the length of the wave to be detected. To reach this same proportion, a parabolic bowl would have to be 60 miles in diameter, or the wires in parallel antennae would have to be 60 miles apart. If the largest bowl built is 250 feet in diameter, the radio astronomer can "see" about as well as an observer with about three fifths of his normal vision and no telescope.

857. Where is the biggest British steerable radio telescope? At Jodrell Bank, in England. This is a 250-foot saucer, which can be turned in any direction, like a telescope. It is administered by the University of Manchester, and is under the direction of Professor Sir Bernard Lovell. It has given invaluable assistance in tracking American and Russian satellites and Moon probes, as well as controlling the radio-transmission of space-probes at fantastic distances from Earth.

858. Are there any radio telescopes in the United States? There are several important installations in operation in the United States and more are being planned and constructed constantly. At present, there are radio telescopes at Harvard College, Cambridge, Massachusetts; at the Naval Research Laboratories in Washington, D.C.; at Ohio State University in Columbus, Ohio; and at the Carnegie Institution in Washington, D.C., among many others.

859. Are there radio receivers in other parts of the world? Yes. There are important receivers in Holland, at the University of Leyden, and many others in Europe, with more being built. Australia seems to have taken the lead in the establishment of large radio telescopes. The receiver at the Radiophysics Laboratory in Sydney is being used to scan the southern skies and the Australian government is financing other establishments.

860. What can radio astronomy tell us that visual astronomy cannot? There are, in space, many sources of intense energy which are not visible to us. These can be located with fair accuracy by means of radio astronomy. Some visible sources which do not appear, visually, to be in a condition to radiate strongly, have turned out to be tremendous radio sources. Radio astronomy can also detect regions in objects which are radiating energy, but which do not appear bright visually. Radio astronomy has also served as a guide and scout to visual astronomy in instances in which a source deemed unimportant or one which was even completely unnoticed has proved to be extremely strong.

861. Who was the first radio-astronomer? The first man to detect and identify electromagnetic impulses from space other than light was Karl Jansky, an engineer from the Bell Telephone Laboratories, at Holmdel, New Jersey, U.S.A. This was in 1931.

862. What sort of an apparatus did Jansky use? Karl Jansky's antenna was an oblong grid which was made to rotate slowly in a horizontal plane.

863. How did Jansky know that the noises he heard were not static? He checked his set very carefully to be sure it was in perfect condition. He found, too, that the source of the noise was not standing still. The noise varied regularly in intensity and was obviously produced by a moving source.

864. What did Jansky receive? Karl Jansky's receiver was arranged to translate the impulses it received into sound. He heard faint hissing noises, very much like ordinary radio static.

865. Where were these first radio impulses originating? From the Milky Way.

866. Who made the first radio map of the Milky Way? Grote Reber, of Wheaton, Illinois, an enthusiastic radio amateur, became interested in Janksy's experiments. He studied at the Illinois Institute of Technology, in Chicago, Illinois, and in 1937 he built a 30-foot saucer. In October, 1938, he was successful in receiving impulses and began then to make a radio map of the Milky Way.

867. What does radio mapping of the Milky Way tell us? The radio map of the Milky Way confirms, in general, the visual picture.

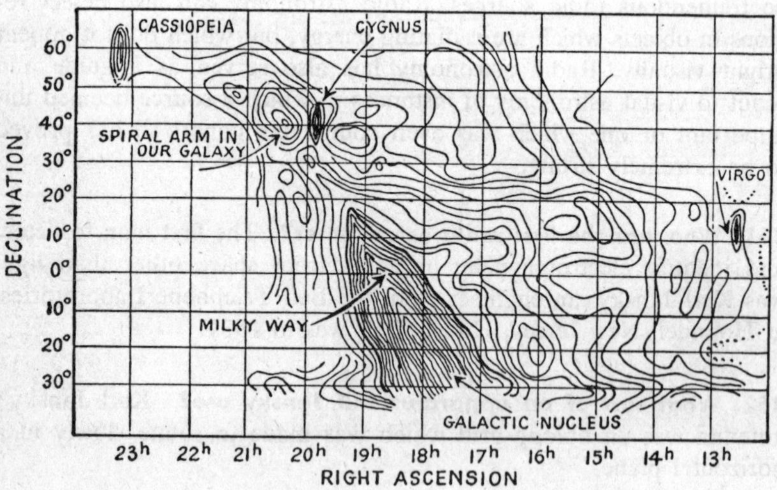

Radio map of the Milky Way showing several of the strong sources

Strong radio waves come from the direction of the centre of the Galaxy, in the area marked by the constellation Sagittarius.

868. Are all radio sources contained in our galaxy? No. Most of them are, in fact, extra-galactic.

869. What are the most famous radio sources in our galaxy? The Crab Nebula in Taurus, and the source known as Cassiopeia A. Both are due to the effects of old supernova explosions. Other galactic sources, rather less intense, are also thought to have a similar origin.

870. What is a Black Hole? Theoretically, a very massive star which collapses (hence the term *collapsar*) at the end of its highly-luminous career may continue contracting until its escape velocity is greater than that of light. In this case no light can escape from it, and it will be invisible, though its gravitational effects upon nearby objects will remain detectable. However, not all astronomers believe in the existence of Black Holes of this nature, and the whole question remains very open.

871. Is there any general radio emission in the Galaxy? Yes. The Galaxy contains large clouds of cold hydrogen, optically invisible, but which emits radiation at a wavelength of 21.1 centimetres. By studying this radiation, it has been possible to chart the positions of the hydrogen clouds, and to confirm that the Galaxy is spiral.

872. What are pulsars? Rapidly-varying radio sources, first detected at Cambridge in 1967. They are now thought to be neutron stars—very small, incredibly dense objects representing the last stages in the life of a star. More than 60 are now known. There is one in the Crab Nebula.

873. Do other galaxies emit radio radiation? Yes, in some cases. The Great Spiral in Andromeda, Messier 31, is a radio source, detectable because it is the nearest of the really large external systems. Other more remote systems are known as "radio galaxies," because they are strong radio sources; a famous case is Messier 87, a galaxy in Virgo.

874. Why are some galaxies particularly strong radio emitters? This is a question which cannot at present be answered. Formerly it was believed that some of the external sources, notably Cygnus A in the constellation of the Swan, were due to two galaxies colliding; there would be few head-on collisions between stars, but the gas and dust spread between the stars would be colliding all the time, so producing the radio emission. This theory was widely accepted, but has now been rejected, because the process could not provide as much energy as is required. So far, we have to admit that we are very uncertain as to the true cause.

875. What are the most distant radio sources known? Probably some of the strange objects known as Quasi-Stellar Objects, or quasars. The most distant of these is thought to lie at over 6,000 million light years, and to be receding from us at 90% the velocity of light.

876. What is a quasar? It is a relatively small object, looking virtually stellar, and yet radiating with immense energy. A strong quasar may be 250 times as powerful as an ordinary system such as our Galaxy.

877. How do the quasars radiate? This is a complete mystery at the moment. All known processes, such as nuclear reactions, are inadequate to explain the colossal energy of the quasars. It has been suggested that the energy is drawn from gravitational collapse, while according to another theory a quasar is due to large amounts of matter having been spontaneously created. There can be little doubt that the identification of quasars is the most important development in cosmology and astrophysics for many years.

878. Why were quasars not discovered before? They have been known for a long time, and some of them are not particularly faint; one shines as an object of the 12th magnitude. However, their true nature was not recognized until the early 1960's, when their spectra, obtained with the Palomar 200-inch reflector, were studied.

879. Are radio emissions detectable over a longer distance than ordinary light? Yes. Some radio sources observable by the new methods are so faint optically that even the Palomar reflector will not show them. Therefore, radio astronomy gives us the best chance of finding out whether the universe is finite or infinite in extent.

880. Are there any radio sources in the solar system? Yes. The Sun is a powerful source; radio waves have also been observed from Jupiter.

881. If the Sun is a radio source, why cannot we detect radio waves from other single stars? Because they are too far away. If the Sun lay at the distance of Alpha Centauri, its radio emissions would be too weak to be detected from Earth by our present instruments.

882. What is the radio emission from Jupiter? It was discovered, quite unexpectedly, in 1955, and takes the form of short bursts. Its origin is still not certainly known. The original idea of thunderstorm activity in Jupiter's atmosphere seems to be wrong; it is, however, undoubtedly linked with the powerful Jovian magnetic field, and is affected by Io, the innermost large satellite. There are very strong radiation zones round Jupiter, as was proved by the Pioneer probes sent past the planet in 1973 and 1974.

883. Is radio astronomy the same as radar astronomy? No. Radio astronomy depends on our picking up radiations from space. In radar astronomy, a pulse of energy is sent out, reflected from another body, and the "echo" detected.

884. What bodies have been contacted by radar? The Moon, the Sun, and the planets Mercury, Venus, Mars, Jupiter and Saturn, as well as the asteroid Icarus and several comets.

885. What have we learned from these radar measures? Information has been gained about the surface of the Moon. Moreover, the time taken for an energy pulse to travel to Venus and back gives us a very accurate value for the distance of Venus; this value may be used to improve our estimate of the astronomical unit, or distance between the Earth and the Sun.

886. Do radio waves and radar pulses travel at the same velocity as light? Yes: 186,000 miles per second. All three are electromagnetic vibrations.

887. Are there any serious amateur radio astronomers in Britain? Yes. The British Astronomical Association has a special section for this work, and there are several valuable amateur installations.

888. When did the space age begin? It depends upon what definition is adopted. One widely-accepted date is October 4, 1957.

889. Why is this date so regarded? Because it was on that day that the Russians launched the first artificial satellite, Sputnik I.

890. What were the details of Sputnik I? According to Russian statements it weighed 184 pounds; it was spherical in shape, with a diameter of 22.8 inches. The second Sputnik, launched on November 3, 1957, was more massive—it was a cone, 19 feet long and 4 feet at its greatest diameter, with a weight of 1,118 pounds.

891. What does "sputnik" mean? It is Russian for "satellite."

892. Did the early Russian satellites stay in orbit permanently? No. Their orbits were elliptical, and near perigee (closest approach to the Earth) they moved in atmosphere of appreciable density. Therefore, they were affected by friction, and came gradually down, destroying themselves in the lower air in the same way as a meteor. Sputnik I descended in the first part of January 1958.

893. Which was the first American satellite? Explorer I, launched by the fourth stage of a Jupiter-C vehicle on January 31, 1958.

894. What do artificial satellites look like? They look like slowly-moving stars. No satellite yet launched shows an appreciable disk to the naked eye. Many satellites are telescopic; a few, such as the American Echo I and Echo II, were bright enough to be really conspicuous.

895. Why were the Echo satellites so bright? Because they were balloons, relatively large in size, and coated with highly-reflective material.

896. What was the use of the Echo satellites? They were passive radio reflectors—that is to say, they could bounce back radio waves sent up from ground stations. They were used to help in communications.

897. Are there any active radio satellites? There have been many, the first of which, Telstar, resulted in the pioneer direct television contact between Europe and America. In 1964, an active satellite was used to relay television broadcasts of the Tokyo Olympic Games all over the world. Satellite transmissions are now common.

898. What are other uses of satellites? There are many uses. Some satellites are designed to photograph whole weather-systems from above; this series of satellites was known as Tiros. Satellites have sent back information about the upper atmosphere, cosmic radiation, meteors, micro-meteorites, solar and stellar ultra-violet rays, and many other topics.

899. Do satellites have any light of their own? A few have carried signal beacons, but most shine only by reflected sunlight. When they enter the Earth's shadow, they are eclipsed, and vanish from sight.

900. What keeps a satellite in its orbit? A combination of its velocity, and gravitational force. Let us take the Moon as an example. The Moon moves round the Earth at 3,350 feet per second; each second, gravitation pulls it towards the Earth by about $\frac{1}{18}$ of an inch. It may be said that the Moon is "falling" all the time, but it never comes any closer to the Earth. At a height of 300 miles, an artificial satellite will be pulled towards the Earth by about 14 feet each second, but it will also travel 5 miles in its orbit. The combination of fall and forward motion will balance each other, and keep the satellite in a stable path. It will come down only if part of its orbit lies low enough for atmospheric resistance to become appreciable.

901. Which was the first successful Moon rocket? Lunik I, sent up by the Russians in January 1959. It passed within 4,000 miles of the Moon, and sent back valuable information. Earlier American attempts to send probes to the Moon had been unsuccessful.

902. Where is Lunik I now? It is still going round the Sun, and has become a tiny artificial planet. All track of it has long since been lost, and it is most unlikely that we shall ever see it again.

903. Which was the first rocket to land on the Moon? Lunik II, sent up by the Russians in September 1959. On September 13 it hit the Moon and destroyed itself.

904. Where did Lunik II land? Probably in the general region of the large lunar crater Archimedes. Its landing was so violent that the whole probe must have been smashed to fragments.

905. Which was the first rocket to go round the Moon? Yet another Russian vehicle: Lunik III, of October 1959. It photographed the reverse side of the Moon, and sent back its pictures by means of television techniques.

906. What happened to Lunik III? Nobody knows. Contact with it was suddenly lost, and was never regained. The Soviet authorities think that it may have been struck by a meteor, so that its transmitters were put out of action. At some time or other it probably re-entered the Earth's atmosphere, and was destroyed.

907. Have the Russians sent up other Moon rockets? Yes, many. Some have even been to the Moon and returned with rock samples, though as yet no Russian cosmonaut has reached the Moon.

908. What was the first real American success with moon probes? Ranger VII, launched on July 31, 1964. It hit the Moon in the Mare Nubium, not far from the Guericke group of craters, and sent back close-up pictures just before impact.

909. Have automatic probes landed gently on the Moon? Yes, both the Russian Luna vehicles and the American Surveyors. During the landing mission of Apollo 12, in 1969, the U.S. astronauts Conrad and Bean walked over to Surveyor 3, a soft-landing probe which had been sent to the Moon previously, and brought parts of its home.

910. Which probes first obtained detailed photographs of all the lunar surface? The American Orbiters. There were five of these, all launched in the 1960s, and all successful. They obtained many thousands of close-range pictures. Without them, the Apollo landings could not have been attempted.

911. Who were the first Russian and the first American in space? The first of all space-travellers was Yuri Gagarin, of the U.S.S.R. On April 12, 1961, he made a complete circuit of the Earth. The first American to enter space, shortly afterwards, was Alan Shepard. The first American to make a complete orbit of the Earth was John Glenn.

912. Who were the first men on the Moon? Neil Armstrong and Edwin Aldrin, the astronauts of Apollo 11, who landed in July, 1969.

913. What are the effects of space-travel on astronauts? In free fall, an astronaut is under conditions of zero gravity—that is to say, weightlessness. The condition is not uncomfortable, and produces no ill-effects over a limited period, but there is still some doubt as to whether it is harmful over a longer period. This is an important factor when we consider a journey to Mars, which will take months—as against less than two weeks for a trip to the Moon and back.

913a. Can an astronaut leave his space-craft when in space? Yes. First man to do so was Alexei Leonov, during an orbital flight in 1965. Since then "space-walks" have been performed on many occasions; for instance, during the return flight of Apollo 16 from the Moon, in April 1972, Astronaut Ken Mattingly went outside the vehicle to retrieve instruments which had been attached to the outer casing. Since the vehicle and the astronaut are moving through space in the same path at the same rate, there is no tendency for them to move apart—though, needless to say, safety lines are carried as a precaution! "Working" in space tends to be exhausting, however, as all the American and Russian astronauts have commented.

913b. What was important about the Gemini flights? The Gemini flights formed a prelude to the Apollo programme, which culminated in 1969 with the landing of the first astronauts on the surface of the Moon. Before anything of the sort could be attempted, much preparatory work had to be undertaken, and the Gemini programme was an essential part of it. The flights were made, and were carefully analysed. Major Virgil I. Grissom and Lieutenant-Commander John W. Young, who piloted the first Gemini spacecraft, shifted their capsule from one orbit to another during their flight. They thus became the first spacemen in history to accomplish such a manoeuvre. Major Grissom fired thruster rockets over central Texas, near the end of the first orbit, lowering the two-man capsule into an orbit ranging from 97 to 105 miles. Previously it had been racing in a more elliptical orbit ranging from 100 to 140 miles above the Earth.

One of the chief purposes of the Gemini flights was to achieve rendezvous between two orbiting vehicles. After some initial difficulties, this was successfully accomplished (the Russians succeeded at about the same time). Following the twelfth and last Gemini mission, the Apollo programme could be brought fully into operation.

914. Who was the first woman in space? This is yet another Russian "first"; the first woman astronaut was Valentina Tereshkova. Subsequently she married another astronaut, Andriyan Nikolayev. They now have a baby daughter.

915. Have any astronauts died on space missions? Yes. V. Komarov was killed when he crash-landed from Soyuz 1 in 1967; later the three cosmonauts of Soyuz 11 (Dobrovolsky, Volkov and Patsayev) were found to be dead after landing. Three American astronauts—Grissom, White and Chaffee—also lost their lives while rehearsing in the spacecraft during a simulated lift-off for the first manned Apollo flight.

915a. What was the Apollo programme? This was the project of sending men to the Moon. The first manned Apollo (No. 7) was an Earth-orbiting vehicle, Apollo 8, carrying Astronauts Borman, Lovell and Anders, went round the Moon in December 1968. Apollo 9, in early 1969, tested the lunar module in Earth orbit. Apollo 10 (Astronauts Stafford, Cernan and Young) also in 1969 tested the lunar module within ten miles of the Moon. Then, in July 1969, came the first actual landing with Apollo 11; Armstrong and Aldrin explored the surface, while Astronaut Collins remained in the orbiting command module. Since then there have been other successful Apollo missions: Numbers 12 (Conrad, Bean, Gordon), 14 (Shepard, Mitchell, Roosa), 15 (Scott, Irwin, Worden), 16 (Young, Duke, Mattingly), 17 (Cernan, Harrison, Schmidt). The unsuccessful mission was Apollo 13 (Lovell, Haise, Swigert).

915b. What experiments led up to the first manned landing on the Moon? The American Rangers, which crash-landed but which sent back close-range photographs before being destroyed; the Orbiters, which provided full photographic coverage of the Moon; and finally the preliminary Apollo vehicles. All these missions played essential rôles leading up to the eventual triumph of Apollo 11.

The Orbiter vehicles, all American, were launched between August 1966 and August 1967. All five were successful. The primary aim was to explore possible landing sites for manned craft—the Apollo programme—but this was enlarged to a photographic coverage of almost all the Moon. The pictures taken were amazingly detailed, and showed features of very small size. Previously, maps of the Moon were not so satisfactory as might have been hoped, particularly with the limb regions and, of course, with the part of the Moon that is always turned away from us. (The Lunik III photographs of 1959, magnificent though they were under the circumstances, were very rough by modern standards.) With the completion of the Orbiter programme, it may be said that charting the Moon's surface was to all intents and purposes complete. In the final maps, some Russian results were also used.

Brilliant as these results were, the soft-landing probes were even more spectacular. The problem here is to bring the vehicle down on to the Moon so gently that its instruments will not be damaged; this involves the use of retro-rockets, and is a very delicate operation indeed. The Russians tried, and failed, five times. Then, with Luna IX, they succeeded. The vehicle came down in the grey plain of the Ocean of Storms (Oceanus Procellarum), and almost at once began to transmit television pictures back to Earth.

One major problem was cleared up at once. There had been a theory, originated by T. Gold, that the main lunar craters were of meteoritic impact origin and that the so-called "seas" were made up of very deep, soft dust into which any space-craft would sink. Luna IX did not sink; it remained on a firm surface, and the dust-theory was immediately discredited. If extra proof were needed, it was given shortly afterwards by the first of the seven American Surveyors, which also came down in the Ocean of Storms and began to transmit. Another important vehicle of the same type was the Soviet probe Luna XIII. Again, the Russian and American results were essentially similar.

One probe of particular interest was Surveyor V, which carried out an on-the-spot chemical analysis of the Moon's crust, and showed that the material was very much like the common volcanic rock called basalt. This strengthened the growing belief that lava-flows were very common on the Moon, and that volcanic action had played the main part in moulding the surface features. It now seems that the Moon used to be very hot—and deep inside the globe it still may be hot.

As with the Orbiters, the Surveyor craft were connected with the Apollo project for landing men on the Moon. Not all were successful; Surveyors II and IV were total failures. However, by the end of the series all the needed information had been collected, and Surveyor VII, the last of the series, was diverted to purely scientific ends. In January 1968 it landed on the rough highlands around the edge of the famous crater Tycho, and sent back pictures of a fantastically rock-strewn surface.

By 1968 the soft-landing unmanned probes had more or less done their work, and it was clear that the first manned expedition was imminent. When Apollo 11 took Armstrong and Aldrin to the Moon, in July 1969, the gap between the two worlds had been bridged at last. This first trip was more or less in the nature of a reconnaissance, but the rock samples brought back caused a complete revolution in all our ideas of the Moon. The landing site was in the Mare Tranquillitatis or Sea of Tranquillity; the surface material was found to be essentially basaltic. There was no measurable atmosphere, and the magnetic field was found to be negligible.

The second landing, with Astronauts Conrad and Bean, took place in the following November. This time parts of the old automatic probe Surveyor 3 were collected and brought home. Then, in 1970, came near-disaster. Apollo 13, with Astronauts Lovell, Haise and Swigert, suffered an explosion on the outward trip. The lunar landing was abandoned, and it was only by a combination of skill, bravery and luck that the astronauts returned unharmed. However, Apollo 14, in 1971, was again successful; the commander was Alan Shepard—who, ten years earlier, had been America's first man in space. The lunar module landed near Fra Mauro, and again samples were collected and experiments set up on the lunar surface.

Since then there have been Apollos 15 (1971), 16 and 17 (1972), near Hadley Rill, Descartes and Littrow respectively. Immense information has been collected, and our knowledge of the Moon has grown out of all recognition. The Apollo programme ended with Apollo 17. As yet the Americans have announced no further plans for manned lunar missions.

916. Why was the Moon the first world to be visited? Because it is much closer than any other natural body in the sky. Venus, the nearest planet, is always at least 100 times as remote; Mars, about 150 times. Moreover, the Moon keeps pace with the Earth as the pair travels round the Sun, whereas Mars and Venus do not. This is the only reason why the Moon was our first space-target; in itself, it is much more hostile than Mars at least. Venus is known to be very hot.

916a. Will a base be set up on the Moon? America is considering this possibility. Studies carried out by the National Aeronautics and Space Administration suggest that a base on the lunar surface is feasible. Such a base, in fact, may be operational before 1990. At first the crew members of the base would probably be limited to three, and they would need something between 9½ and 13½ tons of protective covering equipment and supplies to maintain themselves for a 90-day tour of duty. Later the base would be expanded to house six or nine men. Power might be supplied from a small nuclear reactor. A lunar base would provide a centre from which explorations of the Moon's surface could be carried out. No definite dates have been announced by the Americans as yet, but of course it is known that the Russians also plan to set up bases on the Moon in the foreseeable future.

916b. What vehicles can be used on the Moon's surface? Wheeled vehicles have been found to be perfectly suitable. With Apollos 15, 16 and 17 the astronauts were able to drive across the landscape, using special Moon-cars or Lunar Roving Vehicles (LRV). After the astronauts departed, the vehicles were left behind. With Apollo 15, Scott and Irwin explored the foothills of the Lunar Apennines, and drove right up to the edge of the huge chasm known as the Hadley Rill; with Apollo 16, Young and Duke drove over the Descartes highlands, reaching an altitude of 700 feet up a peak known unofficially as Stone Mountain.

Meanwhile, the Russians had developed an automatic mobile vehicle. Lunokhod 1, launched in 1969, was taken to the region of the Sinus Iridum in a lunar probe, and for ten months it explored the area, being guided and controlled from Earth. It sent back many valuable photographs, and carried out analytical surveys of the surface. This was, perhaps, the greatest Russian achievement in space research up to that time. Lunokhod finally came to the end of its working life, but it is still on the Moon, and no doubt it will one day be examined. In January 1973 the Russians sent up a second Lunokhod vehicle which is at present operating in the area of the Sea of Serenity.

When full-scale lunar colonies are established, there is no reason to suppose that land transport will present any major difficulties.

916c. Have any vehicles been planned? Four types of vehicle are under consideration. The first is known as an advanced surveyor. It is an unmanned robot vehicle designed to travel between 25 and 30 miles in a spiralling direction to make sure the terrain is safe for astronauts to land on. Two prototypes have already been built. Another is a single-seat vehicle in which the astronaut sits in the open. It has a strictly limited range of about three miles. The third type has a pressurised cabin and can travel nine miles. The fourth is a folding vehicle weighing three tons. It has beds, cooking facilities and carries a fully-equipped laboratory.

917. What was the first planetary probe? The Venus probe, launched by the Russians on February 12, 1961. It was a failure, since contact with it was lost at an early stage and was never regained.

918. Which have been the most successful planetary probes? To date, the American Mars probes Mariner IV (1965), Mariners VI and VII (1969) and Mariner IX (1971–2), the soft-landing Venera probes from Russia, and Mariner X to Venus and Mercury.

920. What have the probes told us about Venus? The surface temperature is extremely high (hundreds of degrees F.); the atmosphere is much denser than that of Earth, and is made up chiefly of carbon dioxide. Venus is hopelessly hostile to life.

921. Can a probe from Earth to a planet go by the shortest route? No. This is one of the main problems of astronautical planning. To go on what may be described as a "straight-line" course would mean using power all through the journey, and no rocket could possibly carry enough fuel. The probe must be put into a transfer orbit, so that most of its trip may be carried out without extra power.

922. Have any rockets landed on Mars? Yes. In 1971 the Russian vehicle Mars-3 deposited a capsule on the surface, but its transmitters failed only 20 seconds after arrival. Slightly earlier, the vehicle Mars-2 had deposited a Soviet pennant on the surface.

923. How long does a Mars probe take to reach its target? Over seven months.

924. Have any probes been sent out to more distant planets? Yes. Pioneer 10 by-passed Jupiter in 1973. Pioneer 11 by-passed Jupiter in December 1974, and should encounter Saturn in 1979.

925. Is it necessary to decontaminate probes? Yes, with probes designed to land on their target worlds. Effective sterilization is essential, because any types of bacteria carried from Earth might well produce wide effects on another world, particularly in the case of a world which has an atmosphere. The effects would be irreversible, and would be hard to sort out later, so that priceless scientific information would be lost.

926. Would it be necessary to decontaminate a probe returning from another world? Very definitely yes. Our knowledge of other planets is still very limited, and to risk bringing back unfamiliar bacteria would be the height of folly. This does not apply to the Moon, which is completely sterile, but it has to be borne closely in mind with Mars and Venus, both of which have atmospheres.

927. Can the planetary probes be put to any further use? Yes. For instance, even the unsuccessful Russian Mars probe of 1962 sent back valuable information about conditions in the space region between Earth and Mars; data were obtained about topics such as cosmic radiation and the frequency of meteoric particles.

928. Will it soon become possible to send men from Earth to another planet? This is a matter for debate. The chemically-propelled vehicles which have taken men to the Moon would be unsuitable for a much longer voyage to Mars, and better methods must be found, involving nuclear motors. It may be possible to solve these problems within the next few decades, and we cannot discount the possibility of reaching Mars before the year 2000, but as yet there are many problems to be solved. In particular, it is still uncertain how well an astronaut will stand up to a really prolonged journey through space.

929. Which planet will be the first to be visited? Mars, without a doubt. Though hostile, it is much less unfriendly than Venus, the only other planet within reasonable range.

930. What is Skylab? This was, in effect, the first American space-station, operational in 1973 and 1974. It was constructed from what may be called 'Apollo hardware', and during the mission nine astronauts spent over 170 days on board. There were three crews, each of three astronauts. All kinds of scientific experiments were carried out, as well as medical research and investigations with regard to Earth resources; one minor but notable point of interest was Kohoutek's Comet, which was well observed from the Skylab station.

Perhaps the most important result of Skylab was the proof that astronauts can exist for long periods under conditions of zero gravity, and return unharmed. None of the Skylab crew members felt any ill-effects, even though the final mission lasted for almost three months. The mission ended on 8 February 1974, with the return of the last crew: Carr, Gibson and Pogue.

The success of Skylab paved the way for many other projects, such as the "Space Shuttle" and the Russo-American link-up in space. Had Skylab failed, the progress of space research would have been held up for many decades.

931. Are there prospects of using nuclear-powered vehicles? Yes; in the future this will be done, probably in the 1980s. In an atomic rocket, nuclear power is used, but the final result is much the same; material is expelled from the rocket, and the vehicle moves by reaction in the same way as an ordinary firework rocket. The material expelled would be radioactive, and therefore dangerous, so that it may be necessary to build a compound vehicle, with its first steps chemically powered; in this case the nuclear motors would be brought into action only when the Earth had been left far behind.

932. Will bases be built on other worlds? Unmanned automatic bases on the Moon should become practicable very shortly. If no unexpected hazards are found, manned bases should follow in due course; they have been pictured as domed structures, made of plastic and kept inflated by the pressure of air inside them, but it is quite possible that dangers from meteoric particles and short-wave and cosmic radiations will make it necessary to construct the bases underground. So far as Mars is concerned, some kind of base will be essential at once, since the pioneer voyagers will have to spend some time in or near Mars before the planet and the Earth are suitably positioned for the return journey to be begun. Venus, unfortunately, now seems to be totally unsuitable as a site for a manned base—or even a manned landing. It has been suggested, too, that other worlds may be given breathable atmospheres, but this project will be so difficult that for the moment, at least, it may be entirely discounted.

933. How far will rockets take us? Unmanned rockets have been sent to Mars, Venus and Jupiter; manned craft may follow. Eventually, it is possible that probes from Earth will reach to the most remote worlds in the Solar System, though not all of these worlds are suitable for landings (the gas-giants such as Jupiter are obviously unfavourable). For journeys beyond the Moon, nuclear power is essential.

Interstellar travel is an attractive idea, but even a probe moving at near the velocity of light would take years to reach a planet moving round another star. At the moment, therefore, interstellar travel is hopelessly beyond our capabilities. Whether it will ever be achieved remains to be seen.

933a. Have any of the space-probes given indications of life on any other world? No. In fact, it may be said that the reverse is true.

There was never any real prospect of finding life on the Moon, which is so obviously a sterile world, but the initial outlook for Mars and for Venus was rather brighter. Venus, however, proved to be a disappointment. The probes sent up from 1962 onward showed that it is fiercely hot, and that the atmosphere is very dense; also, the atmosphere is almost pure carbon dioxide. It now seems definite that there is no life on Venus—at least of the kind we can understand

Mars remained, and has always been regarded as the most Earth-like of the planets. In July 1965 Mariner IV passed within a few thousand miles of the Martian surface (the minimum distance between the planet and the probe was 5,400 miles) and transmitted the famous pictures showing that Mars, like the Moon, is covered with large craters. This project was an outstanding triumph. At the time of photography, Mariner was more than 130,000,000 miles from the Earth—and yet communications were still good.

The pictures showed that Mars was more lunar than terrestrial in type, but this would not immediately negative the idea of plant life there. However, an ingenious experiment, whereby signals from Mariner were made to pass through the Martian atmosphere just as the probe entered occultation behind Mars, showed that the planet's atmosphere must be much thinner than had been thought.

This was fully confirmed by the later Mars probes of 1969 and 1971. We are now certain that the Martian atmosphere is extremely tenuous, with a surface pressure of less than 10 millibars, and that it is ineffective as a screen against harmful radiations from space.

On the other hand, the Mariner IX information is slightly more encouraging. A certain amount of water-vapour exists, and at least Mars seems to be geologically active; the volcanoes there may not be dead, though as yet it is too early to come to any definite conclusions. We cannot discount the possibility that low-type organic material survives there. Most astronomers are somewhat sceptical, but the whole problem remains open.

The remaining planets in the Solar System are even more forbidding. Mercury, lacking atmosphere and with extremes of temperature, can be ruled out at once—and can we seriously picture life on a gas-giant such as Jupiter or Saturn?

It seems, then, that in the Sun's family the Earth is the only planet suited to our kind of life. The rest are sterile—apart, just possibly, from Mars, where there is a chance of living material of some sort. This problem should be cleared up in 1976, when the American Viking probe is scheduled to make a soft landing there and to send back information from the surface. Yet we can hope for nothing in the way of advanced life-forms, and most astronomers are doubtful if any organic material will be found. Finally, it must be borne in mind that so far as we are concerned the Martian atmosphere is of very limited use. It is so tenuous that fully pressurized equipment will have to be used all the time, just as on the Moon. We can never turn Mars into a second Earth.

XII. INSTRUMENTS

934. What is a telescope? A telescope is an instrument for collecting light. Its name is Greek in origin and means "to see far off." It enables us to see objects which are too faint to be seen without its aid, and since objects are, as a rule, faint because they are far away, it lives up to its name in this respect.

935. How many different kinds of telescopes are used in astronomy? Basically, there are two types of telescopes used by astronomers, the refracting telescope and the reflecting telescope. There are a number of varieties of each kind used for special purposes.

936. How does a refracting telescope work? A refracting telescope has a large lens, called the objective, at one end of a tube. At the other end of the tube is a smaller lens, the eyepiece or ocular. The

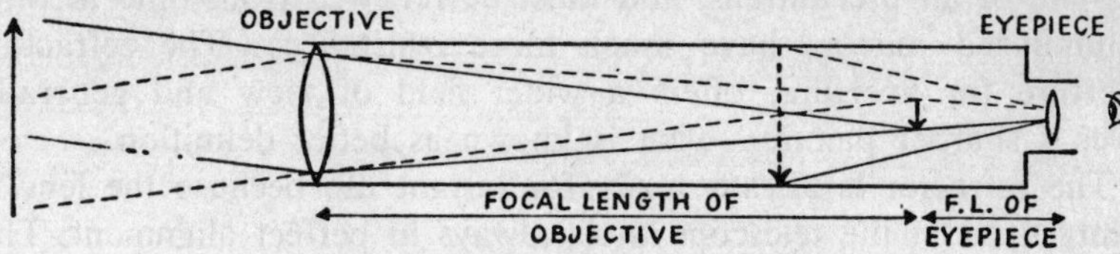

Optical diagram of refracting telescope

objective collects the light from the object viewed and, because of the curvature of its surfaces, forms an image of whatever the telescope is pointed at. This image is formed in the air inside the telescope tube at the point where the objective lens brings the light rays from the object to a focus. The image formed in this way is called the "first image." An eyepiece is placed in the telescope at the opposite end from the objective in such a way that the focal length of the eyepiece—the point where the lens in the eyepiece brings the light rays from whatever object it may be aimed at to a focus (which is very short)—extends just beyond the first image. If this is properly done, the first image will be magnified by the eyepiece and this magnified image is called a "final image." The final image is what the viewer sees.

937. How does a reflecting telescope work? In a reflector, the light from the object is collected by a parabolic mirror at the lower end of the telescope tube. This light is reflected from the mirror back up the tube where it will eventually come to a focus because of the shape of the mirror. Before the focal point, however, the light from the mirror is caught by a plane mirror, set in the tube at an angle of 45°. This mirror reflects the light out through a hole in the side of the tube near its upper end. An eyepiece is set here, arranged so that the "first image" formed by the mirror, in much the same way as by the objective lens in a refractor, is made to fall inside the focus of the eyepiece. This is a Newtonian reflector; there are also other methods of intercepting the light from the parabolic, or primary, mirror.

938. What are the advantages of a refractor over a reflector? The refractor is less likely to suffer damage from mishandling or neglect. The lens of a refractor can, of course, be scratched, but the surface of a silver mirror will deteriorate through atmospheric action in spite of all precautions, and must be renewed from time to time (aluminized mirrors have much more resistance). The refractor, aperture for aperture, offers a wider field of view and generally gives a sharper picture—what is known as better definition.

The refractor is always ready for instant use because the lens is firmly fixed in the telescope tube, always in perfect alignment. The mirror of a reflector is loosely held in a container, or cell, and must be occasionally adjusted for alignment.

939. What are the advantages of a reflector over a refractor? The reflector, size for size, is less expensive than the refractor. The mirror of a reflector may be made of material of any sort that will take and hold a surface without too much expansion or contraction because of temperature changes. The lens of a refractor, however, must be ground from the finest optical glass, free from flaws of any kind. The mirror is likely to be free from colour aberration, which is often found even in excellent lenses. This is an advantage in determining star colours, in lunar work and in planetary detail.

In small telescopes, particularly in visual observing, the reflector is easier to handle and more comfortable to use. The arrangement of the optical system in a reflector permits the visual observer to stand in a more comfortable position regardless of the direction in which the telescope is pointed. A small refractor, even with a reflecting prism at the eyepiece, often requires tiring stooping.

The reflector has, theoretically, no limit in size. The lens of a refractor will sag under its own weight if it is more than about 40 inches in diameter because it can be supported only at its circumference. A mirror, on the contrary, may be supported at many points from underneath. The largest refracting telescope is the 40-inch refractor at Yerkes Observatory. This is, undoubtedly, the largest that will ever be built. All telescopes of greater size will be reflectors. There are several reflectors of 60-inch diameter; there is the 100-inch at Mount Wilson; a reflector of 120 inches' diameter is now in use at Lick Observatory and the great 200-inch Hale Reflector, is in use at Mount Palomar in California. It is physically possible to build telescopes larger than these, but the cost would be tremendous. The Russians are working on a 236-inch reflector, and this should be in full operation soon.

940. What is chromatic aberration? Chromatic aberration is the result of the inability of a simple lens to bring the various wave lengths of light to a common focus. Thus, light passing through a simple lens will produce a blurred image, usually with many spurious colours in it. This effect is chromatic aberration. This fault can be corrected by adding to the simple lens a second lens, usually of flint glass, with one side plane and the other concave. If the concave side of the second lens is turned toward the simple lens and if the second lens is placed behind the simple lens, the chromatic aberration is usually eliminated.

In a similar manner, the mirror of a reflecting telescope, if ground to a spherical surface, will not reflect the various wave lengths of light to a single point, and spherical aberration will result. If a mirror is ground to a parabolic surface, however, this aberration will be overcome.

941. What is the best telescope for amateur use? Either a refractor or a reflector with an aperture of at least three inches. If purchased, a three-inch refractor and a six-inch reflector would cost about the same. Reflectors are relatively easy to make and are often made by amateurs, who grind the mirrors themselves. A refractor, however, requires the knowledge and skill of a professional lens maker and the material for a lens is much more expensive than the material for a mirror, which is usually Pyrex. Either type of telescope, if it is larger than six inches' aperture, should have a mechanical or electronic control to compensate for the earth's motion during observation.

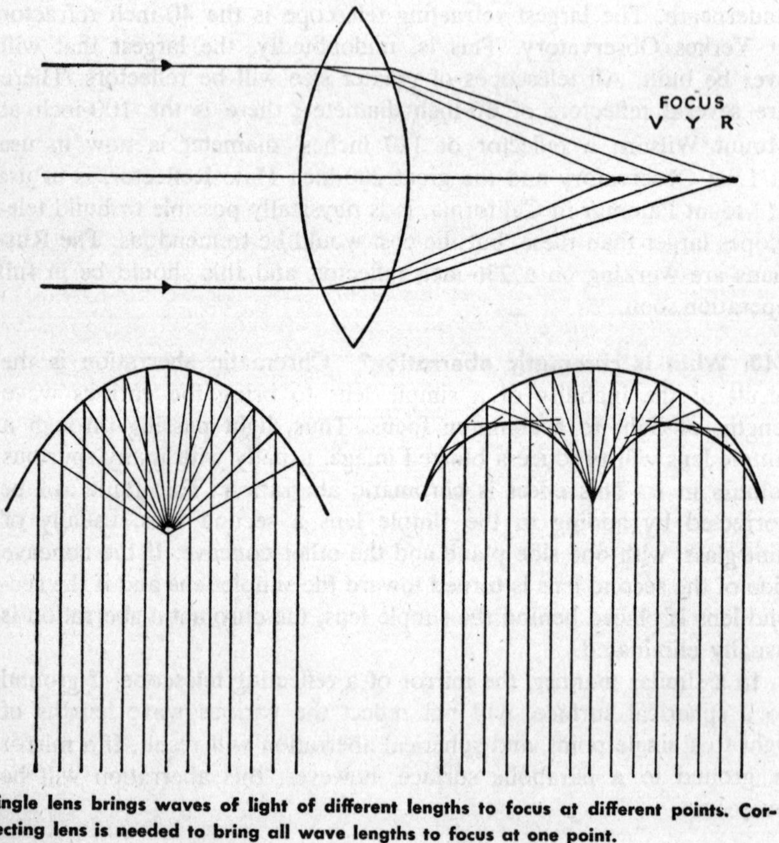

Single lens brings waves of light of different lengths to focus at different points. Correcting lens is needed to bring all wave lengths to focus at one point.
Spherical mirror, on right, brings waves of light of different lengths to focus at different points, while parabolic mirror brings waves of light of all lengths to focus at one point.

942. How many types of telescope mountings are there? There are two basic telescope mountings, but each of them has several variations. The two basic types are the altazimuth and the equatorial.

943. What is an altazimuth mounting? This is the most common kind of mounting. It usually consists of a universal joint which permits the moving of the telescope tube either in altitude—up and down—or in azimuth—horizontally. It is not as good as an equatorial mounting for following a star as the telescope must either be moved in two directions, one after the other, or along a line which is the resultant of the two directions, in order to keep the star in the field. This is not easy to do, as the position of the star must be relocated each time it moves out of the field. This double motion cannot be done mechanically, nor is it easy to do manually.

944. What is an equatorial mounting? An equatorial mounting is equipped with two axes at right angles to each other. One of

Typical equatorial mounting for a small telescope

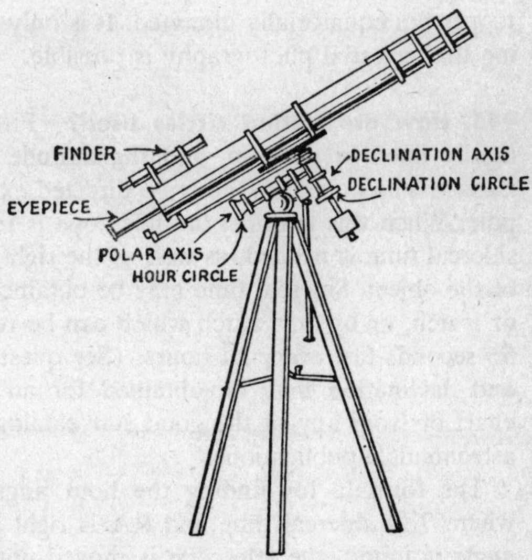

these is the polar axis, which is adjustable and is pointed precisely at the north celestial pole. The angle of the polar axis to the horizon, then, must be the same as the latitude of the telescope. There is usually a graduated circle, marked off in hours and minutes, attached to the polar axis. This is an hour circle and the figures on it correspond to the co-ordinates of right ascension of celestial objects. The polar axis must be capable of rotation.

At right angles to the polar axis, and usually fastened to its upper end, is the declination axis, which must also rotate. The telescope tube is attached to the upper end of the declination axis in such a manner that it too will rotate. Attached to the declination axis is the declination circle, marked off in degrees, minutes and seconds. These figures correspond to the co-ordinates of declination of celestial objects.

With the equatorial mounting, once an object is located and in the field of view, it can be kept in the field by just one movement of the telescope. The object is found in declination and the declination axis is then fastened in position while the telescope is rotated in hour angle about the polar axis. The rotation of the telescope is opposite to the rotation of the Earth, so that the object in view can be followed easily and smoothly, either mechanically or manually. All professional astronomical telescopes and most amateur astronomical telescopes are equatorially mounted. It is only with an equatorial mounting that celestial photography is possible.

945. How are setting circles used? First, the telescope must be rigidly fixed in position, and the latitude of its location accurately known. The polar axis must be directed exactly at the north celestial pole. When this is done, the telescope is ready. To locate an object, sidereal time is needed, as well as the right ascension and declination of the object. Sidereal time may be obtained either by a special clock or watch, or by any watch which can be regulated to run 3 minutes, 56 seconds fast every 24 hours. (See question 423.) Right ascension and declination may be obtained for an object from a good star chart or from any of the good star catalogues, nautical almanacs or astronomical publications.

The formula for finding the hour angle is T—RA=hour angle, where T is sidereal time and RA is right ascension. When the hour angle is found, the telescope is moved until the fiducia on the hour circle marks the hour angle desired. Then the telescope is rotated until the figures on the declination circle read the same as the declination co-ordinates of the object. When this has been done, the object should be in the field of the telescope.

946. What is a finder? A finder is a small telescope with a low power and a wide field of view which is mounted on the outside of the tube of a large telescope in such a way that its eyepiece is convenient to the eyepiece of the larger telescope. The finder and the telescope should be exactly aligned and should have the centres of their fields of view precisely the same. The finder is often equipped with cross hairs or a reticle for locating an object or a part of a field of view with precision. The large field of view of the finder will show the configuration of the brighter stars in the field and will aid materially in locating exactly the particular object to be observed.

947. What is meant by the field of view of a telescope? The amount of sky that can be seen through the eyepiece of a telescope is its field of view. It is smaller for higher powers than for lower.

948. How can the diameter of the field of view be determined? The telescope should be focused upon a star on or near the celestial equator. Move the telescope so that the star is just entering the field. Now let the star move across the field, as it will because of the rotation of the Earth. The number of seconds of time it takes the star to cross the field of the telescope, multiplied by 15, will give the diameter of the field of view in seconds of arc.

949. Is there any difference between the kind of eyepieces used in refractors and in reflectors? No. There are various types of eyepieces which have advantages, one over the other, for different kinds of observing, but they can be used with either type of telescope.

950. How many different kinds of eyepieces are there? There are many varieties of eyepieces, but there are six basic types:

1. The Huygenian (pronounced High-*gee*-nee-an), a negative eyepiece—one in which the final image is formed inside the eyepiece—is made with two lenses, convex on one side and plane on the other. These are mounted in the eyepiece so that the flat sides of the lenses are toward the eye. The Huygenian ocular has a wide angle of view and is inexpensive.
2. The Ramsden eyepiece also has two plano-convex lenses, but with their curved surfaces toward the eye. The Ramsden gives a field of view which is not blurred around the edges, as is often the case with the Huygenian.
3. The Tolles solid ocular is a single glass cylinder with much the

same optical features as the Huygenian eyepiece. It gives good definition and more light than the Huygenian.

4. The Orthoscopic eyepiece is a complicated system with a triple component of two double convex lenses and a double concave lens cemented together. Near to this system, on the side toward the eye, is a plano-convex lens with its flat side toward the eye. The Orthoscopic is the best overall eyepiece, giving a flat field of view, free from any distortion, and is the best eyepiece for medium and high powers.
5. The Kellner eyepiece is a combination of a convex lens and a plano-convex lens. It gives a large field of view and is good for comets and scattered objects.
6. The Monocentric eyepiece has a triple cemented lens and is best for the study of detail in observing the Moon and the planets. It has a small field, but it gives the best definition of any.

951. What is a Barlow lens? A Barlow lens is a supplementary lens which is a concave or a concave-meniscus lens mounted in a short tube and inserted between the telescope and the eyepiece. It increases the focal length of the lens or mirror and almost doubles the size of the final image. It results in some loss of light, but it also doubles the usefulness of any eyepiece.

952. Is there any criterion as to how much magnification a telescope will support? Yes. Telescopes are supposed to work satisfactorily with a power up to about 100 times the diameter of the object lens or mirror. The limit of a three-inch telescope would be 300; that of a six-inch, about 600. As the power is increased, however, the field of view becomes smaller, less light is therefore admitted and the image is fainter. The objects observed pass much more rapidly across the field of view with higher powers and atmospheric disturbances are magnified. Actually, conditions under which powers to the theoretical limit of a telescope may be used are very rare.

953. How many different eyepieces should an observer have? It is as easy to go mad buying accessories for a telescope as for a camera or a fishing rod. If you have money enough for one eyepiece, get one that gives you about 60 power. That will provide excellent viewing of many fascinating objects. From that beginning, get a long-focus, and consequently low-power, Huygenian eyepiece of about 30 power. For the higher powers, it is best to get orthoscopic eyepieces. Don't neglect the formula of about 100 times the aperture in powers, if you want to be able to see anything, for powers higher than that ratio will not admit enough light. Then, double your eyepiece inventory by buying a Barlow lens.

954. What is meant by the power of a telescope? The magnifying power of a telescope is the ratio between the focal length of the primary objective lens or mirror and the focal length of the eyepiece. The focal length of a lens or mirror is usually, but not always, about eight times the diameter of the objective. A six-inch mirror will have a focal length, then, of 48 inches. In such a telescope, an eyepiece with a focal length of one inch will have a power of 48; one with half an inch of focal length will have a power of 96, and so on.

955. Should an observer always use the highest possible magnification? No. There are many occasions when a lower power will give better results. The rule is to use the lowest power that will give the enlargement desired. A high power means a small field and the admission of less light. The small field also means that the object viewed will move across the field much more quickly than with a lower power. All motions of the telescope, due to wind, for example, will be exaggerated through the use of a high power. Experience will tell the observer under what seeing conditions certain powers are best, and for what purposes.

956. What is meant by the resolving power of a telescope? Resolving power is the ability of a telescope to show fine detail or to separate close double stars, and is limited by the diameter of the objective lens or mirror. In general, the larger the objective, the greater is the resolving power. Many other factors, such as the quality of the instrument and its accessories, enter the picture. The resolving power of a telescope is determined by the angular distance between the two components of a double star which the telescope will separate. The formula is that the resolving power of a telescope is equal to 5 seconds of arc divided by the aperture of the telescope.

Thus, the resolving power of a six-inch telescope should be 5 seconds of arc divided by 6, or 0.83″, so that the telescope should separate a double star whose components are 0.83 seconds of arc apart. If, however, the two components in question are of unequal brightness, the formula will not hold, and it often falls down for very faint stars.

957. What are the faintest stars that can be seen with telescopes of various sizes? Here we are dealing with three factors. First, what kind of a night is it? Next, how good are your eyes? Third, how big is the telescope? The only stable factor, of course, is the last, and there is a formula for that. The formula is $m = 9 + 5 \log A$, or "The faintest magnitude visible is equal to 9 plus 5 times the logarithm of the aperture in inches of the telescope." A 4-inch telescope should show stars as faint as twelfth magnitude, if all else is well; a 6-inch telescope should bring in stars to magnitude 12.9; an 8-inch, 13.5; a 10-inch, 14.0; a 12-inch, 14.4, and so on. But don't count on it.

958. What are some of the things that can easily be observed by telescope? Look in the questions about the various constellations for the Messier objects and for wide double stars. The Messier objects are, generally, the brightest and most spectacular of the deep-sky objects, and they are not too difficult to find. Do not neglect the Moon and the other planets. Get a good star atlas—Norton's *Star Atlas and Telescopic Handbook* is one of the best. It gives detailed charts of the sky and lists important and interesting things to see in every part of the heavens. Keep a notebook.

959. How can the faint objects be found without setting circles? That is half the fun of observing. Locate the most prominent object near the particular cluster, nebula or star that you want to see. Know the diameter of your field of view with various powers and—some good advice—start out with a low power. Calculate the direction and distance from the prominent naked-eye object to your goal and march through the sky with your telescope, field by field, until you believe you are near the object you are looking for. Now, start sweeping slowly, back and forth, raising your telescope just a bit after each sweep so that the fields of one sweep will slightly overlap the fields of the last one. With luck you will soon hit the object you want, and seeing the tiny, faint wonder slide into the field of view is one of the thrills that can only be compared with the hole in one or hitting a six.

960. What is a Schmidt telescope? A Schmidt telescope is a reflecting telescope that makes use of a spherical mirror and a correcting lens, or plate, which is placed at the radius of the curve of the sur-

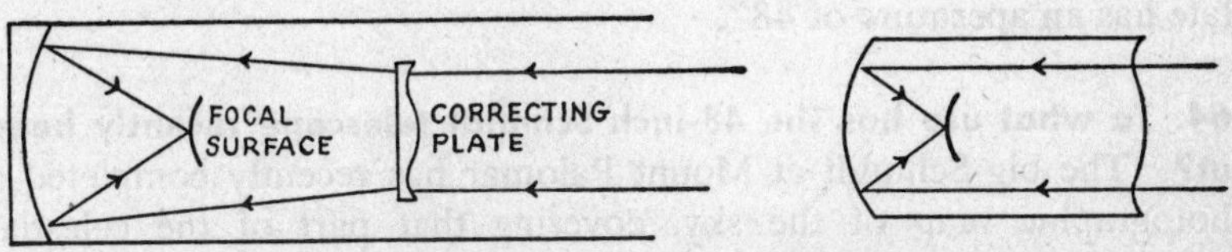

Optical diagram of Schmidt telescope

face of the spherical mirror. The correcting plate is ground to a complicated curve which compensates for the difference between a spherical and a parabolic surface in the primary mirror. This difference is so small that the eye alone would not be able to detect any curve in the correcting plate, any more than the eye could detect the difference between a spherical or a parabolic mirror surface. Schmidt telescopes can be used only as cameras and cannot be used for visual observing.

961. What are the advantages of a Schmidt telescope? The Schmidt telescope has a very wide field, whereas the conventional reflecting telescope has a very small effective field. The 48-inch Schmidt telescope at Mount Palomar can photograph the entire area of the quadrilateral of the Plough, for example, clearly and sharply, to the very edge of the photographic plate, while a comparable power with the 200-inch Hale Reflector would have an effective field about as large as one third the area of the full moon.

962. What are the disadvantages of a Schmidt telescope? The tube of a Schmidt telescope must be about twice as long as that of a conventional reflecting telescope of comparable capacity. The correcting plate is much more difficult to make than a parabolic mirror, but experience is overcoming that difficulty. Schmidt telescopes can be made with much shorter focal lengths than can parabolic reflectors, so that neither of the two difficulties is insurmountable. Because of the impossibility of visual observing with a Schmidt telescope, its use is confined to professional astronomers and to a few greatly advanced amateurs.

963. Where is the largest Schmidt telescope? The largest Schmidt telescope now in use is at the Mount Wilson–Palomar Observatory. It has a concave spherical mirror 72″ in diameter and its correcting plate has an aperature of 48″.

964. To what use has the 48-inch Schmidt telescope recently been put? The big Schmidt at Mount Palomar has recently completed a photographic map of the sky, covering that part of the celestial sphere which is visible from the latitude of Mount Palomar. Two sets of photographs were taken, one on plates sensitive to blue light and one on plates sensitive to red light, so that stars which were either very red or very blue and might fail to register properly on ordinary plates might be pictured. While several years were consumed in making this atlas, it has been estimated that the time required for a comparable job by the 200-inch reflector, with its small field, would have been in the neighbourhood of 2,000 years. The big Schmidt is used as a celestial scout for the Hale Reflector. On the beautifully clear plates of the Schmidt, there appear many faint and tiny objects which are considered worthy of further exploration. The mighty mirror of the 200-inch is then aimed at these places and detailed pictures taken which have turned out to be some of the most exciting and interesting deep-sky objects man has ever encountered.

965. What are some of the variations in the optical systems of the larger telescopes? There are three possible variations in the optical systems of the larger telescopes of the reflecting type. The Newtonian system is the simplest, using an objective parabolic mirror whose cone of reflection is intercepted by a plane mirror inside the focus of the primary mirror so that observing may be done conveniently by an eyepiece set in the side of the telescope tube near the upper end.

The Cassegrainian system requires that a circular hole be bored in the centre of the primary mirror. A convex secondary mirror is mounted above the primary mirror and the rays from the primary are

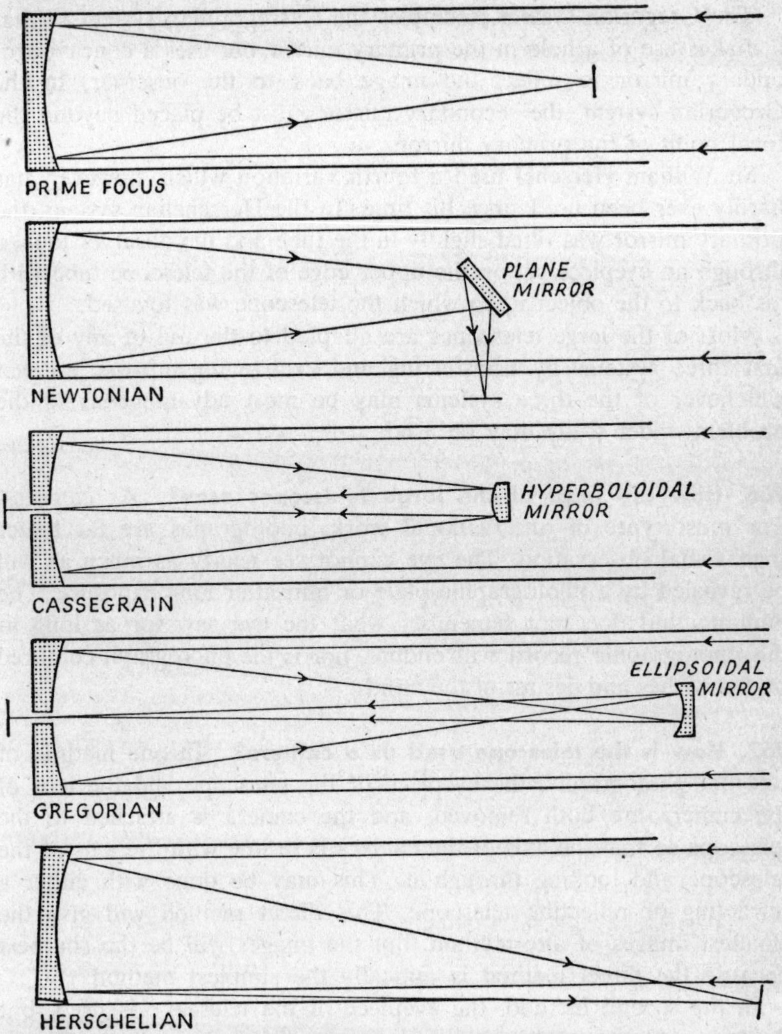

Optical systems of reflecting telescopes

reflected again from the convex mirror through the hole in the primary mirror, where the image is observed through an eyepiece mounted in the centre of the lower end of the telescope. The Cassegrainian system gives a longer focus than is possible with the Newtonian system.

The Gregorian system resembles the Cassegrainian system in that it makes use of a hole in the primary mirror, but uses a concave secondary mirror to reflect the image back to the observer. In the Gregorian system, the secondary mirror must be placed beyond the focal point of the primary mirror.

Sir William Herschel used a fourth variation which, however, has hardly ever been used since his time. In the Herschelian system, the primary mirror was tilted slightly in the tube and the observer looked through an eyepiece set on the upper edge of the telescope tube with his back to the object upon which the telescope was focused.

Most of the large telescopes are adapted to the use of any of the first three systems by transferring and exchanging mirrors, so that whichever of the three systems may be most advantageous to the problem under study may be used.

966. How are most of the large telescopes used? As cameras. For most types of observational work, photographs are far better than visual observation. The eye cannot see nearly as much as will be revealed by a photographic plate or film after long exposure. The human mind does not remember what the eye saw for as long as the photographic record will endure, nor is the photograph coloured by the wishes and desires of the mind.

967. How is the telescope used as a camera? In one method of celestial photography, the eyepiece of the telescope and the lens of the camera are both removed, and the camera is attached to the telescope so that the axis of the camera is in line with the axis of the telescope and looking through it. This may be done with either a refracting or reflecting telescope. This direct method will give the smallest images of any method, but the images will be the sharpest because the direct method is optically the simplest method.

In the second method, the eyepiece of the telescope is used, but the lens of the camera is removed. This method gives a larger image than the first method, but extreme care must be used in order to obtain a sharp, clear image. Longer exposure times are needed than with the first method.

The third method uses both the eyepiece of the telescope and the lens of the camera. Like the second method, the third requires extreme delicacy of operation and longer exposures. In each case, it is preferable to use a camera with a ground-glass focusing plate so that the image of the object to be photographed can be seen while adjustments are being made, and a clear, sharp focus may be obtained.

968. How long an exposure is needed for celestial photography? That depends upon the brightness of the object to be photographed. The Sun requires only minute fractions of a second; the Moon, about half a second and longer, and faint nebulae, clusters and galaxies may need hours of exposure.

969. How is a telescope kept fixed upon distant objects during photography? For exposures of less than half a second, the telescope may be moved by hand to follow the apparent motion of the object being photographed. This tracking should be done mechanically, however, for any exposure longer than half a second. All the large telescopes are equipped with drives which move them counter to the motion of the Earth, so that the object in question may be held in precisely the same position in the field of the telescope all during the exposure. Since the motion of the Earth is not uniform and since the driving mechanism of the telescope is not likely to be uniform in its motion, the telescope must be carefully guided during the photographing.

The astronomer sights through a finder, a small telescope which is exactly aligned with the larger telescope. He watches some bright and easily seen object in the same field with that object being photographed, choosing, if he can, a bright object which can be placed comfortably in a junction of the cross hairs in the finder. He keeps his guide object in the same position all through the operation by speeding up or slowing down the motion of the driving mechanism of the telescope. This can be done by controls which are conveniently placed near the operator's seat.

Astronomical exposures may be started on one night and resumed on subsequent nights in order to get the definition and contrast desired, so precise and exact is the mechanism and the control of these great camera-telescopes.

970. How are astronomical photographs examined? The less handling an astronomical photographic negative receives, the better it is. For that reason, most astronomical plates are left in the negative condition, to avoid the handling which would be needed to make a positive print. The astronomer works with the negative. This is usually mounted on a frame and back-lighted very much in the way an X-ray photograph is mounted for examination in a hospital. The astronomer uses a strong magnifying glass, such as a jeweller's loupe. If the photograph is to be used to detect the presence of a faint moving object, a blink comparator is sometimes used.

971. What is a blink comparator? A blink comparator is an instrument which mounts side by side two photographs taken of the same portion of the sky at different times. In this double frame, first one photograph and then the other is strongly illuminated. This method utilizes the same principle upon which motion pictures are based, and it makes any object whose position in the two pictures is not exactly the same, seem to jump a tiny bit. Thus, an object which moves can be found and studied. This method was used to discover the planet Pluto, and has been used in finding asteroids.

972. For what other purposes are celestial photographs used? The magnitudes of stars can be determined by comparing the sizes of the images they form on photographic plates with the size of the image formed under exactly the same conditions of exposure, telescope size, photographic equipment and so on, by a star of known magnitude. The brighter the star, the larger the image it will produce upon the negative. Careful comparison, by micro-measuring devices, will give the magnitudes of stars to a very small fraction of each step.

Examination of photographs of the same region of the sky taken six months apart are used to determine the parallax of nearer stars by showing their apparent displacement against the background of more distant stars. By this means, stellar distances of these nearer stars may be determined.

Photographs of the Sun, the Moon and the planets are being taken constantly, in order to detect any activity or possible change. Most of the larger observatories maintain nightly photographic patrols covering the entire visible sky. From photographs taken this way, many comets, novae, variable stars, asteroids and meteors have been and are constantly being discovered.

973. What is a spectroscope? A spectroscope is an instrument which makes use of the dispersive properties of a prism or finely ruled grating. Light from any object is admitted to the spectroscope through a very narrow slit which prevents the various colours in the resulting spectrum from overlapping. (See question 477.) Directly

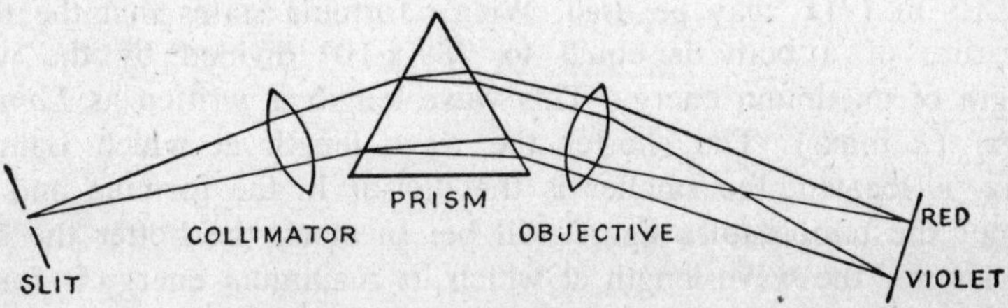

Schematic drawing of a spectroscope

behind this slit is the collimator, which is a lens designed to place in parallel paths the rays of light admitted through the slit. Each ray of light then passes through the prism or across the surface of the ruled grating, and is thus broken into its various wave lengths so that it is seen as one of the band of colours which is the spectrum. A telescope is placed at the far side of the prism or grating to collect and focus the rays of light admitted through the narrow slit. The eye—or the camera—when placed at the eyepiece of this telescope, sees the spectrum, which is really the image of the narrow slit made by each ray of light all focused by the telescope. It is in this spectrum that the lines appear which identify the various elements in the object which is the source of the light used.

974. What is a spectrograph? A spectrograph is essentially the same as a spectroscope, but is designed to photograph the spectrum instead of presenting it visually. The spectographs of the large observatories make use of many prisms or gratings and the images thus produced are captured upon photographic plates. A comparison spectrum of some common element, such as iron, produced by laboratory methods, is often photographed upon the same plate so that its bright emission lines may be compared for position with the dark lines in the spectrum of a star.

975. How will a spectrum give the temperature of a star? The various colours in a spectrum do not appear at the same intensity. There is always one wave length somewhere in the spectrum of a star at which the energy of the star producing the light is at a maximum. When this maximum point is determined, a formula derived by Wilhelm Wien (1864–1928), who received the Nobel Prize for physics in 1911, may be used. Wien's formula states that the temperature of a body is equal to 289×10^5 divided by the wave length of maximum energy. This wave length is written as *Lambda Max.* (λ max.). The shorter the wave length at which Lambda Max. is located, the smaller is the divisor in the formula and the higher the temperature figure will be. In stars, the hotter the star, the shorter the wave length at which its maximum energy is found. That is why the hottest stars look blue, because the wave length of electromagnetic radiation, which is short, produces a blue light. When a star is red, its maximum energy is being produced in a long wave length, which appears red, and its temperature is low. Thus, the colour of a star is an indication of its temperature.

976. What is a photometer? A photometer is an instrument for measuring the intensity of light. Because stellar magnitudes are comparative, astronomical photometers are usually a means of measuring the difference in intensity of two sources of light. One of the light sources will be a star whose brightness is to be determined; the other will be a star of known magnitude or an artificial star of known magnitude. The Pole Star, Polaris—Alpha (α) Ursae Minoris —is used whenever possible, though it is slightly variable.

The light of both stars is admitted to the field of a telescope at the same time. The light of the brighter of the two is then diminished by a system of prisms which can be adjusted to cut down light. The amount of adjustment needed is calibrated so that it will be an accurate measure of the brightness of the star to be determined. Other photometers use the photoelectric cell. The photoelectric cell is made of a substance which will absorb light and eject electrons in proportion to the amount of light absorbed. The ejected electrons can be counted and an extremely accurate measure of the light of the object can be made.

977. What is an interferometer? An interferometer is an instrument for measuring the angular diameters of stars. It is essentially two mirrors placed on the ends of a crossbeam mounted on a telescope in such a way that two beams of light, one from each mirror, are reflected to the eyepiece of a telescope. Even though all stars, except the Sun, are too far away to show any surface disks at all, it is still possible to obtain, by means of an interferometer, a beam of light from each edge of some of the nearest or largest stars. Since light is a wave motion, the two separate beams will either reinforce or cancel out each other according to whether the crests or troughs of the waves from each mirror coincide. The distance between the two mirrors at the ends of the crossbar when such cancellation occurs can be translated into the angular diameter of the star. Knowing the angular diameter and the star's distance, the size of the star may be determined.

978. What is a thermocouple? A thermocouple is an instrument for measuring the temperature of an object. It is usually made of two fine wires of different metals which are joined by having one end of each welded together. At this joint, a tiny disk of blackened tinfoil is soldered. This disk is about the size of the telescopic image of a star. When the light of a star is focused upon the disk, an electric current is generated which can be measured by a delicate galvanometer to which the free ends of the two wires lead. This electric current will vary with the amount of heating of the tinfoil disk. The amount of current, registered by the motion of the hand of the galvanometer across its dial, will—after certain corrections have been made—give the temperature of the star.

979. How is a coelostat used? A coelostat is a mirror so arranged that it will reflect the light of stars or the image of a celestial object constantly into the tube of a fixed telescope. Coelostats are used, as a rule, to study the Sun and are attached to telescopes whose extremely long focal length makes them impossible to move. Such long-focus telescopes will provide a greatly enlarged spectrum of the Sun and make the study of such a spectrum possible in a temperature-controlled room. Coelostats are made to move in such a way that they follow the Sun against the motion of the Earth and provide an opportunity for long, uninterrupted study or long-exposure photography.

980. What is an orrery? An orrery is a model which demonstrates mechanically the motions of the Earth, and sometimes of the Moon and some of the planets, about the Sun. The first such model was built by George Graham early in the eighteenth century. It was built for Prince Eugene of Savoy in order to demonstrate the causes of day and night, the seasons, eclipses, the Moon's phases and so on. Some years later, John Rowley built a more elaborate model, based on the original, for the Earl of Orrery. It is from this model that the name now applied to such devices came. The first orrery built in America was made by David Rittenhouse in 1770.

981. Does *bolometric* have anything to do with *bolide*? No. *Bolometric* refers to the measurement of the total radiation of a body and *bolide* is a meteoroid that explodes with a loud noise. The prefix *bolo* is Greek and means "to throw." The bolometer was invented by Samuel Pierrepont Langley and is used to measure the total amount of radiation which a body produces (or throws) as contrasted to the visual radiation which is called light. Obviously, the total amount of radiation which a star produces is greater than the portion of it which lies within the range of visibility, and bolometric magnitude is always greater than visual magnitude, which is visual radiation only. (See question 451.)

XIII. MECHANICS AND PHYSICS

982. What is meant by the terms *apastron, aphelion, apogee, periastron, perihelion, perigee*? The orbits of planets moving about the Sun, of satellites moving about planets, and of stars moving about each other are never circular, but are always elliptical. Because of this, there will be one point in the orbits of all these bodies when they are farther apart than at any other time in their orbits. In the case of two stars, this point of greatest distance is called *apastron,* from two Greek words *apo,* "far," and *astron,* "star."

Aphelion can refer only to planets and other objects which have orbits about the Sun, and is derived from *apo* and *helios,* "Sun." The opposites of *apastron* and *aphelion* are *periastron* and *perihelion,* from *peri,* "near," and *astron* and *helios.*

In the case of the Moon and artificial satellites sent out from the Earth, the word *apogee* is used to denote the point at which the Moon or satellite is farthest from the Earth and *perigee* for the nearest point. The words use the prefixes *apo* and *peri* and the root *gee,* which means "earth."

983. What is a calorie? A calorie is the amount of heat required to raise the temperature of one gramme of water from 15° Centigrade to 16° Centigrade.

984. What is meant by the solar constant? The solar constant is a figure which represents the rate at which energy is received from the Sun. To avoid all the adjustments which would be needed for atmosphere, clouds, oceans and the like if it were calculated at the actual surface of the Earth, the solar constant is figured as the amount of energy received on the surface of a hypothetical sphere which surrounds the Earth outside of the Earth's atmosphere. The Sun's energy which strikes this hypothetical surface at right angles to the Sun is 1.94 calories per minute per square centimetre, and that is the solar constant.

985. If outer space is emptiness, how can it be either hot or cold? Space itself has no temperature. Any object in space will assume a temperature depending upon its ability to absorb or radiate energy. The human body, for example, would instantly become unbearably hot on the side turned toward the Sun, if it were in space, while the side turned away from the Sun would be at a temperature of almost absolute zero.

986. How long is a cosmic year? A cosmic year is the period of time required for the Sun to be carried completely around the center of the Milky Way Galaxy by the rotation of the Galaxy. It has been calculated that the cosmic year is about 225 million Earth years. Apparently, from the general conception of the age of the Milky Way Galaxy, our Sun is just entering manhood, for its cosmic age is estimated to be between 20 and 21 cosmic years.

987. What is an ellipse? The dictionary defines an ellipse as "a plane curve such that the sum of the distances from any point of

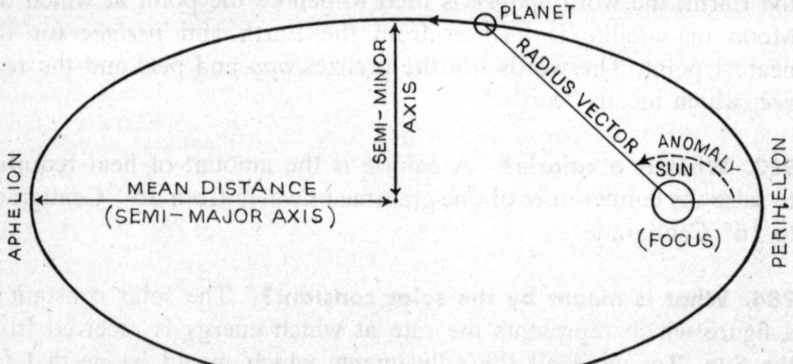

The elements of an ellipse

the curve of two fixed points called foci is a constant, that is, the same; an oval." An ellipse is not an oval in the strictest sense of the term. An oval, from its very name, is egg-shaped—larger at one end than at the other, while an ellipse is uniform. An ellipse is one of the figures that are called conic sections. Imagine a cone standing upon its base. If you make a cut, or section, through the cone parallel to its base, the shape of the surface this cut lays bare is a circle. A circle is a special form of ellipse. If another cut is made through the cone—a cut which is not parallel either to the base of the cone or to either of its sides—the shape revealed will be an ellipse.

The best way to draw an ellipse is to stick two pins firmly into a piece of paper on a drawing board. Then make a loop in a piece of string, and have the loop longer than the distance between the two pins. Lay the loop of string over the pins. Place a pencil inside the loop and draw a line in such a way that the pencil always pulls the loop of string out to its greatest possible distance from the pins. The line drawn will be an ellipse. An easier way is to buy a template of ellipses and run your pencil around inside one of the cut-outs.

988. What is the meaning, astronomically, of *gravitation* and of *gravity*? Loosely, they mean the same thing. Gravity is defined as the gravitational acceleration of a body toward the centre of the earth. Gravitation is the power of a body to attract any other body.

989. What is the difference between the meaning of *weight* and *mass* as applied to some body? Mass is the amount of material in a body, and is usually expressed in terms of comparison, as having so many times the mass of the Earth or the Sun. Weight is the effect of the Earth's gravity upon a body—or the effect of the gravity of any body upon another. Upon the surface of the Earth, a mass of one pound would have a weight of one pound; upon the Moon, the same mass would have a weight of $\frac{1}{6}$ of a pound.

990. Is there any difference between the *speed* and the *velocity* of a heavenly body? *Speed* and *velocity* mean essentially the same thing—motion through space expressed in units of distance covered in units of time, as ten miles per second. Velocity is generally considered to mean speed in a given direction as related to a specific object. Objects speeding away from the Earth are said to have a plus radial velocity and those speeding toward the Earth have a minus radial velocity.

991. How can there be an east or west direction in the heavens? East and west are directions on earth only and are extended into space to indicate motions or locations of objects as seen from the Earth. Absolute celestial directions are given by indicating the co-ordinates of right ascension and declination for a series of dates so that the motion of an object may be plotted against the celestial sphere. For fixed objects, the positions are given again by the celestial co-ordinates of right ascension and declination.

992. In astronomy, what is meant by absorption? Under certain conditions, the electrons of an atom are capable of absorbing energy. The relatively cool gases in the outer regions of the Sun, for example, absorb some of the tremendous energy that comes from the extremely hot centre of the Sun. Absorption manifests itself in producing darker lines across the spectrum of the Sun—and other stars—which identify the elements whose atoms and electrons cause the darker lines. Such a spectrum is called an absorption spectrum in contrast to the spectrum which produces bright lines in the same places, which comes directly from a highly compressed gas and which is called an emission spectrum.

The effect of the vast amount of cosmic dust in interstellar space which absorbs certain wave lengths of energy from objects embedded in this dust or lying beyond it, causes a reddening of the light from these objects, and is spoken of as absorption. This absorption effect must be taken into consideration in any calculations based upon the colour of stars or other distant objects.

993. What is azimuth? Azimuth is distance around the horizon, measured in degrees, minutes and seconds of arc. In astronomy, azimuth is measured from the south point of the horizon westward, while in navigation, azimuth is measured from the north point of the horizon eastward. Actually, both astronomer and navigator are going in the same direction, although they started from different points.

994. What is meant in astronomy by perturbations? Any disturbance in the regular or calculated motion of a satellite, planet, or star in its orbit is a perturbation. Perturbations indicate some outside influence which causes irregularities in the motion of one body in its orbit about another. By perturbations in the path of Uranus, for example, the presence of a body farther from the Sun was disclosed. As a result, a search was made for the body causing those perturbations and the planet Neptune was discovered.

995. What is meant by astrophysics? Astrophysics is that part of the general science of astronomy that deals with the physical properties of the stars. Descriptive astronomy studies the positions and appearance and arrangement of the stars as they are seen from the Earth. Celestial mechanics covers the factors of mass and motion and embraces orbits and the applications of gravitational and other laws on stars and planets. We know now that the stars are more than just large and fairly massive bodies shining in the skies; they are extremely complicated structures of gas and are in all stages of development.

Astrophysics is the application to the stars of the laws governing the behaviour of atoms and molecules, the laws of turbulence, of hydrodynamics, the use of the spectrum—in fact, the probing into the ultimate particles and forces which form the stars, cause their growth and development and final decay. The stars are now regarded as great laboratories in which experiments in nuclear physics are being conducted on a cosmic scale under conditions of temperature and pressure—or lack of it—which could never be duplicated in any terrestrial laboratory. Astrophysics seeks to take advantage of these conditions, not only to find out about the stars themselves, but also to gain some insight into the physical properties of matter and of atomic structure.

996. What is the "Purkinje effect"? The human eye does not see light of all colours with equal intensity. Blue is much easier to see, for example, than red. For that reason, if two points of light, one blue and one red, from two different sources, are both diminished by the same amount, the red object will appear fainter than the blue. This is known as the "Purkinje effect." It was first explained by Johannes Evangelista Purkinje (pronounced *poor*-kin-yay), a Czech physiologist of the nineteenth century.

997. What are co-ordinates and how are they used? Co-ordinates are imaginary lines in the sky which mark off the division of the celestial sphere into north and south or east and west. The co-ordinates which indicate east and west are called hour lines of right ascension and correspond roughly to longitude lines on the surface of the Earth. They converge at the celestial poles and are most widely separated at the celestial equator. There are 24 of these hour lines and the series begins—and ends—at the vernal equinox, that point in the sky at which the Sun crosses the celestial equator on its way north in the spring.

The co-ordinates that mark the north and south distances are called parallels of declination. They start at zero at the celestial equator and run to 90° at the poles. For northern latitudes—or for northern declination—the degrees are considered plus, and for southern declination, minus.

Declination is measured in degrees, minutes and seconds of arc north and south from the celestial equator. Right ascension is measured in hours, minutes, and seconds eastward completely around the celestial sphere. Thus, the twenty-third hour of right ascension, although it is just one hour west of the vernal equinox, is considered to be 23 hours east of that point. One hour of right ascension is equal to 15 degrees of arc.

Every object in the heavens can be located if its two co-ordinates of right ascension and declination are known.

998. What is the aberration of light? The aberration of light is the apparent displacement of a star from its true position and is caused by the velocity of the Earth's revolution about the Sun. It is most easily explained by likening it to the apparent change in the direction of rain drops in a shower caused by the motion of a person

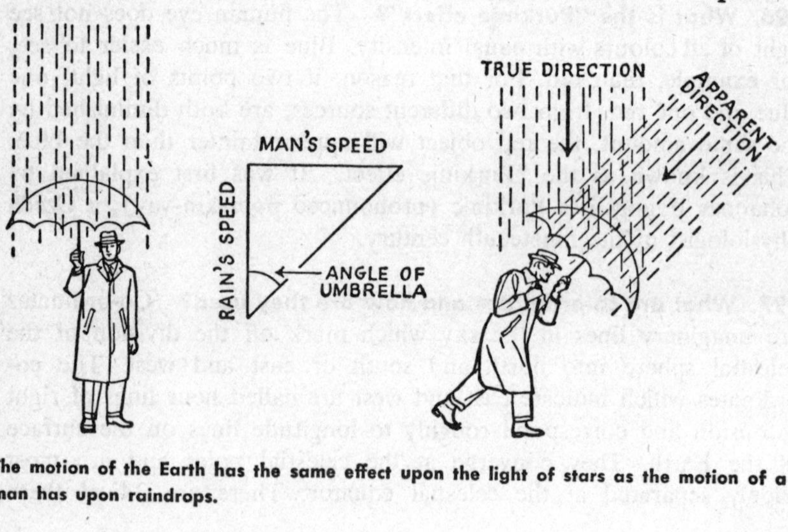

The motion of the Earth has the same effect upon the light of stars as the motion of a man has upon raindrops.

passing through the rain. If the person stands still, the drops fall straight down, and he will get only his hat and shoulders wet. If he moves forward slowly, the drops will seem to come from in front of him, and the faster he moves, the more nearly horizontally will the rain drops strike him.

As the Earth moves about the Sun, the light of stars which are located in a direction perpendicular to the Earth's motion is displaced by a very small amount—20″.5 of arc. Thus, through the year, such stars will appear to move 20″.5 first one way and then 20″.5 the other way. Stars located at or near the zenith will move in tiny circles whose radius is 20″.5, and stars between the zenith and a position perpendicular to the Earth's motion will seem to move in an ellipse whose major axis is 41″, twice the radius of the small circle of the zenith stars.

This apparent motion, which is one of the proofs of the Earth's revolution about the Sun, was discovered in 1725 by James Bradley (1693–1762), and, small though it is, it must be taken into account in determining the precise position of a star.

999. How does aberration of light differ from chromatic aberration? Aberration of light is a constant relationship between the speed of light and the speed of the Earth in its orbit about the Sun. It causes an apparent displacement of the stars from their true positions as observed from the moving Earth. Chromatic aberration is caused by the varying reflection from a mirror of light waves of different length, or their varying refraction when they pass through a lens. Chromatic aberration produces a fuzzy image surrounded by beautiful but annoying bands of colour.

1000. What is the difference between Centigrade and Fahrenheit? Both Centigrade and Fahrenheit are systems for measuring temperature. The Centigrade method arbitrarily sets its zero point at the freezing point of water and calls the boiling point of water 100 degrees. The Fahrenheit system uses 32° as the freezing point of water and 212° as the boiling point of water. Thus there are 100 divisions between those two points on the Centigrade scale and 180 on the Fahrenheit scale.

To translate from Fahrenheit to Centigrade, subtract 32 from the Fahrenheit reading and take 5/9 of the result. If the Fahrenheit reading is below 0, add 32 and take 5/9. To convert Centigrade readings into Fahrenheit, add 32 and take 9/5 of the result.

Absolute zero, the lowest possible temperature, at which all molecular action ceases, is —459.4° Fahrenheit or —273 Centigrade. In the Kelvin system, indicated by the letter K and often used in astronomy, the zero point is absolute zero. 0° K equals —273 C. and —460 F. The Kelvin scale is actually the Centigrade scale plus 273 and has no minus readings.

1001. What is meant by the carbon cycle? The *carbon cycle* is the name for the process which makes a major contribution to the production of energy in stars. It was postulated by Dr. Hans Bethe and appears to satisfy conditions known to exist in the interiors of hot stars.

In this process, the nucleus of a common carbon atom, C^{12}, and the nucleus of a hydrogen atom, H^1, called a proton, collide. The two nuclei combine to form an isotope of nitrogen, N^{13}. The mass of this new nucleus is slightly less than the sum of the masses of the carbon nucleus and the proton and the difference is one gamma ray —a burst of energy. The newly formed isotope of nitrogen is unstable and within a short time after its birth it ejects a positive particle, a positron, and becomes a heavy isotope of carbon, C^{13}. When this heavy nucleus collides with another proton, another isotope of nitrogen, N^{14}, is formed, with the release of another gamma ray. Collision with a third proton transforms the nitrogen to an isotope of oxygen, O^{15}, and produces a third gamma ray. The oxygen nucleus is unstable and ejects a positron, thus transforming itself into still another isotope of nitrogen, N^{15}. A fourth proton splits the heavy nitrogen nucleus into two parts. One of these parts is the nucleus of a helium atom, He^4; the other is the nucleus of the carbon atom, C^{12}, which reverts to its original state to begin the cycle all over again. In stars such as the Sun, the process is slightly different, and the reaction is known as the proton-proton reaction; the carbon cycle applies to hotter stars. However, the end result is the same: hydrogen is converted into helium. In the Sun, 564 million tons of hydrogen are transformed into 560 million tons of helium every second. Most of the four million tons of matter which are lost are converted into energy—of which the Earth receives one part in two thousand million. Large as the tonnage figure for the loss of mass sounds, it represents a very small percentage of the total mass of a star. If an average star like the Sun were to convert its entire supply of hydrogen into helium and energy, it would lose less than 1% of its mass.

The length of time during which the various cycles can continue depends entirely upon the amount of hydrogen available. Since the universe consists largely of hydrogen, there is plenty of fuel. The temperature necessary to start the carbon cycle going is of the order of 15 million degrees Centigrade. This temperature is found in the interiors of most stars. The possibility of the reaction's actually taking place has been proved in laboratory tests.

1002. What is the proton-proton reaction? The proton-proton reaction is a process of creating stellar energy which operates at a lower temperature than that required for the carbon cycle. It is considered to exist in the interiors of stars whose temperatures are of the order of 13 million to 16 million degrees Centigrade. It is now believed to be the main process operating in main-sequence stars such as the Sun.

There are two possible processes by which hydrogen is converted into helium and energy in the proton-proton reaction. One of them begins with the collision and combining of two protons, or hydrogen nuclei. This forms one nucleus of heavy hydrogen, H^2, or deuterium, and releases a positive particle, or positron. The deuterium nucleus is struck by a second proton to form an isotope of helium, He^3. This isotope of helium combines with a stable helium nucleus, He^4, to form an isotope of beryllium, Be^7, which captures an electron and becomes an isotope of lithium, Li^7. The lithium nucleus collides with another proton and forms two nuclei of helium, He^4. In this process, most of a star's store of beryllium and lithium are consumed. This may account for the small quantity of these elements found in many stars.

The second form of the proton-proton reaction begins in the same manner, with the collision of two protons which form one nucleus of deuterium and release a positron. The deuterium collides with another proton and becomes an isotope of helium, He^3. This helium isotope meets another helium isotope just like it and breaks up into one normal helium nucleus, He^4, and two protons. In either case, the resultant release of energy is the same.

1003. How big is an atom? The atom was once considered to be the ultimate, indivisible particle of matter. Its very name comes from two Greek words which mean "unsplittable" (*a,* "not"; *tomos,*

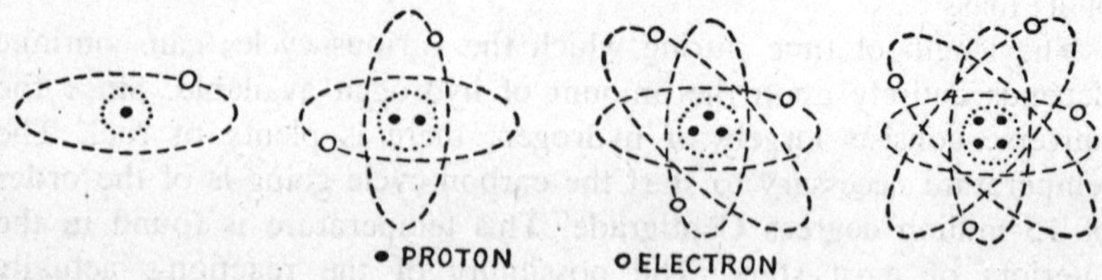

Atoms. Each atom has as many electrons moving in orbits about its nucleus as there are protons in the nucleus.

"split"). We now know that the atom itself is an exceedingly complicated system composed of two distinct groups: the nucleus, which is made up of protons and neutrons; and the electrons, which move about the nucleus in orbits in such a way that a model of the more complicated atoms shows a tight, closely knit, heavy group of protons and neutrons, surrounded by a haze of speeding electrons.

There are as many protons in the nucleus of an atom as there are electrons in motion about the nucleus, and since each electron has only about one thousandth of the mass of a proton, most of the mass of an atom is in its nucleus. The proton and the neutron in the nucleus of an atom have masses of 1/167,000,000,000,000,000,000,000,000 grams; the mass of an electron is 1/9,110,000,000,000,000,000,000,000,000,000 grams.

1004. What is an electron? An electron is an elemental particle of matter which is part of the structure of the atom. Every atom consists of a small, heavy nucleus approximately 10^{-12} centimeters in diameter, surrounded by a region 10^{-9} centimeters in diameter. In this region, the electrons move about the nucleus somewhat like planets around the Sun. Where the planets move in approximately the same plane, however, the electrons move in orbits at all angles to the nucleus and at various distances from the nucleus in what might be described as a haze. The orbits of the electrons are known as rings or shells. Each electron carries a small negative electric charge. The negative charges of all the electrons of an atom are equal to and balance the positive electric charges carried by the protons in the nucleus, so that the atom as a whole has an electric charge of zero. The number of protons, each with its positive charge, determines the number of electrons around the nucleus and also determines the atomic number of the atom.

All the atoms of a given element have the same atomic number even though there may be differences in the arrangement of their elemental particles. Hydrogen, for example, normally has one proton in its nucleus. Consequently the nucleus has a positive electrical charge. Hydrogen also has one electron in each atom balancing the positive charge of the proton with a negative charge. There can be hydrogen atoms with either two or three neutrons in their nuclei in addition to the proton, but they will still be hydrogen with one electron per atom. Hydrogen atoms with one proton and one neutron in their nuclei are called heavy hydrogen or deuterium, and hydrogen with one proton and two neutrons is tritium. Deuterium is an isotope of hydrogen. There are possible isotopes of most of the elements. An isotope must have the same atomic number and chemical identity as the normal form of the element, but it will have a different mass and will be another species of the same chemical element.

1005. What is a proton? A proton is the nucleus of the hydrogen atom. It carries a positive charge of electricity and is also found in the nuclei of other atoms in combination with neutrons. The number of protons in the nucleus of an atom determines the number of electrons which revolve in orbits about the nucleus of an atom, for the electron carries a negative charge of electricity and the negative charges of the electrons must balance the positive charges of the protons. The hydrogen atom, which has one proton and no neutrons in its nucleus, has therefore only one electron and is the simplest of all atoms. The uranium atom has 92 protons and 146 neutrons in its nucleus and 92 electrons which make a complicated haze of orbits about the complicated nucleus.

1006. What is a neutron? A neutron is a particle of matter found in the nuclei of atoms in combination with protons. The neutron is of greater mass but carries no electrical charge. It is the neutron which is driven from the nucleus of an atom to produce, by collision with the nuclei of other atoms, various atomic reactions.

1007. What is meant by ionization? Ionization is the detaching of one or more electrons from the structure of an atom. This can be done by bombarding the atom with energy. An atom is capable of absorbing a certain amount of energy without losing any of its electrons and will thus remain electrically and chemically stable up to a certain point. If more energy than the critical amount is absorbed, an electron will be detached from the atom. This is ionization. The electron and the rest of the atom are now separate and each has its own electrical charge; negative in the case of the electron and positive in the case of the rest of the atom, because there will be one more proton in the nucleus with a positive charge than there are electrons with negative charges to balance. Thus, the electron is a negative ion and the nucleus and remaining electrons make up a positive ion.

An atom may become ionized as many times as it has electrons in the region surrounding its nucleus. It is possible, then, for a complicated atom to be ionized many times. Each succeeding step in the progressive ionization of an atom requires the absorption of much more energy than the preceding step. It requires, for example, the energy equivalent of a temperature of about one million degrees Centigrade to ionize the iron atom 14 times.

1008. What is Einstein's formula about the relation between energy, mass and motion? Einstein's formula is $E = mc^2$. This formula says that the amount of energy (E) produced in a star because of the operation of the carbon cycle or the proton-proton reaction is equal to the mass (m) of the star's material that is converted into energy, times the square of the speed of light (c^2). In the case of the Sun, for example, 564 million tons of hydrogen are transformed into 560 million tons of helium every second. This is a conversion of four million tons of the Sun's mass every second, into energy, and yet only $\frac{7}{10}$ of 1% of the entire mass of the Sun will ever be involved in this process. Under Einstein's formula, this minute fraction of the Sun's mass is sufficient to account for the vast amount of energy in the form of radiation which the Sun—and other stars—produce and broadcast constantly.

1009. What is light? Visible light is the very short segment of the wave band of electromagnetic radiation which affects our eyes. The waves of all forms of radiation vary in length from about one million inches to about $\frac{1}{100}$ millionth of an inch. The length of the radiation waves which affect our eyes and which we call light runs from $\frac{1}{200,000}$ of an inch to $\frac{1}{300,000}$ of an inch. The longer of these waves we see as red and the shorter as violet. All the waves of intermediate length produce all the other colours and their variations.

1010. Does light weigh anything? Yes. If a delicate scale balance is so arranged that one of the weighing pans is kept dark, and light is allowed to fall upon the other one, the lighted pan will in theory sink slowly.

1011. What is the symbol for the speed of light? The small Roman letter c.

1012. What is the symbol for the wave length of light? The small Greek letter, Lambda (λ).

1013. What is an angstrom unit? The unit of measurement of the length of a wave of light is an angstrom unit, named in honour of Anders Jonas Ångström (1814–1874), a Swedish physicist. The angstrom unit is one ten-millionth of a millimeter, sometimes called a tenth-meter. The visibie range of light runs from wave lengths of 7,600 angstroms (red) to 3,900 angstroms (violet). The symbol for the angstrom unit is A or Å.

1014. How can light be broken up into its various wave lengths? When light passes from a medium of one density into a medium of a different density, it is refracted, or bent. This property of light is utilized by passing light through a prism. In this process, the shorter waves of light are refracted to a greater extent than are the longer waves. For that reason, light emerges from the prism as a band of colours, with the longer waves, which we see as red, at one end of the band, and the shorter waves, which we see as violet, at the other end.

1015. Do waves of light of different lengths travel at different speeds? No. The speed of light of whatever wave length is constant at approximately 186,000 miles per second. Variations in the length of waves of light cause a variation in frequency; that is, in the number of waves that pass a given point in a given time.

1016. What is a wave length of electromagnetic radiation? The distance from the crest of one wave to the crest of the following wave is one wave length.

1017. What is the difference between the waves of energy that carry our radio programmes and those that we call X-rays? The difference is in length only. The radio waves are long—up to one million inches—while X-rays are on the order of one millionth of an inch and shorter.

1018. Is a prism the only means of analyzing light? A finely ruled grating, called a diffraction grating, may also be used. This is usually a piece of glass upon which a great many lines, which must be parallel, have been ruled—the more lines to the inch, the better the grating. The action of such a grating, with as many as 20,000 lines to the inch, is interference. This interference produces a spectrum which is superior to the prism spectrum in that it is more widely spread. The prism spectrum, however, is likely to be brighter than the grating spectrum.

1019. What is a continuous spectrum? A continuous spectrum is a band of all the colours without transverse lines. It is produced by heating a solid, a liquid or a very dense gas to incandescence and passing the light so produced through a prism or over a grating. In a continuous spectrum, the various colours blend into one another without interruption or boundaries.

XIV. ASTRONOMERS

1020. Who was Thales? Thales was one of the first, if not the first, individual whose name is prominent in the science of astronomy. He left no writings of his own, but we can see his form dimly, with blurred outlines, in the writings of Herodotus. Thales of Miletus was the son of two members of the minor nobility of Miletus. Herodotus says he was of Phoenician extraction. Later authorities, however, claim that his family were Boeotians who lived with the Ionians of Asia Minor. At any rate, he was a Greek and not a Phoenician, although he may have lived for a time in Phoenician territory.

Thales was a man of great wisdom and common sense. He was widely famed in his day for his political sagacity rather than for his scientific ability, but this was because his skill in government was more easily appreciated by the world in general than was his wide knowledge in the field of mathematics. His best-known astronomical undertaking was his prediction of the eclipse of the Sun that took place on May 28, 585 B.C. The date of this eclipse has been somewhat disputed, but Herodotus fixed it by recording that it occurred during a battle between the Lydians and the Medes and that it led to a cessation of hostilities and a permanent peace between the warring nations. Herodotus says that Thales also foretold the eclipse to the Ionians and fixed the date on which it was to take place. There is considerable discussion about the possibility of Thales' being able to predict any eclipse. The Chaldeans had discovered the saros long before this time. Thales must have known of the saros and may have been able to use these tables of lunar cycles to make his prediction.

Thales went to Egypt and what he learned there from the Egyptian priests, who had invented and were practising geometry, inspired him to further speculation. Using the knowledge of the priests, he determined and expressed certain abstract mathematical relations which would apply to all geometrical figures and not simply to certain specific areas of land. He calculated the method of determining the height of an object by measuring its shadow. He discovered that the circuit of the Sun between solstices is not always uniform; he calculated that the year contained 365 days, and he found that the angular diameter of the Sun was the 720th part of the zodiac, or 30 seconds of arc. Thales also advised the navigators of his day to adopt the constellation of Ursa Minor as a fixed point in sailing, rather than Ursa Major, because the Little Bear was nearer to the North Pole. In spite of all this, however, Thales was certain that the Earth was a flat disk, and he may have thought that the Sun and the Moon were similarly constructed. Thales was born in 640 B.C. and lived for 94 years.

1021. Who was Pythagoras? Pythagoras was born in 572 B.C., in one of the islands of the Greek archipelago—probably in Samos. He studied and travelled widely and is said to have visited the Brahmins of India and the Druid priests of Gaul. When he was about fifty, he had to leave Greece because of political pressure, and settled in Crotona, a city in southern Italy. Here he worked and taught, gathering about him a group of men who formed one of the schools of philosophy which were common in those days. The basis of their tenets was the doctrine of metempsychosis, or the transmigration of souls.

Aside from the religious mysticism of Pythagoras' philosophy, he is justly remembered for his pioneering in mathematics and astronomy. He maintained that the Earth, and presumably each of the other bodies in the universe, was a sphere and that they all revolved about one central body. He did not elect the Sun to the central position, but he imagined an area of central fire, filling the hub of the universe. Around this central fire there revolved ten bodies: the Sun, the Earth, the Moon, the five planets which were known at that time, heaven with its stars and a newcomer named Antichthon or the "counter-Earth." This counter-Earth could never be seen by man because it was always on the opposite side of the central fire. Earth people could never even see the central fire because the side of the Earth upon which man lived was always turned away from it. The light and heat which reached the Earth from the central fire were reflected to the Earth by the Sun. Each of the bodies in the Pythagorean system was fixed upon the surface of one of a series of concentric spheres which revolved upon each other. The motions of these spheres in contact gave out musical tones. The notes produced by the spheres nearest the central fire were low, and the farther out in the system the spheres were and the swifter was their motion, the higher in pitch their tones were. The whole system produced a harmonious chord which, unfortunately, was not perceptible to man. Man lived too close to it and, like the worker in a boiler factory, was never able to hear the sound. This imperceptible sound was the music of the spheres.

Pythagoras was the first to paint the picture of a system revolving about a central body with its various components rotating upon their axes. From this germ of an idea, Copernicus acknowledges taking the nucleus for his own truer conception.

1022. Who was Meton? Meton was an Athenian astronomer about whom very little is known. He flourished about 432 B.C. and attempted to find a stable basis for the calendar. From a study of the records kept by the Chaldeans and the Egyptians, he suggested a calendar based upon lunations, or complete cycles of the changes of the Moon, somewhat similar to the saros. Meton found that 235 lunations took the Moon through all its possible changes and brought it again to a point which was as near to its starting place as it ever got. He used this cycle as the overall period of his calendar. According to Meton's calculations, 235 lunations equalled 19 years. He was not too far wrong, for 235 lunations is actually only two hours more than 19 years. Meton's overall period is called the Metonic cycle. The present method of determining the date of Easter is based upon the Metonic cycle.

1023. Who was Aristotle? Aristotle was perhaps the greatest name in an age of great men. He was born in Stagira, in Greece, and is often known as "the Stagirite." Aristotle studied under Plato, tutored Alexander the Great and founded a school in the Lyceum, in Athens. Aristotle's studies and philosophies covered the whole range of human knowledge and speculation and it is unfortunate that his astronomical pronouncements were farther from actuality than were the views he held on almost any other branch of science. He supported the concept of Eudoxus who tried to explain the motions of the planets by having them move in a system of 27 concentric spheres with the Earth at the centre. Aristotle elaborated somewhat on this system, requiring 55 spheres. He also declared that the Sun and the Moon were both perfect—divine manifestations of celestial fire. So great was the regard in which Aristotle was held that his espousal of these errors made them persist for 2,000 years. Any expression which ran counter to the dictates of the Stagirite was almost certain to bring the man who had the temerity to utter it under an accusation of blasphemy.

1024. Who was Aristarchus? Aristarchus of Samos lived from 310 to 230 B.C. He was one of the first to apply mathematical principles to astronomy. Only one of the writings of Aristarchus remains, and this work does not go into the mathematical side of the picture, but Archimedes quotes Aristarchus as propounding a heliocentric system. Aristarchus pictured a system many times larger than any that had been previously imagined. In this system, the Sun and the stars were motionless, and the Earth and the other planets revolved about the Sun. Aristarchus also declared that the apparent diameter of the Sun was half a degree, fixed the true length of the tropical year, and he may have invented the armillary sphere, an assemblage of rings representing the positions of important circles of the celestial sphere. In his treatise on the Sun and the Moon, he expressed the belief that the Sun was twenty times as far from the Earth as was the Moon.

1025. Who was Archimedes? Archimedes of Syracuse was the most prodigious mathematical genius of his time. He was born in 287 B.C. and early became an ardent astronomer. He constructed of brass a planisphere—a projection of the celestial sphere—that showed the revolutions of the Sun, the Moon and the five known planets, and which demonstrated the nature of eclipses. He was a mechanical genius and an excellent working physicist. His researches and his ingenious devices had impressed the popular imagination and had made him one of the staff of Hiero II, who was the king of Syracuse.

In the best-known story about Archimedes, Hiero had ordered a crown of gold. When the crown had been delivered, Hiero suspected that a certain proportion of the metal in the crown was rather less than gold, thereby enhancing the profit of the maker. He put the problem of finding out the true nature of this bauble up to Archimedes. There was no thought of melting down and analyzing the crown. Archimedes puzzled over the problem for some time. Deeply engrossed in his problem, as he was during his waking hours, he was in his bath, which he had filled to the brim. He stepped in and watched, with some consternation, probably, the spreading puddle of water on the floor. His practical, mechanical mind went to work. "The tub was full. Therefore, as I got into it, the bulk of water which ran over on to the floor must have equalled the bulk of my body. If I were to put anything, then, into a basin full of water, as much of that water would be displaced as would be equal to the bulk of whatever I put into it. But the water would not necessarily weigh as much as whatever I put into the basin. Surely, I weigh more than the water that spilled on to the floor. Now—suppose I take that crown of Hiero's—that is supposed to be made of pure gold—and put it into a basin of water. I'll catch all the water that runs over and weigh it. Then I'll take a lump of gold—that I'm sure is all gold—that weighs exactly the same as Hiero's crown. I'll put that in a basin of water and catch what overflows. If they weigh the same—those two lots of water—then Hiero has a gold crown. If the water from Hiero's crown weighs more, there's silver in the crown, for silver weighs less than gold and therefore takes up more room for each unit of weight than gold does, and will displace more water. That's it! I've found it!"

Archimedes jumped out of his overflowing tub and ran through the streets of that fortunately liberal-minded town in his buff, shouting the Greek equivalent of "I've found it!" His "Eureka!" has come ringing down the years but the specific result of his experiment—as to whether or not Hiero was cheated—has been lost in the larger implications of the discovery of specific gravity.

During a Roman siege of Syracuse, Archimedes rounded up what was probably the first corps of military engineers, keeping off the attacks of the Roman forces until the armies of that overbearing empire were foaming with rage and frustration. There is a legend of his having devised a gigantic mirror which focused the Sun's rays upon the Roman ships, to their complete discomfiture. At any rate, he did, by his mechanical contrivances, succeed in holding off the ultimate victory of Roman arms for some time. When, by sheer force of overwhelming numbers, hunger and thirst, the Romans had overcome the Syracusan defences, orders were given that Archimedes was not to be harmed, but was to be brought to Rome.

Archimedes was not, at the age of 75, the most tactful of men. His later years had been filled with honours, and he did not tolerate ignorance and brutality. A Roman soldier chanced upon him after the breach of the city's defences. Archimedes, probably unkempt and weary from his labours during the siege, was squatting on the ground, drawing mathematical figures in the dust. The soldier stood over him, casting a shadow over the old man's dusty problem. Archimedes may not have bothered to look up, but he did speak, giving the soldier, no doubt, none the best of it. He is reported to have said, "Get away from my circles, you dog!"

The soldier, a little mad with what must have been a very bloody business, cut down the world's greatest thinker with one more sweep of his red sword. So died Archimedes, who said of the principle of the lever, "Give me where I may stand, and I will move the world!"

His planisphere was taken to Rome and was described by Cicero, 150 years later. It was, unfortunately, based upon the premise that the Earth was the centre of the universe.

1026. Who was Eratosthenes? Eratosthenes, who was born about twelve years after Archimedes, in 275 B.C., and who lived to be six years older, was the successor to Aristarchus as the librarian of the Alexandrine School. This great centre of scholarship was founded by Alexander the Great, who laid the foundations of no less than eleven cities, each of which bore his name. The most famous of these cities was Alexandria in Egypt. It was here that the great library and school were located and it was here that Eratosthenes worked and taught. Eratosthenes did not adhere so slavishly to the empirical methods of Aristotle that he eschewed all experiment, but he did seem to have a profound respect for the theories and dogmas of his great predecessor. He devoted himself to clearing up a few details of the Aristotelian universe.

For example, Eratosthenes had heard that, at a place called Syene which may be the moden Assuan, the Sun, upon a certain day each year and only upon that day, shone down a well so directly that the well was illuminated to the very bottom. The day upon which this invariably happened was the day that the Sun reached its most northern point in the heavens—the summer solstice on the twenty-first of June. Eratosthenes knew that the distance between Syene and Alexandria, where he was, measured 5,000 stadia. The stadium then was about ten to the mile. He measured the angle by which the Sun's position at Alexandria deviated from the vertical on that day. Assuming that the well at Syene was actually vertical, he then had the difference between the direction of the Sun over the two cities at the same time. He found that difference to be $\frac{1}{50}$ of a full circle. The circumference of this circle, then, should be about 50 times the distance between Alexandria and Syene, or about 250,000 stadia. With ten stadia to the mile, he was remarkably close to the correct answer.

Eratosthenes also found that the Sun, at its most northern point in the sky, was $\frac{11}{83}$ of the circumference of a circle above the equator. This works out to an elevation of 23°51′20″. The present value of this figure, which is the inclination of the Earth's axis, is 23°27′8″.26.

1027. Who was Hipparchus? Hipparchus was born at Nicaea, in Bithynia, in 160 B.C., and died in 125 B.C., just 35 years later. In that short time, he gathered together all he could find that was known about the heavens, winnowed it, blew away the chaff, checked and revised the measurements and founded, although he certainly was not conscious of so doing, the science of astronomy. He re-measured and checked the inclination of the Earth's axis by studying the recorded variations in the place of the rising and setting of the Sun at various times of the year and, no doubt, by confirming these variations from his own observations. He measured the length of the year down to the smallest division of time then in existence. He recorded what may have been a nova and made the first comprehensive charts of the stars. He drew up tables for the various phases of the Moon and of its positions for years ahead. He discovered the precession of the equinoxes and, in his spare time, he invented trigonometry.

1028. Who was Ptolemy? The dates of Ptolemy's birth and death are not known exactly. He flourished from about 125 to about 180 A.D. He was a Greek who lived in Egypt and he, too, was associated with the library and school at Alexandria. Ptolemy revised the star catalogue that Hipparchus had compiled 250 years before and brought together all the records he could find which dealt with eclipses.

Ptolemy held to the conventional theory that the Earth must be the centre of the universe and worked out an elaborate scheme to account for the motions of the Sun, the Moon and the planets as seen from the Earth. He modified the Aristotelian concept of concentric spheres and devised cycles, epicycles and deferents—circles rolling around the circumferences of other circles. The strange picture he drew of the solar system was accepted for 1,200 years.

Ptolemy, like Hipparchus, put down all that he knew about astronomy, knowledge that he had gleaned from men long gone as well as some notable additions of his own, into one tremendous work which he called *The Mathematical Syntax*. Those who studied it later called it *The Great Astronomical Syntax*. This was done more from a matter of size than of excellence, and served to distinguish it from a work that was physically smaller—a compendium of the writings of various earlier astronomers, which was called *The Smaller Astronomical Syntax*.

After Ptolemy there was darkness. No other great astronomer appeared before the intellectual fog of the Dark Ages rolled over the world. Alexandria was burned by the Arabs. One of the few scrolls that they saved from the holocaust was the great work of Ptolemy. They admired it excessively and gave to it as a title the superlative of the Greek word for "great"—"the greatest." The best rendering of this word into English letters is *Megiste*. To this title the Arabs added their own definite article *al*. The writings of Ptolemy are still known as *The Almegest*.

1029. Who was Copernicus? *Copernicus* is the Latinized form of *Koppernigk*. Nicolaus Copernicus was born at Thorn, in Prussian Poland, in 1473. He was educated at Cracow and Bologna in canon law, in which he took his doctorate, but at Bologna, he attended the astronomical lectures of Domenico Maria Novara. He also studied medicine at Padua and upon the completion of his formal education, he went to Heilsberg in Poland, where he became physician to his uncle, the Bishop of Ermeland.

When his uncle died, Copernicus took up his duties as canon of the cathedral of Frauenburg and devoted himself for many years to various administrative activities there, besides acting as physician to the community. During this time, he conceived, from a study of the astronomy of the ancients, from his courses under Novara and possibly from some observations of his own, a theory of the operation of the solar system. In 1530, he wrote a treatise in which he put forth his ideas and expressed his dissatisfaction with the Ptolemaic system. The material in this treatise became known through a short popular version of it which he wrote for the edification of a small circle of his friends. It was not long before the wide dissemination of this abridged work led to a demand upon him by high church officials for a publication of the full treatise. Copernicus was reluctant to permit this, however, and it was not until 1540 that he released his manuscript to the printer. Three years later *De Revolutionibus Orbium Coelestium* was printed. Copernicus was on his deathbed when the first copy was put into his hands.

1030. Who was Tycho Brahe? Tycho Brahe was born in 1546 in Knudstrup, in Scania, which was then a part of Denmark. Although his family was in comfortable circumstances, he was adopted by an uncle who undertook to care for his education. Tycho was sent to Copenhagen to begin his studies. While he was there, he was greatly impressed by the occurrence of a total eclipse of the Sun at precisely the time predicted. He bought some books on astronomy, including a Latin version of Ptolemy's works, and took up the science as a hobby. In 1562 he entered the University of Leipzig as a law student, but he continued to pursue his avocation and began to observe by means of a celestial globe, a cross-staff and a pair of compasses. When his uncle died, he left Leipzig and went to Wittenberg, the university at which Hamlet was educated, but soon left that school for Rostock. At Rostock, in 1566, he became involved in what was probably a student duel and lost his nose. He immediately fashioned himself a new nose, painted it and attached it. He wore this artificial nose the rest of his life, and it is quite evident in some of his portraits.

At Rostock, Tycho studied chemistry. In 1571, he returned to Denmark, where another uncle financed the establishment of a laboratory for him, near his birthplace of Knudstrup. Here, on November 11, 1572, Tycho saw a new star which blazed up in the constellation of Cassiopeia. He observed this star with great diligence and wrote an exhaustive account of its appearance and location, minutely describing every phase of its career until, after 18 months, it faded to invisibility. This star was a supernova which has been known ever since as "Tycho's Star." The publication of Tycho's report on this star won him some fame as an astronomer and he travelled and lectured in many places in Europe.

In 1572 Tycho married a girl whose station was considered far below his own and his relations with his family were badly strained. He continued his astronomical work, however, with growing fame and was finally subsidized by Frederick II, king of Denmark. Frederick gave him the tiny island of Hven in The Sound, along with a substantial pension. On this island Tycho built an observatory which became, when it was finished and equipped with the finest astronomical instruments, the best in existence.

Here Tycho worked for 21 years, using instruments which he had purchased and devising and making many of his own. His work as an observer was meticulous in the extreme. He plotted planetary positions night after night, almost hour after hour. His accuracy was phenomenal and he developed a system of averaging the results of his observations in order to eliminate, as nearly as possible, any human errors. It is said that he held his science in such respect that he never entered the observatory, which he named Uraniborg, "The Fortress of Heaven," unless he was dressed in the finest clothes he owned. His overall purpose was the correction of the astronomical theories which existed at that time. No efforts comparable to his had been made since the days of Ptolemy. His observatory became a show place and distinguished guests of the royal house of Denmark were customarily taken there to see the great man at work. James VI of Scotland, who afterwards became James I of England, was one of the great personages who visited him.

Tycho did not exercise a great amount of tact in his contacts with those around him. He was arrogant and difficult. He quickly made a mortal enemy of the chancellor of the court of Frederick. The legend is that Tycho had a small and probably obnoxious dog that he kept always by him. The chancellor had the misfortune to incur the enmity of this dog and, in wrath over a nip at his ankles, he delivered a sound kick where he hoped it would do the most good. Tycho apparently threw the chancellor out of the observatory and, when his king and patron died shortly thereafter, Tycho was unceremoniously ousted from Uraniborg.

After visiting briefly at several places, Tycho finally came to Prague, where he was made welcome by the emperor of Austria, Rudolph II. Tycho's equipment was sent to him from Hven and he seemed again to be firmly established. His protégé and pupil, Johannes Kepler, who had been working with him for several years, joined him there and together they continued their observations and deductions. This was not to continue, however, for in October of 1601, the same year in which he had established himself at Prague, Tycho died, at the age of 55.

Tycho suggested certain modifications to the Ptolemaic system and flatly rejected that of Copernicus. Tycho held that the Earth was stationary in space. He said that the Sun moved around the Earth, but that the rest of the planets moved around the Sun. The stars Tycho relegated to their obscure position in the outermost circle of objects which also moved about the Earth. He compiled a refined and corrected catalogue of stars, a catalogue which contained 777 objects. He published an account of his theory of the state of the universe and of the solar system. In Volume II of this work, which was called *Astronomiae Instauratae Progymnasmata,* he discussed the comet of 1571 and, by observation, proved that it had no perceptible parallax. Thus, it could not have been, as was generally thought, an emanation of the Earth's atmosphere, but must have been a visitor from outer space.

Tycho's invaluable legacy was the tremendous amount of accurate data he had collected over the years about the positions of the Sun, the Moon and the five visible planets. The painstaking and deadly mathematical accuracy of his observations and measurements enabled him to refine and correct all of the known astronomical constants and to provide material for the confirmation of the theory of Copernicus who, ironically, was considered by Tycho to be a presumptuous muddler.

1031. Who was Galileo? Galileo Galilei was born on February 15, 1564, in Pisa, Italy, the oldest of seven children of a charming, accomplished and very poor musician and scholar. Galileo was educated at the monastery in Vallombrosa, near Florence, and early exhibited a tremendous facility for manual skill, giving also every indication of a quick and ingenious mind. For a time, he showed an inclination to join one of the established orders of the church, but his father wished him to become a physician and, at considerable financial sacrifice, placed him in the university at Pisa.

During his stay there, his inborn qualities became even more apparent, for at the age of 17, he is credited with having made suggestions which assisted to eminence one of the great artists of the time, Lodovico Cigoli. Galileo's powers of observation and his insatiable curiosity lead him to the discovery of the isochronism of the pendulum. While he was attending mass in the cathedral at Pisa, he noticed that a lamp, set swinging there, completed its oscillation in the same length of time, no matter how wide its swing. This discovery was used by him during the medical phase of his career in timing pulses, and has since become the basic principle in pendulum clocks. His father greatly wished him to become a physician and kept him away from other subjects of study which might divert him, but the story goes that he managed to attend a lecture in geometry. That subject fascinated him and he finally persuaded his father to allow him to change from medicine to mathematics.

Before Galileo could take his degree, financial stringency forced him to withdraw from Pisa and he returned to Florence, where he lectured and taught privately. He invented a hydrostatic balance and his account of that invention attracted wide attention. He wrote an essay on the centre of gravity of solids and was appointed, because of it, to the post of mathematical lecturer at the University of Pisa. Here, from the famous leaning tower, he conducted the experiments which proved that all bodies are attracted to the Earth so that they fall with the same speed, regardless of their weight. This was contrary to the pronouncements of Aristotle, who declared that heavier bodies fall faster than lighter bodies. Galileo's experiments proved the falsity of this doctrine and lead to a general scepticism, on his part, toward many of the ancient and accepted principles. He discoursed at length and with biting sarcasm on these matters and gained much fame and many enemies. He had to resign his position at Pisa and as his father died at this time, he was required to return to Florence to help care for his family. After a short interval of hardship, he was appointed professor of mathematics at the University of Padua, and here he stayed for 18 years. His tenure was marked by increasing prosperity and honour. He invented a thermometer, while he was at Padua, which did not work very well but which formed the basis for later improvements and through the years was developed into the form of mercury thermometer in common use today.

While he was at Padua, he corresponded with Kepler and his letters indicate that he was ready to accept the Copernican theory of the solar system. He observed the bright new star of 1604, and again this event spurred him to another attack upon the fixed and immutable principles of Aristotle.

In 1609, he heard rumours of a strange discovery that had been made during the previous year by an obscure optician named Johannes Lippershey who lived in Holland. This was a device that increased the apparent size of distant objects. Galileo's interest was inflamed and the legend is that on the very day he first heard of this instrument, he succeeded in duplicating it in his own laboratory. This telescope had the power of magnifying objects three times. He improved upon this first telescope very quickly and at the end of constant experiment during which he personally made hundreds of telescopes, he succeeded in producing one that had about 33 powers of magnification. His telescopes were simple. They consisted of a single fixed tube at one end of which was a plano-convex lens with a plano-concave lens at the other end. Galileo aimed his telescope at the sky and began a new era in astronomy. This was in 1609.

Once Galileo looked up, discovery followed discovery with amazing swiftness. The Moon was found to be another world, mountainous and rugged. Its apparent phosphorescence—the mysterious illumination which crept over its surface during two weeks—was announced to be the same light that differentiates day from night. Venus was found to have a variety of shapes which corresponded to those of the Moon. The faint, distant ribbon of light which was then and which still is called the Milky Way lost its cloudy character and was resolved into untold numbers of stars. Saturn, which was then, unfortunately, at that period of its progress along its orbit when its inclination presents its rings almost edge on to our viewpoint, was shown to be something which looked like a large body with two tiny companions nestling beside it. The Sun—the great, pure, inviolable Sun—was shown to have defects; and the Sun repaid what must have been a horrible ordeal of observation for Galileo by striking him blind in his last years.

The sight which moved Galileo to public pronouncement, however, was Jupiter, the great wandering "star." Here, obviously circling around the central body, were four other tiny bodies—moving in changeless paths. Here was another universe. It was on the seventh of January, 1610, that Galileo first saw this miniature system, and its significance must have burst upon his mind with a light as overwhelming as that of the Sun itself. Here was a demonstration of the possibility that smaller bodies could revolve about larger bodies. The foundation was laid for the struggle between Galileo and the Church.

In Galileo's accounts of his discoveries, he did not intend to raise any issue between himself and the established doctrine, but the implication was there and inescapable. When it was brought to a head, Galileo threw himself into the fray with joyous impetuosity. His views were finally laid before the Inquisition and Galileo was advised by friends in high places in that body to be careful. Galileo was overconfident, however, and two of his propositions were condemned formally. The first of these was that the Sun was immovable in the heavens, which was "absurd and expressly contrary to Holy Writ," and the other, that the Earth rotated daily upon its axis, "was erroneous in faith." Galileo was officially scolded and told to refrain from teaching or defending the heretical doctrines. He agreed and returned to Florence with an official certificate of purgation. He was permitted to express his views in the form of an hypothesis, and he believed he was clever enough to do this and to keep from further clashes with the ecclesiastical authorities. A personal friend of Galileo's was elected to the papacy, and Galileo was unofficially told that his views, while rash, were not really heretical.

With this encouragement Galileo produced his famous *Dialogue on the Two Great Systems of the World*. This was published in 1632, and purported to be a three-sided conversation between "Salviati," who expressed Galileo's own views; "Sagredi," an eager and intelligent listener; and "Simplicio," a rather stupid antagonist. Galileo's enemies started the rumour that "Simplicio" was intended to caricature the Pope himself, and there was no doubt that the whole tenor of the dialogues violated the edict of the Inquisition. The sale of the book was prohibited and Galileo was sternly summoned to Rome. He tried in every way to avoid accepting this invitation. He pleaded his advanced age—he was nearly 70—his poor health and many local difficulties of travel, but the Pope, angered at what he considered Galileo's ingratitude, insisted.

Galileo went to Rome finally, and was treated with great consideration and respect, but he was brought to trial. Under threat of torture, which was never carried out and which was obviously never intended to be carried out, he recanted. It is sad that, like many dramatic moments in the lives of the great, the dramatic moment in which Galileo most often figures did not happen. He probably did not, as mankind fondly relates, stamp his aged foot and mutter "*E pur si muove!*—Nevertheless, it moves!"

The last eight years of Galileo's life were spent in his villa at Arcetri in what amounted to house arrest. He published here his very important work *The Dialogue of the New Science,* which dealt with his early experiments and with some thoughts upon the principles of mechanics. He continued to observe, and his last contribution to the cause of astronomy was his discovery of the Moon's diurnal and monthly libration. He became hopelessly blind in 1637, and was blind when he was visited by Milton in 1638. He died at Arcetri in 1642. He was born in the same year that Shakespeare was born and in the year in which Michelangelo died, and he died in the year in which Isaac Newton was born.

1032. Who was Kepler? Johannes Kepler was born in 1571 in Weil, in Württemberg. His parents were unworthy of the son their marriage produced. His father, after failing at many occupations, deserted his family and was never heard of thereafter. His mother was a strange, self-willed, difficult woman who, in her seventy-fourth year, was tried for witchcraft. Kepler himself spent a miserable boyhood. He survived an attack of smallpox when he was four which left him a cripple with defective eyesight, and weakened his constitution to the point where, for the most of his life, he was a prey to almost every minor ailment known to man.

The family poverty drove him to work as a field hand after elementary schooling, but at the age of 13 he was sent to school at Adelberg and later at Maulbronn. Here his innate mental powers were developed and, in 1588, he was admitted to the University of Tübingen. There he studied astronomy and ardently espoused the Copernican theory. When his studies were complete, he accepted, rather hesitatingly, an appointment to a professorship in astronomy. Kepler soon found that the greater part of his duties was the fashioning of almanacs and the casting of horoscopes. He studied the history and methods of the ancient astrologers and produced, probably with some misgivings, passable horoscopes and almanacs which were filled with ambiguous predictions. He kept up a study of the planets and their motions and believed that he had discovered a relationship between the planets. The publication of this theory gained for him a friendship with the two greatest astronomers of his day, Galileo and Tycho.

At this time, Kepler married a wealthy young woman and after considerable vacillating between his scientific, political and religious views, was kept on at Tübingen. The university, however, was almost destitute and Kepler had no pupils. This circumstance, added to his principles, finally cost him his post there, but at exactly the proper moment, he was invited by Tycho Brahe to become Tycho's assistant at Prague. Kepler joined Tycho there on a temporary basis and soon made the association permanent. Tycho, however, died in 1601, a few months after Kepler had joined him.

At Tycho's death, Kepler's appointment as his successor was confirmed, and he completed the lunar tables upon which Tycho had been engaged. These tables were known as the Rudolphine Tables, in honour of Tycho's patron, Rudolph of Austria. Rudolph was a badly balanced man with astrological tendencies and Kepler had again to turn to the expedient of casting horoscopes and making carefully worded predictions. He did have at his disposal, however, the priceless treasure of Tycho's observations. He set himself the task of proving, by means of them, the correctness of the Copernican system, in spite of Tycho's deathbed admonition that their use should be to refute that same theory. Kepler selected Mars as his test object in his attempt to determine the definite relationship between the various objects in the solar system. He chose Mars because of the rather large eccentricity which the orbit of that planet appeared, from Tycho's figures, to possess. After nine years of almost unremitting labour, he published his results which contained the first two of his famous three laws of planetary motion. These two laws were that the orbits of the planets about the Sun are not circles, but ellipses, and that the planets so move that an imaginary line drawn from a planet to the Sun will sweep over equal areas in equal intervals of time.

During the nine years taken up with this work, Kepler had two side interests. He observed and reported upon a nova which blazed up in 1604 to a brightness comparable to that of a star which Tycho saw and reported in 1572. He also observed and speculated upon a triple conjunction of Mars, Jupiter, and Saturn, which occurred in 1603. Kepler calculated the periods of the three planets and found that such an association would take place at intervals of 805 years. Its possible occurrence in 6 B.C. led him to expound the theory that the appearance of these three bright planets in close association was the Star of Bethlehem. This view did not increase his general popularity.

In 1610, he was given a strange and wonderful new instrument, one that had been developed by his correspondent Galileo the year before—a telescope. His delight and interest in this were unbounded and he produced an excellent treatise on the theory of refraction by lenses, suggesting the principal of the inverting, or astronomical telescope. He corresponded with Napier, the inventor of logarithms, and was an ardent advocate of their use. In 1619, the culmination of his labours with the problem of planetary orbits appeared. In this publication, he stated the third and final law which bears his name. This harmonic law says that the squares of the times the planets require to make complete revolutions about the Sun are in proportion to the cubes of their distances from the Sun.

The laws of Kepler are the bases for all determinations of the orbits of bodies in space and are one of the foundation stones of astronomy. Kepler continued a career divided between sound astronomical work in which he was able to predict and see verified several transits of Mercury and one of Venus, and during which he also maintained a steady flow of less reputable astrological works. He died in 1630, at the age of 55, from an illness brought on by a long journey on horseback. He was twice married and was the father of twelve children, only two of whom lived to maturity.

1033. Who was Huygens? Christian Huygens (pronounced *Highgens*) (1629–1695) was a Dutch mathematician and physicist. He greatly improved the telescope lenses of his time, substituting a compound lens of two pieces of glass of different shapes cemented together for the flat lens with an inordinately long focus which was the only known way before his time of overcoming chromatic aberration. Huygens found Titan, the largest satellite of Saturn, and discovered the true nature of Saturn's rings. He was the first man to use a pendulum in a clock, basing his invention in this field upon Galileo's discovery of the isochronism of the pendulum.

1034. Who was James Gregory? James Gregory (1638–1675) was a Scottish mathematician. His theories concerning the possibility of the reflecting telescope led Isaac Newton to build a telescope which embodied many of the principals of modern reflectors. Some of Gregory's ideas are also used in the optical system which today is called the Gregorian. (See question 965.) Gregory was one of the first to attempt to measure stellar distances.

1035. Who was Isaac Newton? Isaac Newton was born in 1642 in Woolsthorpe, Lincolnshire, England. He was a sickly child and was educated, with considerable interruption because of his health, at the Grammar School at Grantham. In spite of his irregular attendance at school, he was admitted at once to Trinity College, Cambridge, upon his graduation from the Grammar School. Before he was 25, he had invented the binomial theorem, the method of tangents and fluxional calculus. His papers on these subjects gained him a professorship at Cambridge. Here he began a study of planetary motions, engaging in a correspondence with John Flamsteed who was, at that time, Astronomer Royal. His association with Flamsteed grew acrimonious at times because of Flamsteed's unwillingness to supply Newton with certain observational data which were necessary to him. It was about at this same time that Edmond Halley came to him for help along the same lines, and Newton's association with Halley and Halley's enthusiastic encouragement led to the publication of Newton's *Philosophiae Naturalis Principia Mathematica,* known as the *Principia,* which contained his formulation of the laws of gravity. This was published in 1687.

Newton was elected to Parliament, but serious illness interrupted his career for several years. In 1694, he was given a sinecure as Warden of the Mint and, in 1697, he was made Master of the Mint. In 1703, he became president of the Royal Society and was re-elected to that office every year until he died. In spite of his mathematical mind and the importance of most of his work, he was tremendously interested in ancient prophecies and theological problems, about which he wrote extensively. He died at Kensington in 1727 and is buried in Westminster Abbey.

1036. Who was Flamsteed? John Flamsteed (1646–1719) was the first Astronomer Royal. He was appointed to that post by Charles II, king of England, in 1675, and carried on astronomical research at the Greenwich Observatory. Flamsteed's salary as Astronomer

Royal was 100 pounds a year; he had taken holy orders, and obtained a post as a clergyman in order to live and hold his position as an astronomer. He was a man of unimpeachable character, but he defeated the purpose of his post by a strange reluctance to give out the information he had obtained through his researches as the official astronomer of England. His reason for this unwillingness to pass on information was that he claimed to feel a need to check and recheck his findings constantly. Indeed, he compiled the first Greenwich Star Catalogue which was published, over his violent protests, in 1712. He succeeded in burning about 300 copies of this catalogue which came into his possession. This passion for withholding the results of his studies led him into many controversies with other scientific men in England, particularly with Sir Isaac Newton, whom he had known for many years. His system for numbering the fainter stars in each constellation is still in use. He was never a vigorous or healthy man; he died in 1719 and was succeeded in his post by Edmond Halley.

1037. Who was Edmond Halley? Edmond Halley was born in London in 1656. His father was a wealthy soap maker who provided his son with an excellent education, sending him to Queen's College, Oxford, where Halley studied mathematics and astronomy. He did not complete his studies at Oxford, but went, with his father's blessing, to St. Helena for the purpose of studying and locating accurately those stars which could not be seen from the latitudes of England. He stayed in St. Helena for a year and, in 1679, he published a catalogue of southern stars, giving the positions of 341 stars. Halley also observed and reported upon a transit of Mercury which occurred on November 7, 1677, while he was in St. Helena.

Upon his return to England, Halley was elected to the Royal Society, which had been recently formed, and began to try his hand at several astronomical problems. Among these was the law of force under which the planets move in elliptical orbits. He was familiar with Kepler's laws of planetary motion, but was unable to carry their principles any farther. He visited Sir Isaac Newton to consult him about the mathematics of the problem, and found that Newton had already given much thought to the same puzzle and had the solution to it lying somewhere about his home. Halley persuaded him to find it, elaborate upon it and present it to the Royal Society. This Newton ultimately did under the title of *Philosophiae Naturalis Principia Mathematica,* which was the famous work in which he formulated the laws of gravitation.

Halley studied the variations of the compass in the Atlantic and published a chart of his findings which was of tremendous value to navigation. He found a method of determining solar parallax by means of the transits of Venus and was a pioneer in the analysis of the proper motion of supposedly "fixed" stars. Upon the death of John Flamsteed, Halley was made the second Astronomer Royal of England, and carried out a complete series of lunar observations covering 18 years. Among his astronomical pursuits was the study of the orbits of comets. Two bright comets had appeared over England in 1680 and 1682, which excited Halley's interest. He collected, as a starting point, records of the appearances of all bright comets from A.D. 1337 to 1698. As he was compiling his lists, he became convinced that the bright comets of 1531 and 1607 were successive appearances of the same object and not two different comets. Again he called upon Newton for the mathematics to prove his point and was so assured by Newton's findings that he declared that this particular comet which, according to him, had appeared approximately every 76 years since 240 B.C., would return again in 1759.

Halley died in 1742, and the comet whose return he had predicted 37 years before did appear again in 1759, establishing the proof that comets are members of the solar system with orbits about the Sun, and not casual visitors from space, as they had been considered before that time. The comet was named for him by public acclaim and is probably the most stupendous monument that any man has ever had.

1038. Who was Charles Messier? Charles Messier was born in France in 1730. He must have had some technical schooling for, as a young man, he went to Paris and became a draftsman for an astronomer there. This aroused his own interest in astronomy and in a few years he had become a full-fledged observational astronomer himself. He was employed by the Observatory of Paris in such prosaic and poorly paid projects as charting the rise and fall of the tidal waters of the Seine, but he devoted his spare time and his enthusiasm to the search for comets, believing that this field would quickly open for him the doors to fame and fortune.

Halley's discovery that comets belonged to the solar system was just beginning to be accepted and comets were held in great esteem in those days. Halley's own comet was about due to make its first predicted return and Messier was searching the skies for it. He did find it, eventually, with the dubious help of some celestial charts that were crowded with errors, but it was discovered upon this historic occasion by a German astronomer on Christmas Day in 1758. Messier was not able to see it until almost a month later, but this adventure stimulated his interest in comets and he became an assiduous pursuer of these strange, elusive objects. Altogether he observed about 50, claimed discovery of 21 and was officially credited with the discovery of 15.

In seeking for comets, Messier used charts, especially drawn, of the regions he was probing with his telescope. From time to time he would encounter objects which were obviously not single stars and which, at times, superficially resembled comets in the early stages of development. He noted these objects very carefully upon his charts, perhaps at first to avoid the repeated and useless reference checking which encounters night after night with the same objects might cause. As he noted their positions, he described these objects as well as he could. Large telescopes are not best suited for comet sweeping and Messier probably used a telescope with an aperture of a little more than six inches and he worked largely with an eyepiece which magnified about 120 times.

He soon realized that while he was smoothing the path for comet observers, he was also making a list of faint and unusual objects which were permanent in space. In February, 1771, he published his first list in the *Journal of the Royal Academy of Sciences.* It is called "A Catalogue of Nebulae and Star Clusters, discovered among the fixed stars, in the skies above Paris."

Messier's first list contained 45 objects, including two obvious groupings, the Pleiades and Praesepe. Later listings brought his total to 103. A few of those originally listed by Messier may have been, ironically enough, comets which eluded the very eye of the man who was called, by Louis XV, the "ferret of comets." Other objects which were later discarded were additions to Messier's list by co-operating astronomers. Messier was not always careful in checking either the position or the description of these contributions. Some duplication and some wishful thinking marred the correctness of his catalogue.

This was, nevertheless, the first attempt to list these mysterious objects.

Messier's memorial exists in the letter M which is prefixed to the catalogue number of each of the Messier objects, and which is still often used to identify them, even though they may also be listed in more up-to-date and complete catalogues under other designations.

Messier survived the French Revolution and the Napoleonic Era and died in 1817.

1039. Who was William Herschel? Sir William Herschel was born Friedrich Wilhelm Herschel in 1738 in Hanover, Germany. He was the son of a musician, became himself a skilled performer on several musical instruments, and rose to the eminence of oboist in the band of the Hanoverian Guards. In 1757 he decided that he would be better off under the benevolent skies of England and migrated to that country, where he worked as an organist, composer and teacher. For a hobby he brushed up on his mathematics and turned to astronomy. This avocation rapidly got the better of him and he built his own telescope and began with it a systematic exploration of the skies. With this telescope, in 1781, he found the first planet ever to be discovered in recorded times, since man had known all of the planets out to Saturn from the very beginning. Herschel picked up an object which, from its observed motion on several occasions, he thought to be a comet. When he realized he had found a planet, he named it, in a burst of patriotism, Georgium Sidus—"George's Star," attempting thereby to honour George III, the ruler of his adopted country. This name did not endure, and Herschel's contemporaries for some time called the new planet "Herschel." Finally, in keeping with the names of the other planets, it was called Uranus, after the Greek deity who had charge of all the heavens.

In 1782 Herschel was appointed private astronomer to the king, and in 1789 he attempted to build the largest telescope in the world. The monster he constructed was 40 feet long and had an aperture of 48 inches. It was a reflector with a mirror made of a combination of tin and copper which was called "speculum metal." Most telescopic mirrors of that time were made of this same unsatisfactory substance, which was so soft and easily scratched that it had to be shaped and polished repeatedly. Herschel mounted his giant on a tremendous scaffolding on a turntable, and it must have given employment to a large crew of men to raise and lower the tube and to turn the great machine in azimuth.

Herschel did build and use an 18.5-inch reflector with a 20-foot focal length to great advantage. The mirror in this telescope was given a slight tilt in the lower end of the tube so that its cone of reflection could be picked up by an eyepiece mounted on the upper edge of the tube. The observer could see the object at which the telescope was pointed by turning his back upon the skies and peering into the eyepiece from the top of the cylinder that housed the mirror. This must have been an excellent mirror, for Herschel could use eyepieces with it that gave him powers up to 2,000.

He made a study of double stars, and catalogued over 800 of them. He was the first to locate the point in space toward which the Sun is travelling—the apex of the Sun's way—and his systematic and detailed exploration of those skies available to him from his observatory resulted in a great catalogue of nebulae. This catalogue was arranged by his sister, Caroline Herschel, who was his devoted helper and no mean astronomer in her own right.

Herschel's catalogue listed over 2,500 nebulae and clusters, a tremendous increase over Messier's 103. Herschel also determined the rotational speed of Saturn and found the sixth and seventh satellites of Saturn. He was a prolific writer and left a remarkable legacy of astronomical notes. He died, full of honours, in 1822, and may be called the first of the great observational astronomers.

1040. Who was William Huggins? Huggins was an English astronomer, born in 1824. He worked from his home in a part of London called Tulse Hill, and it was there, in 1870, that he hit upon the idea of using a spectroscope for the study of the stars. His description of the first sight of spectral lines is thrilling. Spectroscopy had been in use in laboratory work in the analysis of materials for some time, but Huggins was the first to apply it successfully to astronomy. He also made notable improvements in the then new science of astronomical photography and published the first catalogue of stars by spectral classification. He was president of the Royal Society from 1900 to 1906 and died in 1910.

1041. Who was Percival Lowell? Percival Lowell was born in Boston in 1855 to a family that gave Harvard one of its famous presidents and the world a great poetess. A. Lawrence Lowell, president of Harvard for many years, was his brother, and Amy Lowell was his sister. During the first part of his adult life, he visited Korea and Japan, wrote several books about these countries, their problems and customs, and was made a special counsellor to a Korean mission to the United States.

Schiaparelli's discoveries on Mars in 1877 aroused Lowell's interest in that planet and, in 1894, he built the Lowell Observatory at Flagstaff, Arizona, in order to devote all his time and energy to the study of Mars and of astronomy in general. He became a skilled observer and was able to see details upon Mars that were not apparent to many astronomers, but whose existence was supported by collaborative evidence from others whose drawings of Mars matched those of Lowell closely. Lowell's belief that Mars was inhabited by an intelligent race went far beyond the general viewpoint, and he wrote much in support of his theories, presenting his ideas logically and convincingly. His books *Mars and Its Canals* and *The Evolution of Worlds* are famous for their espousal of his theories and for their descriptions of his research.

His theory that a ninth planet existed beyond the orbit of Neptune led him to initiate the research that resulted, in 1930, in the discovery of Pluto. Lowell died in 1916, unaware that his persistence and his faith in his idea of a trans-Neptunian planet would one day be justified.

1042. Who was George Ellery Hale? George Ellery Hale is the man to whom the world owes the existence of three great telescopes, the 40-inch refractor at Yerkes Observatory—the largest of its kind that will ever be built—the 100-inch Hooker Reflector at Mount Wilson—for many years the largest telescope in existence—and the 200-inch Hale Reflector at Mount Palomar—the largest telescope in the world and likely to be so for some years.

Hale was born in Chicago in 1868 and was graduated in 1890 from the Massachusetts Institute of Technology. He went immediately into astronomy and soon became the leading authority on many aspects of solar activity. He invented the spectroheliograph, an instrument which makes possible the photography of the Sun's prominences by using the regions of the spectrum in which these prominences can best be seen. Before his invention of this device, the prominences could be photographed only at times of the total eclipse of the Sun. Hale also studied the mechanics of sunspots.

In addition to his scientific standing, Hale was able to convince men of great wealth of the need for better astronomical instruments. Yerkes, a Chicago streetcar magnate, was persuaded to finance the Yerkes Observatory and the great 40-inch refractor there, which is still in use. Hooker, of California, was convinced by Hale of the need for better observing facilities to the extent of underwriting the construction and establishment of the 100-inch Hooker Reflector at Mount Wilson and, finally, the Rockefeller Foundation supplied, at Hale's urging, the funds to build the 200-inch Reflector at Mount Palomar, which is most fittingly known as the Hale Reflector. Hale was director of the Yerkes Observatory from 1895 to 1905, and director of Mount Wilson from 1904 to 1923. He died in 1938, before his dream of the Glass Giant of Palomar was realized.

1043. Who was Albert Einstein? Albert Einstein was born at Ulm, Germany, in 1879. In 1900, he was graduated from the Federal Institute of Technology in Zurich, Switzerland, and became a Swiss citizen. For seven years he was an examiner at the Patent Office in Berne, Switzerland. During these years he obtained his doctor's degree at the University of Zurich and began work upon his special theory of relativity and the study of atoms. As a result of this work he was made a professor at the University of Zurich, but soon left there to assume the post of professor at the German University at Prague. From there, he went to his old school, the Federal University of Technology, where he held the chair of theoretical physics. His fame grew steadily and, in 1912, he was offered and accepted the post of professor of physics and head of the theoretical physics department of the Kaiser Wilhelm Institute.

In 1921, Einstein was awarded the Nobel Prize in physics. When Hitler became chancellor of Germany, Einstein was deprived of his citizenship and property and left Germany, never to return. He was invited to the Institute for Advanced Study at Princeton, New Jersey, in 1933, and remained there until his death in 1955.

Einstein was never an astronomer in the strictest sense of the word, but his work posed problems for astronomers, and astronomy was called upon to confirm many of his theories, notably that of the effect of gravitation upon light. (See question 391.) The field of nuclear physics claimed much of his knowledge and skill. His formula explaining the conversion of matter into energy—$E = mc^2$—is the result of his facility in unlocking the mysteries of the universe. His kind, gentle nature and his unassuming character endeared him to the world. One anecdote illustrates this. He and his wife were guests of honour at the dedication of a new astronomical observatory. One of the astronomers had been assigned to escort Mrs. Einstein about the splendid building and had finally taken her into the dome of the great new telescope. He pointed to it with justifiable pride and said, "With this instrument, Mrs. Einstein, we expect to solve many of the mysteries of the universe." Mrs. Einstein nodded and smiled. "Yes," she said, "my husband does that on the back of an old envelope."

XV. TABLES AND BIBLIOGRAPHY

1044. Where are the largest refracting telescopes? Here is a list of refractors with objective lenses of more than 25 inches in diameter. This list is taken from *Telescopes and Accessories,* by Dimitroff and Baker.

Diam.	*Focal Length*	*Date*	*Maker*	*Location*
40.0 in.	63.5 ft.	1897	Alvan Clark	Yerkes Observatory, Williams Bay, Wisconsin
36.0 in.	57.8 ft.	1888	Alvan Clark	Lick Observatory, Mount Hamilton, California
32.7 in.	53.0 ft.	1889	Henry Bros.	Observatory of Paris, Meudon, France
31.5 in.	39.4 ft.	1899	Steinheil	Astrophysical Observatory. Potsdam, Germany
30.0 in.	46.3 ft.	1914	Brashear	Allegheny Observatory, Pittsburgh, Pennsylvania
30.0 in.	52.6 ft.	1886	Henry Bros.	Bischoffsheim Observatory, University of Paris, Nice, France
30.0 in.	46.3 ft.	1885	Alvan Clark	Central Astronomical Observatory, Poulkovo, U.S.S.R.
28.0 in.	27.9 ft.	1893	Howard Grubb	Royal Observatory, Greenwich, England
27.0 in.	....	1927	McDowell	Lamont-Hussey Observatory, University of Michigan, Bloemfontein, South Africa
26.8 in.	68.9 ft.	....	Steinheil	Treptow Observatory, Berlin, Germany
26.8 in.	34.4 ft.	1880	Howard Grubb	University Observatory, Vienna
26.5 in.	35.1 ft.	....	Howard Grubb	University Observatory, Johannesburg, Transvaal, Union of S. Africa.
26.0 in.	32.5 ft.	1873	Alvan Clark	U. S. Naval Observatory, Washington, D.C., U.S.A.
26.0 in.	32.8 ft.	1883	Alvan Clark	Leander McCormick Observatory, University of Virginia, Charlottesville, Virginia, U.S.A.
26.0 in.	22.5 ft.	1897	Howard Grubb	Royal Greenwich Observatory, Herstmonceux, England
26.0 in.	36.1 ft.	....	McDowell	Yale Observatory, Johannesburg, Transvaal, Union of S. Africa
25.6 in.	34.1 ft.	....	Zeiss	Berlin-Babelsberg University Observatory, Neubabelsberg, Berlin, Germany
25.6 in.	....	....	Zeiss	Imperial University Observatory, Mitaka, Tokyo, Japan
25.6 in.	34.1 ft.	....	Zeiss	University of Belgrade, Yugoslavia

1045. Where are the largest reflecting telescopes? Here is a list of reflectors with mirrors of more than 60 inches in diameter. (The Russian 236-inch is not yet operational.)

Diam.	*Focal Length*	*Date*	*Observatory*
200	55 ft.	1948	Mount Palomar (U.S.A.)
120	50	1959	Lick Observatory (U.S.A.)
102	33	1960	Crimea (U.S.S.R.)
98	40	1967	Herstmonceux Castle (England)
84	18	1961	Kitt Peak (U.S.A.)
82	39	1939	McDonald Observatory, Texas (U.S.A.)
76	32	1958	Haute-Provence (France)
74	30	1960	Helwan (Egypt)
74	30	1958	Radcliffe, Pretoria (S. Africa)
74	31	1955	Mount Stromlo (Australia)
74	30	1960	Okayama (Japan)
74	31	1935	Dunlap, Toronto (Canada)
72	30	1918	Dominion, Victoria (Canada)
69	25	1932	Perkins, Flagstaff (U.S.A.)
61	25	1942	Bosque Alegre (Argentina)
61	25	1934	Harvard (U.S.A.)
60	25	1908	Mount Wilson (U.S.A.)
60	26.5	1930	Boyden, Bloemfontein (S. Africa)

1046. What are the elements of the solar system?

Planet	*Mean Distance from Sun*	*Period of Revolution*	*Eccentricity of Orbit*	*Inclination to Ecliptic*	*Longitude of Node*	*Longitude of Perihelion*
Mercury	36,000,000	88.0 days	0.206	7°.0	47°.8	76°.8
Venus	67,200,000	224.7 days	0.007	3°.4	76°.3	130°.9
Earth	92,900,000	365.3 days	0.017	—	—	102°.2
Mars	141,500,000	687.0 days	0.093	1°.8	49°.2	335°.2
Jupiter	483,300,000	11.86 yrs.	0.048	1°.3	100°.0	13°.6
Saturn	886,000,000	29.46 yrs.	0.056	2°.5	113°.3	92°.2
Uranus	1,783,000,000	84.01 yrs.	0.047	0°.8	73°.8	169°.9
Neptune	2,791,000,000	164.8 yrs.	0.009	1°.8	131°.3	44°.2
Pluto	3,671,000,000	248.4 yrs.	0.249	17°.1	109°.6	223°.2

1047. What are the physical characteristics of the planets?

	Sun	*Mercury*	*Venus*	*Earth*	*Mars*
Diameter (miles)	864,000	2,900	7,700	7,927	4,140
Mass (Earth = 1)	332,000	0.0543	0.8136	1.000	0.1069

	Sun	*Mercury*	*Venus*	*Earth*	*Mars*
Mean Density (Water = 1)	1.41	5.46	5.06	5.52	4.12
Surface Gravity (Earth = 1)	27.9	0.38	0.88	1.00	0.39
Albedo		0.058	0.76	0.39	0.148
Inclination of Equator to Orbit	7°10′	—	—	23°27′	24°
Magnitude at Greatest Brilliancy	—26.8	—1.9	—4.4	—	—2.8
Oblateness	0	0	0	1/296	1/192
Velocity of Escape (Miles per Second)	383	2.0	6.3	6.95	3.1

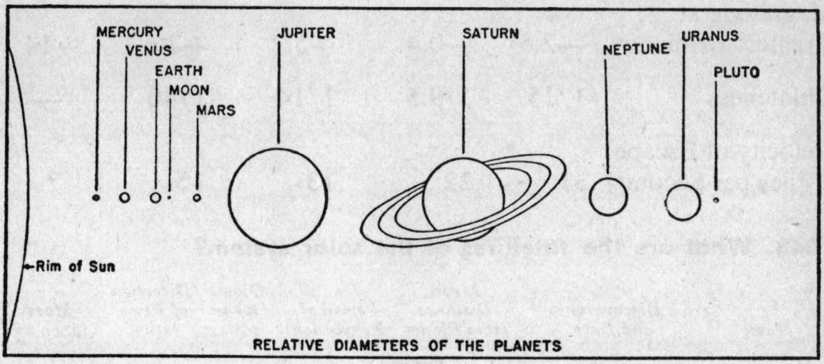

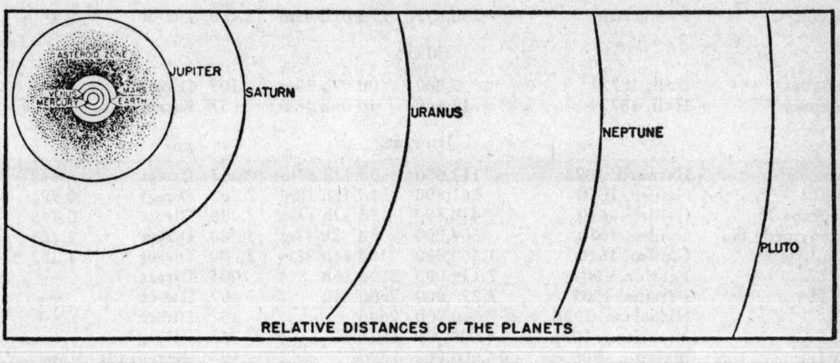

The relative sizes of the planets against a segment of the Sun. The planetary orbits shown approximately to scale.

	Jupiter	*Saturn*	*Uranus*	*Neptune*	*Pluto*
Diameter (miles)	86,900	75,100	29,300	31,200	3,700?
Mass (Earth = 1)	318.35	95.3	14.54	17.2	0.033?
Sidereal Rotation Period	9h 50m	10h 14m	10h 45m	About 14h	6d 9h?
Mean Density (Water = 1)	1.35	0.71	1.56	2.47	2?
Surface Gravity (Earth = 1)	2.65	1.17	1.05	1.23	0.16?
Albedo	0.51	0.50	0.66	0.62	0.16 ?
Inclination of Equator to Orbit	3°7′	26°45′	98°	29°	—
Magnitude at Greatest Brilliancy	−2.5	−0.4	+5.7	+7.6	+14
Oblateness	1/15	1/9.5	1/14	1/40	—
Velocity of Escape (Miles per Second)	37	22	13	15	?

1048. What are the satellites of the solar system?

Name	*Discoverer and Date*	*Mean Distance from Planet*	*Period of Revolution*	*Diameter in Miles*	*Direction of Revolution*	*Mass* (Moon = 1)
		EARTH				
Moon	Prehistoric	238,857	27d 07h 43m	2,160	Direct	1
		MARS				
Phobos	Hall, 1877	5,800	0d 07h 39m	10?	Direct	—
Deimos	Hall, 1877	14,600	1d 06h 18m	5?	Direct	—
		JUPITER				
V	Barnard, 1892	112,600	0d 11h 57m	100?	Direct	—
Io I	Galileo, 1610	261,800	1d 18h 28m	2,300	Direct	0.991
Europa II	Galileo, 1610	416,600	3d 13h 14m	2,000	Direct	0.645
Ganymede III	Galileo, 1610	664,200	7d 3h 43m	3,200	Direct	2.123
Callisto IV	Galileo, 1610	1,169,000	16d 16h 32m	3,200	Direct	1.322
VI	Perrine, 1904	7,114,000	250d 16h	100?	Direct	—
VII	Perrine, 1905	7,292,000	260d 1h	40?	Direct	—
X	Nicholson, 1938	7,300,000	260d	15?	Direct	—
XI	Nicholson, 1938	14,000,000	692d	15?	Retrograde	—
VIII	Melotte, 1908	14,600,000	739d	40?	Retrograde	—
IX	Nicholson, 1914	14,900,000	758d	20?	Retrograde	—
XII	Nicholson, 1951	—	—	15?	Retrograde	—

Name	*Discoverer and Date*	*Mean Distance from Planet*	*Period of Revolution*	*Diameter in Miles*	*Direction of Revolution*	*Mass* (Moon = 1)
		SATURN				
Janus	Dollfus, 1966	98,000	0d 17h 58m	150?	Direct	?
Mimas	W. Herschel, 1789	115,000	0d 22h 37m	400?	Direct	0.00048
Enceladus	W. Herschel, 1789	148,000	1d 8h 53m	500?	Direct	0.00195
Tethys	Cassini, 1684	183,000	1d 21h 18m	800?	Direct	0.00887
Dione	Cassini, 1684	234,000	2d 17h 4m	700?	Direct	0.01440
Rhea	Cassini, 1672	327,000	4d 12h 25m	1,100	Direct	0.03108
Titan	Huygens, 1655	759,000	15d 22h 41m	2,600?	Direct	1.9297
Hyperion	G. Bond, 1848	920,000	21d 6h 38m	300?	Direct	—
Iapetus	Cassini, 1671	2,210,000	79d 7h 56m	1,000?	Direct	—
Phoebe	W. Pickering, 1898	8,034,000	550d	200?	Retrograde	—
		URANUS				
Miranda	Kuiper, 1948	81,000	1d 9h 56m	—	Retrograde	—
Ariel	Lassell, 1851	119,000	2d 12h 29m	600?	Retrograde	—
Umbriel	Lassell, 1851	166,000	4d 3h 28m	400?	Retrograde	—
Titania	W. Herschel, 1787	272,000	8d 16h 56m	1,000?	Retrograde	—
Oberon	W. Herschel, 1787	364,000	13d 11h 7m	900?	Retrograde	—
		NEPTUNE				
Triton	Lassell, 1846	220,000	5d 21h 3m	3,000?	Retrograde	1.80
Nereid	Kuiper, 1949	3,460,000	359d	200?	Direct	—

1049. What books are there for further reading on astronomy?

There are many books on astronomy for the layman. Here are a few of them.

Abetti, Giorgio. *History of Astronomy.* Sidgwick and Jackson, 1954.

——. *Exploration of the Universe.* Faber and Faber, 1968.

Baker, R., and Fredric, L. *Astronomy.* Van Nostrand, 1971.

Beet, E. A. *Astronomy Old and New*. Bell, 1966.

Bok, B. J. *The Astronomer's Universe.* Cambridge University Press, 1958.

Davidson, M. *Elements of Mathematical Astronomy.* Hutchinson, 1948.

Firsoff, V. A. *The World of Mars.* Oliver and Boyd, 1969.

——. *The Interior Planets.* Oliver and Boyd, 1968.

——. *Life, Mind and Galaxies*. Oliver and Boyd, 1968.

Hey, J. S. *The Evolution of Radio Astronomy.* Elek, 1973.

Howard, Neile E. *Telescopic Handbook and Star Atlas.* Crowell, 1967.

King-Hele, D. *Observing Earth Satellites.* Macmillan, 1966.

Lovell, B. *The Story of Jodrell Bank.* Oxford University Press, 1968.

——. *Out of the Zenith.* Oxford University Press, 1971.

Moore, Patrick. *Astronomy*. Macdonald, 1974.
——. *Space*. Lutterworth Press, 1974.
——. *The Amateur Astronomer*. Lutterworth Press, 1974.
——. *The Amateur Astronomer's Glossary*. Lutterworth Press, 1967.
——. *Naked-Eye Astronomy*. Lutterworth Press, 1973.
——. *Guide to the Planets*. Lutterworth Press, 1974.
——. *Guide to the Stars*. Lutterworth Press, 1974.
——. *The Sky at Night*. B.B.C., 1972.
Motz, L., and Duveen, A. *Essentials of Astronomy*. Blackie, 1966.
Norton, A. P. *Star Atlas and Telescopic Handbook*. Gall and Inglis, 1972.
Payne-Gaposchkin, C. *Introduction to Astronomy*. Eyre and Spottiswoode, 1967.
Robinson, J. Hedley. *Astronomy Data Book*. David and Charles, 1973.
Ronan, C. A. *Their Majestie's Astronomers*. Bodley Head, 1967.
Roth, G. D. *The Amateur Astronomer and his Telescope*. Faber and Faber, 1971.
Schatzman, E. *Origin and Evolution of the Universe*. Hutchinson, 1966.
Various. *Astronomical Telescopes and Observatories for Amateurs*. David and Charles, 1973.
Wright, H. *The Great Palomar Telescope*. Faber and Faber, 1953.

The following books are suitable for the reader who wishes to read further in astronomy.

Abetti, Giorgio. *The Sun*. Faber and Faber, 1962.
Alexander, A. F. O'D. *The Planet Saturn*. Faber and Faber, 1962.
——. *The Planet Uranus*. Faber and Faber, 1964.
Baxter, W. M. *The Sun and the Amateur Astronomer*. David and Charles, 1972.
Baade, W. *The Evolution of Stars and Galaxies*. Oxford University Press, 1963.
Cross, C. A. *Mars*. Mitchell Beazley, 1973.
Ellison, M. A. *The Sun and its Influence*. Routledge and Kegan Paul, 1968.
Fielder, G. *Lunar Geology*. Lutterworth Press, 1962.
Glasby, J. S. *Variable Stars*. Constable, 1968.
Kahn, F., and Palmer, H. *Quasars*. Manchester University Press, 1967.

Moore, Patrick, and Cattermole, P. J. *The Craters of the Moon.* Lutterworth Press, 1967.

Moore, Patrick. *The Comets.* Keith Reid, 1974.

Newton, H. W. *The Face of the Sun.* Penguin Books, 1959.

Nicolson, I., and Moore, P. *Black Holes in Space.* Orbach and Chambers, 1974.

Peek, B. M. *The Planet Jupiter.* Faber and Faber, 1961.

Richter, N. *The Nature of Comets,* Methuen, 1963.

Smart, W. M. *Spherical Astronomy.* Cambridge University Press, 1967.

Størmer, C. *The Polar Aurora.* Oxford University Press, 1955.

Various. *Practical Amateur Astronomy.* Lutterworth Press, 1973.

INDEX

References are to question numbers